Delta Sustainability

Weiguo Zhang · Huib de Vriend
Editors

Delta Sustainability

A Report to the Mega-Delta Programme of the UN Ocean Decade

Springer

Editors
Weiguo Zhang
State Key Laboratory of Estuarine and Coastal Research
East China Normal University
Shanghai, China

Huib de Vriend
Faculty of Civil Engineering and Geosciences
Delft University of Technology
Delft, The Netherlands

ISBN 978-981-97-7258-2 ISBN 978-981-97-7259-9 (eBook)
https://doi.org/10.1007/978-981-97-7259-9

This Springer imprint is published by the registered company Springer Nature Singapore Pte Ltd.
The registered company address is: 152 Beach Road, #21-01/04 Gateway East, Singapore 189721, Singapore

Acknowledgements

This book is an outcome of the research consortium programme 'Mega-Deltas: Seeking Solutions for Sustainable Issues', which is adopted by the UN Ocean Decade in 2021. The authors of the 15 deltas are detailed as follows:

Amazon River Delta

Xiaomei Nian, Weiguo Zhang
State Key Laboratory of Estuarine and Coastal Research, East China Normal University, China
Marcelo Cancela Lisboa Cohen
Graduate Program of Geology and Geochemistry, Federal University of Pará, Brazil
Department of Oceanography and Coastal Sciences, Louisiana State University, USA
Dailson José Bertassoli Júnior, **André Oliveira Sawakuchi**
Institute of Geosciences, University of São Paulo, Brazil
Janice Muriel-Cunha
Institute of Coastal Studies, Federal University of Pará, Brazil
Henrique Oliveira Sawakuchi
Department of Thematic Studies-Environmental Change, Linköping University, Sweden

Changjiang River Delta

Leicheng Guo, Linjing Ren, Haifei Yang, Weiguo Zhang, Lei Gao, Xiuzhen Li, Ben Li, Wenzhen Zhao, Ben Li, Wenzhen Zhao
State Key Lab of Estuarine and Coastal Research, East China Normal University, China

Danube River Delta

Liliana Ene, Gabriel Lupu, Adrian Burada, Marian Mierlă, Mihai Doroftei, Silviu Covaliov, Iuliana-Mihaela Tudor, Orhan Ibram, Aurel Năstase, Alexandru Doroşencu, Lucian-Eugen Bolboacă, Mihai Marinov, Vasile Alexe, Marian Tudor
Danube Delta National Institute for Research and Development, Romania

Fly River Delta

Fan Xu, Qinzi Cheng
State Key Laboratory of Estuarine and Coastal Research, East China Normal University, China

Ganges-Brahmaputra-Meghna River Delta

Sheikh Aftab Uddin
Institute of Marine Sciences, University of Chittagong, Bangladesh
Lijun He
State Key Laboratory of Estuarine and Coastal Research, East China Normal University, China
Md. Jaker Hossain
Institute of Marine Sciences, University of Chittagong, Bangladesh
State Key Laboratory of Estuarine and Coastal Research, East China Normal University, China
Nabila Nusrat
Institute of Marine Sciences, University of Chittagong, Bangladesh
Milki Debi
Institute of Marine Sciences, University of Chittagong, Bangladesh
State Key Laboratory of Estuarine and Coastal Research, East China Normal University, China

Indus River Delta

Waqar Ahmed, Samina Kidwai
National Institute of Oceanography, Pakistan

Irrawaddy River Delta

Qinzi Cheng
State Key Laboratory of Estuarine and Coastal Research, East China Normal University, China
Xing Li
School of Geography, Geomatics and Planning, Jiangsu Normal University, China

Lena River Delta

Sergey Chalov, Kristina Prokopeva, Dmitry Magritsky
Faculty of Geography, Lomonosov Moscow State University, Russia

Mekong Delta

Cong Mai Van
Faculty of Civil Engineering, Thuyloi University, Vietnam
Faculty of Civil Engineering and Geosciences, Delft University of Technology, The Netherlands
Tri Cao Mai
Faculty of Coastal and Offshore Engineering, Hanoi University of Civil Engineering, Vietnam
Xuefei Mei
State Key Laboratory of Estuarine and Coastal Research, East China Normal University, China

Mississippi River Delta

Y. Jun Xu
School of Renewable Natural Resources, Louisiana State University, USA
Coastal Studies Institute, Louisiana State University, USA
Nina S. N. Lam
Department of Environmental Sciences, Louisiana State University, USA
Kam-biu Liu, Kehui Xu
Department of Oceanography and Coastal Sciences, Louisiana State University
Coastal Studies Institute, Louisiana State University, USA

Nile River Delta

Alaa Salem
Faculty of Science, Kafrelsheikh University, Egypt
Yan Liu
State Key Laboratory of Estuarine and Coastal Research, East China Normal University, China

Pearl River Delta

Haixian Xiong, Jie Ren, Huan Liu, Junjie Deng, Leiping Ye, Jiaxue Wu
School of Marine Sciences, Sun Yat-sen University, China
Southern Marine Science and Engineering Guangdong Laboratory, China
Benwei Shi
State Key Laboratory of Estuarine and Coastal Research, East China Normal University, China

Rhine-Meuse-Scheldt-Delta

Huib de Vriend
Faculty of Civil Engineering and Geosciences, Delft University of Technology, The Netherlands
Zhengbing Wang, Bas van Maren
Faculty of Civil Engineering and Geosciences, Delft University of Technology, The Netherlands
Deltares, The Netherlands
Zhong Peng
State Key Laboratory of Estuarine and Coastal Research, East China Normal University, China

Río de la Plata Estuary

Natalia Venturini, Ernesto Brugnoli, Ofelia Gutiérrez, Pablo Muniz, Gustavo J. Nagy, Daniel Panario, José E. Verocai
Instituto de Ecología y Ciencias Ambientales (IECA), Facultad de Ciencias, Universidad de la República, Uruguay

Xianye Wang
State Key Laboratory of Estuarine and Coastal Research, East China Normal University, China

Yellow River Delta

Houjie Wang, Xiao Wu, Naishuang Bi, Peng Li
College of Marine Geosciences, Ocean University of China, China

Special thanks go to Prof. Qing He, Director of the State Key Laboratory of Estuarine and Coastal Research, East China Normal University (ECNU), and Prof. Xiuzhen Li, Prof. Ying Wu, Prof. Yunxuan Zhou, Prof. Ning Zhao, Prof. Zhixuan Feng, Dr. Weiming Xie, Dr. Ying Huang, Mrs. Li Tan from ECNU, and Dr. Donald L. Forbes from Geological Survey of Canada for their support during the preparation of the book. This work is supported by the Science and Technology Commission of Shanghai Municipality (23590780200).

Contents

Introduction 1
Weiguo Zhang and Huib de Vriend

Amazon River Delta 5
Xiaomei Nian, Marcelo Cancela Lisboa Cohen, Weiguo Zhang, Dailson José Bertassoli Júnior, Janice Muriel-Cunha, Henrique Oliveira Sawakuchi, and André Oliveira Sawakuchi

Changjiang River Delta 33
Leicheng Guo, Linjing Ren, Haifei Yang, Weiguo Zhang, Lei Gao, Xiuzhen Li, Ben Li, and Wenzhen Zhao

Danube River Delta 57
Liliana Ene, Gabriel Lupu, Adrian Burada, Marian Mierlă, Mihai Doroftei, Silviu Covaliov, Iuliana-Mihaela Tudor, Orhan Ibram, Aurel Năstase, Alexandru Doroşencu, Lucian-Eugen Bolboacă, Mihai Marinov, Vasile Alexe, and Marian Tudor

Fly River Delta 73
Fan Xu and Qinzi Cheng

Ganges-Brahmaputra-Meghna River Delta 89
Sheikh Aftab Uddin, Lijun He, Md. Jaker Hossain, Nabila Nusrat, and Milki Debi

Indus River Delta 117
Waqar Ahmed and Samina Kidwai

Irrawaddy River Delta 133
Qinzi Cheng and Xing Li

Lena River Delta 145
Sergey Chalov, Kristina Prokopeva, and Dmitry Magritsky

Mekong River Delta 165
Cong Mai Van, Xuefei Mei, and Tri Cao Mai

Mississippi River Delta 187
Y. Jun Xu, Nina S. N. Lam, Kam-biu Liu, and Kehui Xu

Nile River Delta 217
Alaa Salem and Yan Liu

Pearl River Delta 237
Haixian Xiong, Jie Ren, Huan Liu, Junjie Deng, Leiping Ye,
Benwei Shi, and Jiaxue Wu

Rhine-Meuse-Scheldt Delta 263
Huib de Vriend, Zhengbing Wang, Bas vanMaren, and Zhong Peng

Río de la Plata Estuary 293
Natalia Venturini, Ernesto Brugnoli, Ofelia Gutiérrez, Pablo Muniz,
Gustavo J. Nagy, Daniel Panario, José E. Verocai, and Xianye Wang

Yellow River Delta 319
Houjie Wang, Xiao Wu, Naishuang Bi, and Peng Li

Summary 343
Weiguo Zhang and Huib de Vriend

Introduction

Weiguo Zhang and Huib de Vriend

Deltas, located at the interface of land and sea, are dynamic regions heavily influenced by terrestrial and marine factors. However, these areas are facing increasing environmental pressures due to factors such as population growth, urbanization, and climate change. These pressures cause coastal erosion and subsidence, flooding, freshwater shortage, pollution, and loss of biodiversity. These issues hinder the achievement of the United Nations' sustainable development goals (SDGs) in deltaic regions, such as eradicating poverty (SDG 1), ensuring clean water and sanitation (SDG 6), and promoting sustainable cities and communities (SDG 11).

In 2018, a workgroup entitled "Mega-Delta" was established under the umbrella of the Future Earth Coasts Programme. After a series of discussion, the workgroup developed the initiative "The Mega-Delta Programme", in its full name as "Mega-Deltas: Seeking Solutions for Sustainability Issues". This initiative was adopted by the UN Ocean Decade in 2021. The main goal of the programme is to develop a revised development strategy that considers the unique challenges and varying levels of socio-economic development present in different types of deltas within the context of global change. This will be achieved through collaborative efforts across multiple disciplines and sectors. Ultimately, the program seeks to protect the well-being and long-term stability of these critical deltaic environments.

The Mega-Delta Programme is organized around three primary themes, aligning with those of Future Earth Coasts (formerly IGBP/IHDP LOICZ, Land–Ocean Interactions in the Coastal Zone):

W. Zhang (✉)
State Key Laboratory of Estuarine and Coastal Research, East China Normal University, Shanghai 200241, China
e-mail: wgzhang@sklec.ecnu.edu.cn

H. de Vriend
Faculty of Civil Engineering and Geosciences, Delft University of Technology, Delft 2628 CN, The Netherlands

W. Zhang and H. de Vriend (eds.), *Delta Sustainability*,
https://doi.org/10.1007/978-981-97-7259-9_1

Theme 1: Dynamics of coupled human-nature systems, which aims to understand the evolution of deltaic systems across multiple timescales.
Theme 2: Forecasting the evolution of deltaic systems, which focuses on predicting how these systems will respond to global climate change and social development.
Theme 3: Solutions for sustainable development, which aims to provide pathways towards sustainability by addressing both global and local challenges.

In the initial phase (2021–2023), the programmes focus on 16 representative deltas worldwide, which reflects the diversity of physical settings and socio-economic developments (Fig. 1). The partners agree that an assessment report on deltaic issues and management challenges is necessary. This report will provide researchers and the public at large with general information on the most pressing sustainability issues in each of the selected deltas. For each delta, the chapter generally comprises five sections, namely physical setting, socioeconomic characteristics, current issues, management measures and perspective for future pathway. At the end of this report, a synthesis chapter summarizes the general facts of the deltas and efforts needed to address the sustainability issues. In November 2023, we finished the report and submitted it to Springer in May 2024. Due to a copyright issue, the chapter on Mackenzie River Delta written by Dr. Donald F. Forbes is not included in this book. We will seek publication elsewhere. We extend our appreciation to all the authors for their contribution.

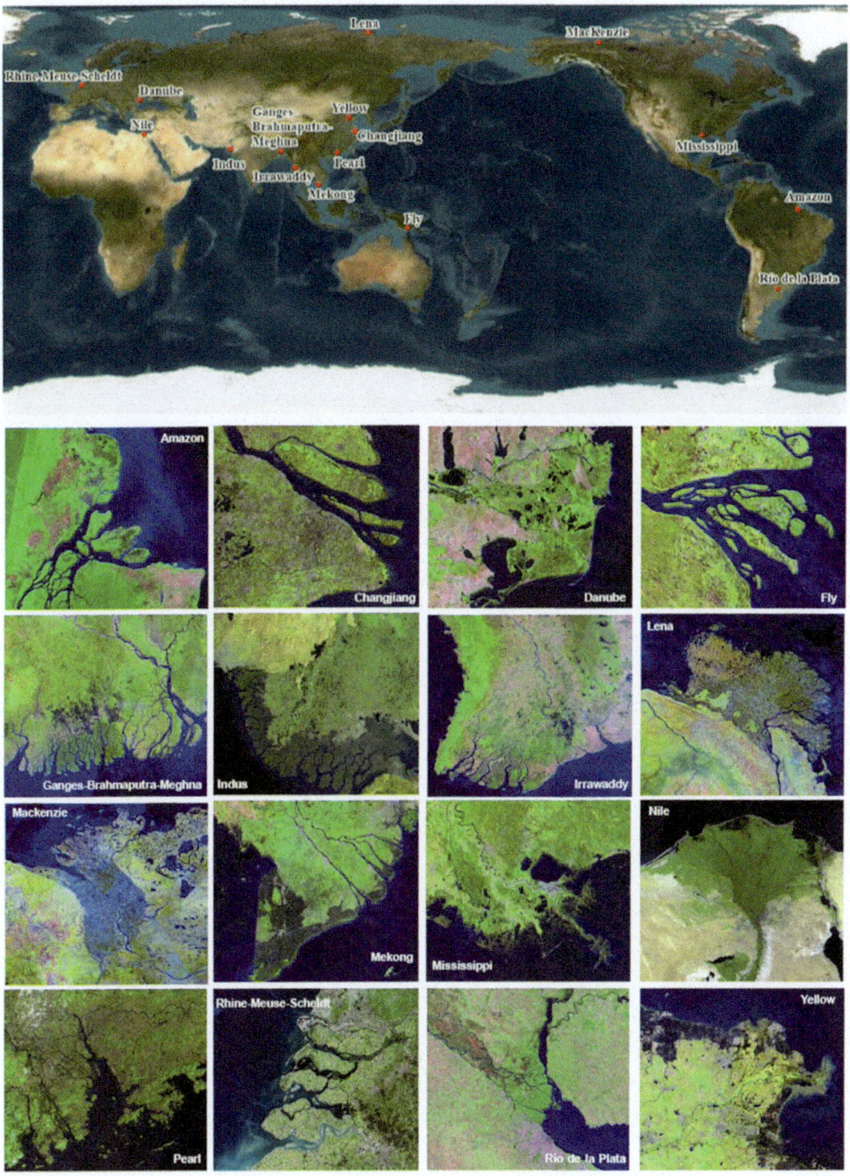

Fig. 1 Selected deltas in the UN Ocean Decade Mega-Delta Programme

Amazon River Delta

Xiaomei Nian, Marcelo Cancela Lisboa Cohen, Weiguo Zhang, Dailson José Bertassoli Júnior, Janice Muriel-Cunha, Henrique Oliveira Sawakuchi, and André Oliveira Sawakuchi

1 Physical Setting

1.1 The Amazon River

The Amazon River basin (Fig. 1), with its headwater located in the Andes Mountains in Peru, covers approximately 40% of the Brazilian territory and 75% of the Peruvian territory, as well as significant parts of Colombia, Ecuador, and Bolivia. It represents the Earth's largest drainage system (6.15×10^6 km^2) and the second longest river (~6,437 km). The combination of the wet tropical climate, widespread rainforests across the basin and the steep relief of the Andean headwaters produces a large load of dissolved and particulate inorganic and organic materials that the Amazon River transports, processes and delivers to the Atlantic Ocean (Gibbs, 1967; Ward

X. Nian (✉) · W. Zhang
State Key Laboratory of Estuarine and Coastal Research, East China Normal University, Shanghai 200241, China
e-mail: xmnian@sklec.ecnu.edu.cn

M. C. L. Cohen
Geology and Geochemistry, Federal University of Pará, Belém 66077-530, Brazil

Department of Oceanography and Coastal Sciences, Louisiana State University, Baton Rouge 70803, USA

D. J. B. Júnior · A. O. Sawakuchi
Institute of Geosciences, University of São Paulo, São Paulo 05508-080, Brazil

J. Muriel-Cunha
Graduate Program in Environmental Biology, Institute of Coastal Studies, Federal University of Pará, Bragança 68600-000, Brazil

H. O. Sawakuchi
Department of Thematic Studies-Environmental Change, Linköping University, Linköping 58183, Sweden

W. Zhang and H. de Vriend (eds.), *Delta Sustainability*,
https://doi.org/10.1007/978-981-97-7259-9_2

et al., 2013). The mean water and suspended sediment discharge at Óbidos station in central-eastern Amazonia (upstream limit of tidal influence) are 178,000 ± 65,000 $m^3\ s^{-1}$ (AD 2000–2012) and 45–80 mg L^{-1} (data compiled by Ward et al. (2015)), respectively. These values are in the range of the combined water and suspended sediment discharge measured in the north and south channels of the Amazon River mouth at Macapá (Fig. 2), respectively showing water discharges of 110,000 ± 35,000 $m^3\ s^{-1}$ and 73,000 ± 32,000 $m^3\ s^{-1}$, and suspended sediment concentrations of 51.9 ± 30.9 and 38.8 ± 18.8 mg L^{-1} (Ward et al., 2015). The suspended load of the Amazon River discharged into the Atlantic Ocean is largely composed of a mixture of silt and clay (up to 90%) (Milliman & Meade, 1983) and particulate organic components (Ward et al., 2013, 2015). The Solimões and Madeira Rivers (Fig. 1) are the main western Andean tributaries of the Amazon River, contributing around 90% of the total suspended load transported by the Amazon River (Martinelli et al., 1989).

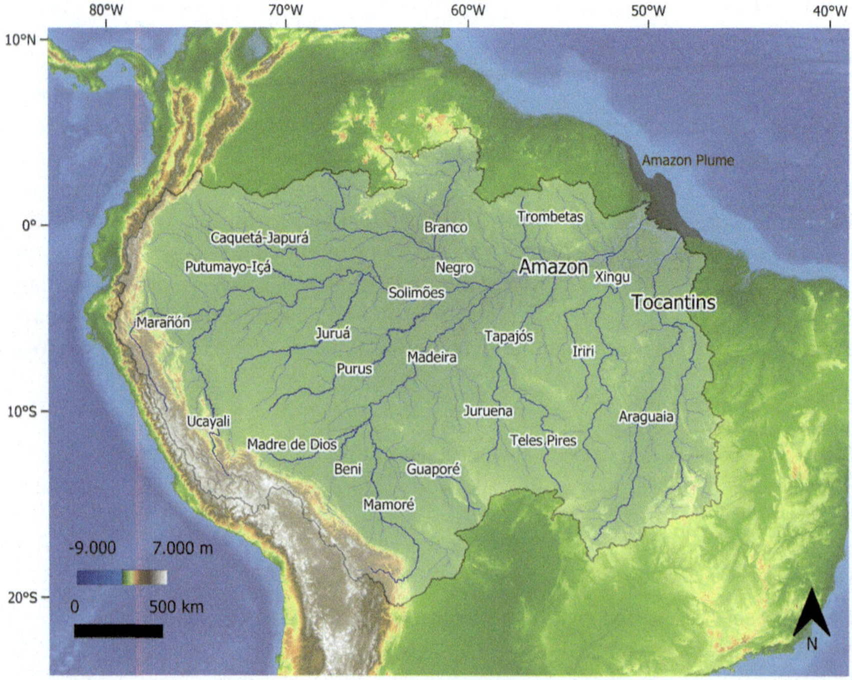

Fig. 1 Northern South America showing the Amazon River and the Tocantins River basins (whitish area) representing the major watersheds draining the Amazon biome and discharging into the Atlantic Ocean (geospatial data from Venticinque et al., 2016; NOAA National Centers for Environmental Information, 2022)

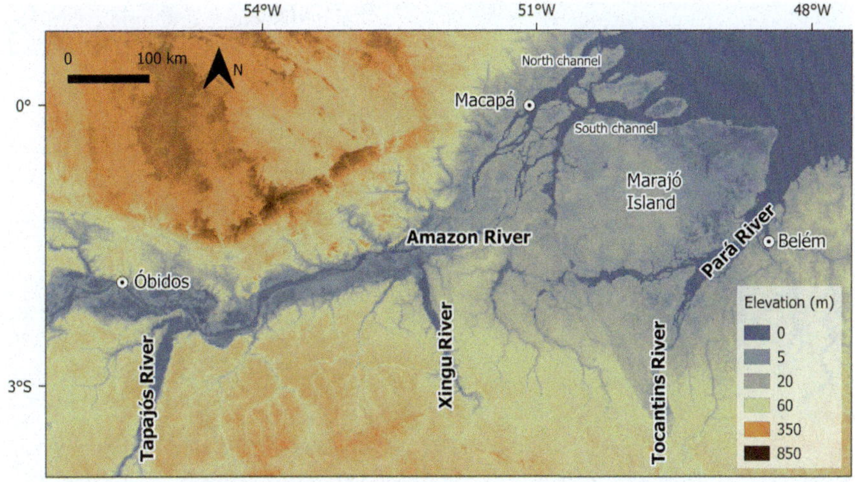

Fig. 2 Elevation map of the onshore portion of the Amazon River Delta and the downstream sector of the Amazon River showing the Óbidos locality (tidal limit) and the estuary of the Pará River bounding the east side of the delta (NOAA National Centers for Environmental Information, 2022)

1.2 The Amazon River Delta

The Amazon River Delta (ARD) is a tide-dominated setting with strong tidal current speeds of 200 cm s^{-1} nearshore. The tidal range varies from 4 to 8 m at the river mouth and progressively decreases upstream up to the Óbidos fluviometric station (Fig. 2; ANA, 2023). The delta plain of the ARD reaches a peak elevation of 11 m above sea level due to the macrotidal setting. According to definition by Edmonds et al. (2020), the delta plain area of the ARD is estimated to be 85,667 km^2. The delta has a typically hot, humid and tropical climate that is characterized by high temperatures and high levels of precipitation throughout the year. The average temperature in the region is around 25–30 °C, and the average annual rainfall is around 2,000–3,000 mm. The maximum water discharge measured at the Óbidos station occurs in May–June with over 250,000 m^3 s^{-1} and the minimum discharge of ca. 85,000 m^3 s^{-1} in November–December (Ward et al., 2015). Water level shows a ca. 8 m seasonal variation at the Óbidos station, and peak sediment discharge occurs in February–March (Armijos et al., 2020). The Amazon conducts to the ocean one-fifth of all the freshwater on the Earth's surface, forming a large river plume (Fig. 3) and diluting the ocean salinity up to 160 km offshore (Goulding et al., 2003).

Numerous distributary and tidal channels traverse the landscape in the delta plain of the ARD, with intervening islands, mangrove swamps, intertidal mudflats, sandy beaches, brackish lagoons and marshes, palm swamps and seasonally flooded grasslands and forests (Scott & Carbonell, 1986). Near the coast, the ARD is connected with the estuary of the Pará River through tidally influenced channels, supplying water and sediments to the Marajó Island (Fig. 2), a large coastal wetland

Fig. 3 Amazon River plume extending alongshore (From ISS, by ESA/NASA)

with seasonally-tidally flooded vegetation representing most of the intertidal and supratidal zones of the ARD. The Pará River estuary is connected with the Tocantins River watershed (Fig. 2), which is also draining the Amazonian biome and influencing the eastward portion of the ARD. The ARD and its watershed provide important habitats to a wide range of species, including fish, birds, and mammals (Best, 1984; Remsen & Parker, 1983; van der Sleen & Albert, 2017). The wetlands and ecosystems in the ARD are not only of ecological significance but also provide vital services to local communities, such as food resources, water purification, flood control, and climate regulation (Magris & Barreto, 2010).

The offshore progradation of sediments delivered by the Amazon River to the equatorial Atlantic Ocean forms the Amazon subaqueous delta. This feature can be divided into topset, bottomset and foreset regions and extends to water depth of ca. 60 m. It transits to an outer-shelf sand belt and to the Amazon submarine fan (Fig. 4). In addition to developing the subaqueous delta, sediments are also transported northwestward along the coast by longshore current, which forms a 1600 km-long mud bank up to the Orinoco River delta (Fig. 5; Nittrouer et al., 2021).

1.3 Mangrove

Mangroves occupy ~9,900 km^2 of the Brazilian coast, with around 30% of this area represented by the coasts of the states of Amapá and Pará encompassing the Brazilian

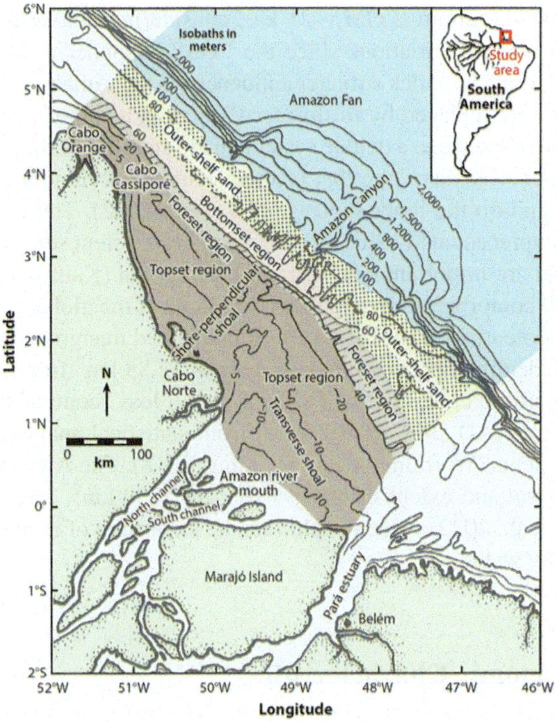

Fig. 4 Subaqueous portion of the Amazon River Delta and its deep water submarine fan (From Nittrouer et al., 2021)

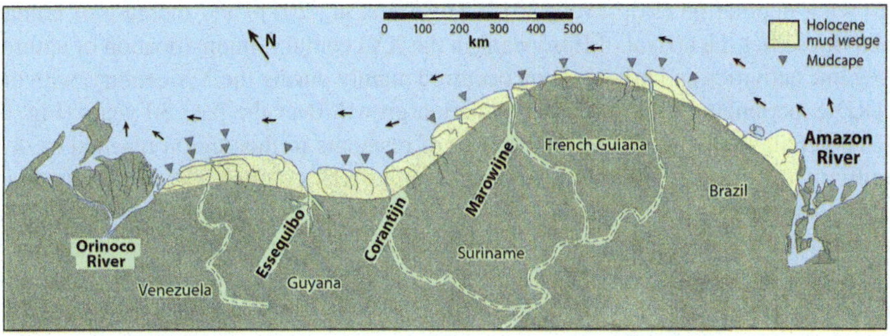

Fig. 5 Westward longshore transport of the Amazon River sediments (From Nittrouer et al., 2021)

segment of the ARD (Diniz et al., 2019). Mangroves were wider than today on the Amazon River mouth between 8800 and 2250 cal yr BP (Cohen et al., 2012), forming a continuous mangrove belt along the ARD. The establishment of this mangrove strip would be a direct consequence of the marine incursion caused by sea-level rise after the Last Glacial Maximum (LGM, ~21 ka), considering that water and sediment discharges had marked variations since the LGM (Höppner et al., 2018). In the late Holocene, in areas under stronger influence of the Amazon River discharge, the mangroves were replaced by freshwater vegetation and the coastal morphology evolved from an estuarine to a deltaic system due to stability of the relative sea-level (RSL) and higher river discharge that favored sediment accumulation at the mouth of large rivers and on the inner shelf (Cohen et al., 2012). The mangroves of the ARD have an unprecedented value as climate and ecosystem sentinels (Bernardino et al., 2022) and are important reservoirs of blue carbon (Kauffmann et al., 2018). The ARD could comprise an even higher percentage of the global mangrove extent, as there is still potential for the discovery of unmapped mangroves in the area. The mangrove loss identified by satellite data is about 12.85 km^2 from 2010 to 2016 in the ARD (Bernardino et al., 2022). This is a "small loss" compared to the net loss of over 6,000 km^2 (AD 1996–2016) found from statistical analysis of global data on mangrove extents (Worthington and Spalding, 2018). The ARD has exhibited an increased tidal wetland extent from 1999 to 2019 (730 km^2, ratio of loss to gain 0.98) (Murray et al., 2022), which could fuel the expansion of mangroves under the influence of salty water.

2 Socio-economic Characteristics

2.1 *Population*

Human occupation of the ARD began at least around 3,000 years ago, followed by growing population (Roosevelt, 2013; Silveira et al., 2011) and increase of human activities after the arrival of Europeans in the XVI century. Intensification of anthropogenic activities and urbanization occurred mainly during the XX century, with the ARD experiencing rapid human population growth over the past 30 years (Fig. 6; Szabo et al., 2016). In 2020, the number of residents in this region reached ca. 4.3 million, marking a remarkable growth of around 70% compared to the figures from 1990. The majority of the population (ca. 80%) is concentrated in urban areas along the region's coast and major rivers (IBGE, 2020). The rural population of the ARD decreased slightly from 1990 to 2000, with a minor increase from 2000 to 2020, reflecting growing demand in forests- and agroforestry-related goods (Szabo et al., 2016). The population density in the ARD is generally lower than in other deltas around the world (Syvitski et al., 2009), due to the distance from major migratory routes from southern Brazil, and the vast size of the Amazon area and its wetlands. Human occupation and deforestation therefore focused on the southern edge of the Amazon River basin.

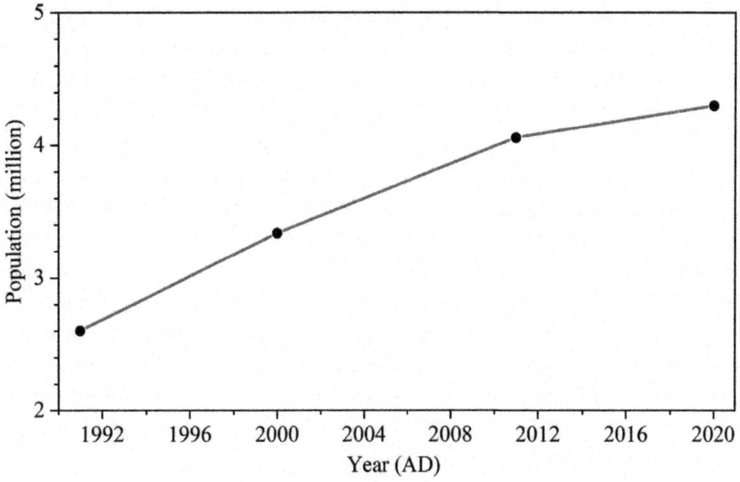

Fig. 6 The population growth in the Amazon River Delta (modified after Szabo et al., 2016). The data were collected from Instituto Brasileiro de Geografia e Estatística (IBGE, 2020)

2.2 *Major Industries*

Fishing is a significant economic activity in the ARD, where communities rely on both subsistence and commercial fishing, targeting a variety of fish species found in the Amazon River and its tributaries (Almeida et al., 2001). Additionally, there are initiatives to develop sustainable aquaculture practices, including fish, shrimp and crab farming, to support the growing demand for seafood and increase economic development (Tenório et al., 2015).

Agriculture is also an important industry in the ARD (Fearnside, 1985), and rice and jute cultivation are primary agricultural activities in the delta (Kato et al., 1999; WinklerPrins, 2006). Meanwhile, more small-scale farming takes place with the emergence of new collaborative arrangements in the region (Futemma et al., 2020). Livestock rearing, particularly cattle ranching, also occurs in certain regions (Veiga et al., 2002; da Silva et al., 2021).

Logging in the ARD began in the context of the last several centuries and has gradually become a major industry in the region since then (Johns et al., 1996; Ometto et al., 2011). Large-scale commercial logging operations, although subject to regulations, have contributed to deforestation and habitat destruction in some areas (Skole & Tucker, 1993).

Mining activities in the ARD region include bauxite and kaolinite, but inland areas hydrologically connected with the ARD host large deposits of iron, copper, manganese, nickel and gold, which makes the Pará State responsible for around 45% of the Brazilian ore metal production (ANM, 2023). There is no oil and natural gas production in the ARD region (ANP, 2022). Its offshore area, however, including the subaqueous delta, hosts main oil and gas exploration prospects. Recent discoveries of deep-water giant oil and gas fields extend westward in the offshore areas of French Guiana, Suriname and Guyana, as well as in the mirrored continental margin of west Africa (Cruz et al., 2021). The oil and gas exploration on the offshore of the ARD is considered a major environmental threat, with an ongoing debate about its benefits and drawbacks (Araujo et al., 2021; Marques et al., 2023). The ARD is a major transportation route for goods and people in the region, with barges and boats being the most common form of transportation (Vogt et al., 2016). The Amazon River also serves as an important transportation route for the oil and gas produced at Urucu field in central-western Amazonia (ANP, 2022) and for soybean produced in the deforestation edge of southern Amazon and in the central Brazil savanna (Cerrado).

The unique socioecological features and biodiversity of the ARD attract tourists and nature enthusiasts (Pereira et al., 2018). Ecotourism is an emerging industry, with visitors exploring the delta's natural habitats, engaging in wildlife viewing, birdwatching, boat tours, and cultural experiences with indigenous and river dwelling communities (Ohl-Schacherer et al., 2008).

3 Current Environmental and Climate Change Issues

Although the ARD is better preserved than most other large river deltas (Besset et al., 2019; Syvitski et al., 2009), it is still facing many challenges with the growth of the human population (IBGE, 2020), enhanced human activities (Giam, 2017), and climate change (Giosan et al., 2014). Environmental changes affecting the ARD strongly rely on human activities occurring in the Amazon River and its tributaries, including channel impoundment for hydropower, deforestation, mining and water pollution. Climate change threats include vulnerabilities to sea level rise and intensification of extreme climate events leading to flooding and droughts. These issues are discussed in the following subsections.

3.1 Damming and Deforestation

Anthropogenic activities such as channel impoundment for hydropower (Fig. 7) and deforestation in the Amazon River´s tributaries can also shift the deposition-erosion balance in downstream reaches of large rivers through changes in discharge regime and sediment supply, with complex effects propagating to the ARD. The Santo Antônio and Jirau hydroelectric dams caused a 20% decrease for the mean annual surface suspended sediment concentration in the Madeira River (Fig. 8; Latrubesse et al., 2017). This tributary delivers a suspended load of 280×10^6 t yr^{-1} to the Amazon River, representing around 45% of the Amazon river suspended load at Óbidos (Filizola & Guyot, 2009). In addition, some hydroelectric plants, such as Belo Monte, the world's 4th largest hydroelectric plant in terms of installed capacity (11,233.1 MW), have modified the seasonal flood pulse, the carbon budget (Bertassoli

Jr. et al., 2021) and the connectivity of seasonally flooded environments (Kalacska et al., 2020) of the lower Xingu River, the fourth-largest tributary of the Amazon basin. Thus, hydropower expansion across the Amazon basin can lead to abrupt and steep declines in inland river connectivity and sediment transport to the ARD (Flecker et al., 2022).

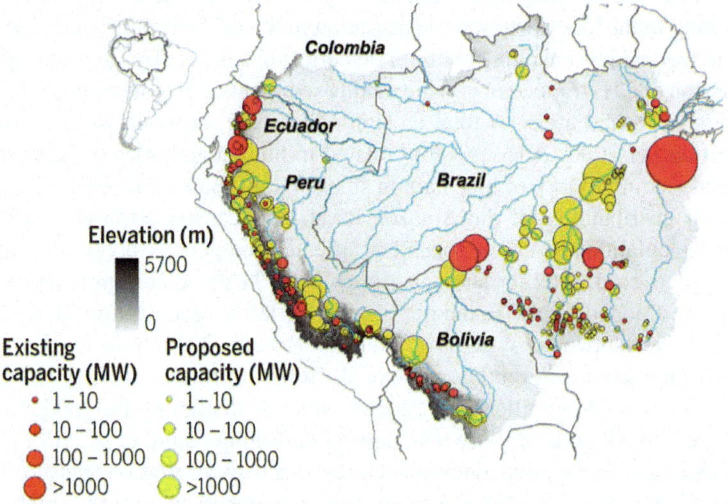

Fig. 7 Spatial distribution of 158 existing hydropower dams in the Amazon basin and 351 additional proposed dams (From Flecker et al., 2022)

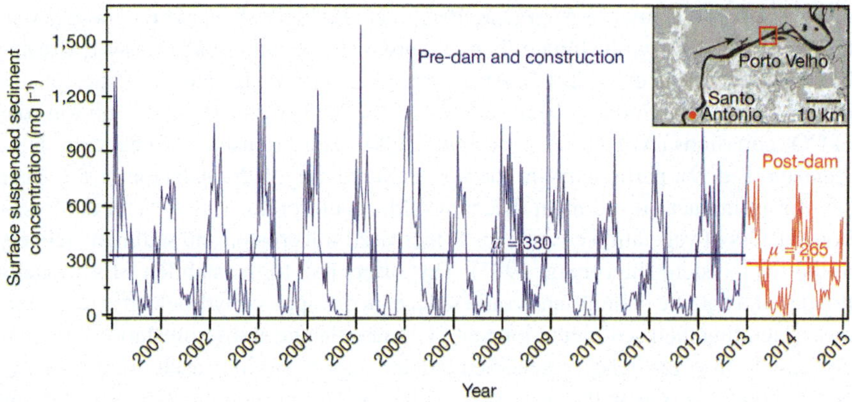

Fig. 8 Changes in surface suspended sediment concentration in the Madeira River downstream of the Santo Antônio dam (8°48′06″ S, 63°57′03″ W) for pre-and post-dam construction periods. Horizontal coloured lines indicate mean surface suspended sediment concentrations for each period: pre-dam construction (2001–2013), and post-dam construction (2014–2015) (From Latrubesse et al., 2017)

Although hydropower is a potentially clean source of renewable energy, some Amazon lowland dams may emit more greenhouse gases per unit energy than coal-fired power plants (Almeida et al., 2019; Kemenes et al., 2011), especially due to the increase in the emission of methane (CH_4) from reservoirs (Bertassoli Jr. et al., 2021). In 1990, hydroelectric dams in the Brazilian Amazon emitted about 0.26 million tons of CH_4 and 38 million tons of CO_2 (Fearnside, 1995). Damming is not only a major contributor to greenhouse gases emission in the Amazon River Basin, it is also contributing to the loss of carbon sequestration in the ARD through the reduction of sediment load and its ensuing erosion (Forsberg et al., 2017). This potentially favors the emission of CH_4 trapped in organic muddy sediments and reduces the blue carbon reservoirs in coastal areas. A total erosion of 130 km^2 occurred on the east coast of Brazil during 2000–2016, possibly related to human activities (Goldberg et al., 2020) and leading to loss of blue carbon stocks (Kauffman et al., 2018). Moreover, dams built in tributaries of the Amazon River have caused displacement of river dwelling populations to urban zones and affected indigenous peoples (Fearnside, 2015; Pezzuti et al., 2022). They also impact the health of the population living close to the flooded areas (Franco et al., 2018), led to degradation of freshwater and seasonally flooded ecosystems (Castello & Macedo, 2016; Resende et al., 2019) and stimulated deforestation (Laurance et al., 2002). Furthermore, they blocked the migration of fish (Winemiller et al., 2016), which is disrupting the food chain and affecting the livelihoods of those who depend on fishing (Baird et al., 2021),

The rapid loss of the Amazon tropical forest due to deforestation leads to changes in water discharge and sediment balance within watersheds, posing a high risk of irreversible changes to biodiversity and ecosystems relying on seasonally flooded and aquatic habitats. This is exacerbated by climate change (Giam, 2017; Nobre et al., 2016). The Amazon rainforest is being rapidly cleared for agriculture, mining, and other human production activities, which has led to the destruction of habitats and a decline in biodiversity (de Area Leão Pereira et al., 2020). Meanwhile, these activities, in combination with climate change, cause shifts in rainfall patterns and threaten biodiversity (Röpke et al., 2017). An estimated 2.6 million hectares of undesignated public forests have already been deforested by 2018, resulting in 1.2 billion tons of CO_2 emissions (Azevedo-Ramos et al., 2020). The extensive deforestation in the Amazon (Fig. 9), particularly in the upstream regions of the Amazon basin, has a significant impact on sediment discharge (Martinelli et al., 1989). With deforestation, the loss of vegetation cover leads to increased soil erosion and sediment delivery to the river system (Sternberg, 1987). This offers risk to the stability of seasonally flooded and aquatic habitats and could reduce the lifespan of hydroelectric reservoirs by accelerating their sedimentation rates. Climate change and deforestation have also decreased forest resilience, intensified the dry season and increased fire occurrence and carbon emissions in the Amazon over the last decades (Gatti et al., 2021; Zemp et al., 2017).

Suspended sediment discharge of the Amazon River at Óbidos increased by about 20% from 1995 to 2007 (Martinez et al., 2009), and the north shelf mud bank of the Amazon at Cabo Norte was found to be accreting at a rate of 2 km yr^{-1} (Gensac et al., 2016). However, the increasing trend reversed, and sediment loading decreased

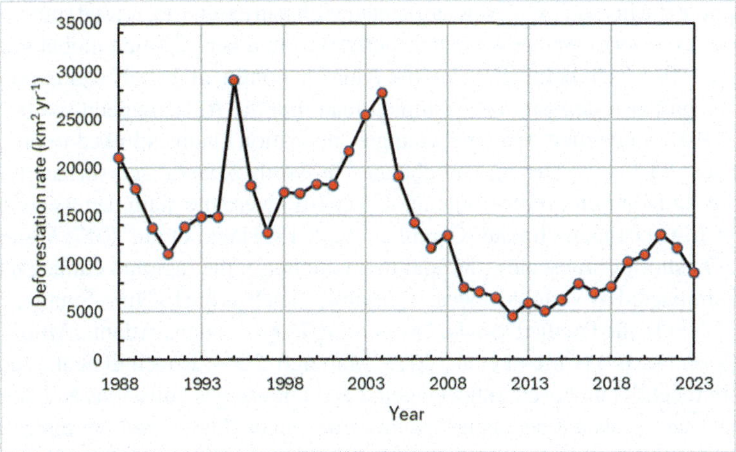

Fig. 9 Official deforestation rates for the legal Brazilian Amazon in the period 1988–2023. Data from PRODES, the deforestation monitoring program carried out by the Brazilian Institute for Space Research (INPE). The Brazilian government's Action Plan for the Prevention and Control of Deforestation in the Legal Amazon (PPCDAm) was implemented in 2004 and interrupted in 2014 (Silva Junior et al., 2021)

during 2007–2018 (Li et al., 2020), despite a steady trend of water discharge over the whole period. The study by Montanher et al. (2018), based on satellite imagery and in situ data, also showed that suspended sediment transport at the Óbidos station intermittently increased and decreased without a systematic unidirectional bias. The increasing trend reported in previous studies may be caused by the internal variability from seasons to decades (Montanher et al., 2018). Thus, it seems plausible that sediment discharge at the Óbidos station has not shown any trends associated with deforestation and river damming yet.

3.2 Climate Trends in Amazonia

The rainfall determining the water discharge of the Amazon River and the outflow of the ARD is driven by the South American Monsoon system, which varies at a range of timescales. Paleoclimate proxy records have revealed insight into the links between long-term climate variability in Amazonia and forcings such as insolation, tropical Atlantic sea surface temperatures and regional atmospheric circulation (Cheng et al., 2013; Vuille et al., 2012; Wang et al., 2017). Glacial-interglacial cycles (ca. 40–100 kyr cyclicity) have profoundly changed the Amazonian climate during the Quaternary period (Baker et al., 2020). For instance, during the LGM, climatic and environmental conditions in the ARD markedly diverged from those observed today. Despite the marked drop in temperatures, increased summer insolation in the southern hemisphere during the LGM was associated with increased (decreased) precipitation in western (eastern) Amazon. Concurrently, during a period when global sea levels were ca. 120–130 m below present, the Amazon continental shelf was exposed and the Amazon River deposited its sediment load into the Amazon deep-sea fan (Nace et al., 2014). Since then, sea level changes, insolation changes linked to precession cycles (ca. 23 k yr cyclicity) and changes in North Atlantic sea surface temperatures during Heinrich events (millennial cyclicity) were the main factors controlling long-term Amazonian climate (Cheng et al., 2013; Häggi et al., 2017; Wang et al., 2017). On shorter timescales (decadal to interannual), the regional climate has been notably impacted by various modes of variability such as the El Niño-Southern Oscillation (ENSO), the Pacific Decadal Oscillation (PDO), and the Atlantic Multidecadal Oscillation (AMO) (Chiessi et al., 2009; Marengo, 2004; Ronchail et al., 2002).

More recently, however, anthropogenic greenhouse gas emissions and deforestation have led to substantial changes in the Amazonian climate. Studies have revealed an increase in the mean air temperature in the Amazon region of 0.6 °C from 1973 to 2013, with 2016 and 1998 being the warmest years (Almeida et al., 2017; Jiménez-Muñoz et al., 2016; Marengo et al., 2018). Although precipitation records show no consistent long-term trend across the entire Amazon Region, significant regional variations have been identified in previous works. The northern and northwestern parts of Amazonia exhibited an increase in precipitation during the last decades (Gloor et al., 2013). Conversely, a drying trend has been observed in eastern Amazonia, coinciding with widespread deforestation (Espinoza et al., 2019). For instance, the watershed of the Tocantins River, an important source of freshwater input to the ARD region, has had a significant increase in the number of dry days during both wet and dry seasons (Haghtalab et al., 2020).

The frequency and intensity of extreme droughts and floods during strong El Niño and La Niña events in multiple parts of Amazonia have also increased over the past 100 years (Marengo & Espinoza, 2016). Climate change caused by anthropogenic emissions of greenhouse gases and the natural variability in climate-ocean interactions are important components influencing the hydrological cycle in the Amazon basin (Machado et al., 2018; Soares & Marengo, 2009). On the other hand, widespread deforestation and fire events cause forest fragmentation, increasing the vulnerability of the Amazon forest (Brando et al., 2014; Aragão et al., 2018). The forest is critical in supporting rainfall and sediment fluxes in the Amazon River and its tributaries (Armijos et al., 2020), and in recycling and transporting moisture from Amazonia to Southeastern South America (Arraut et al., 2012).

The combination of climate change with deforestation poses a major threat to the Amazonian ecosystem stability (Spracklen & Garcia-Carreras, 2015; Aragão et al., 2018), including the widespread wetlands in the ARD. Modeling efforts have shown that if deforestation extends beyond a ca. 40% threshold, a decrease in regional evapotranspiration would significantly reduce precipitation in Amazonia and promote a severe dieback throughout the entire biome (Boers et al., 2017; Nobre et al., 2016). In fact, deforestation and severe drought events have already affected the resilience of the Amazon Forest, posing a severe threat to climate stability in the area (Boulton et al., 2022).

3.3 *Relative Sea-Level Rise*

Sea-level variations have significantly affected the distribution and extent of coastal environments in the Amazon region during the Holocene (Cohen et al., 2021). The post-glacial sea-level rise favored landward transgressive expansion of coastal wetlands and stabilized near its current level between ~8000 and ~5000 cal yr BP (Behling et al., 2001; Cohen et al., 2005, 2021). Regarding the Anthropocene (Zalasiewicz et al., 2018), the RSL in Belém, near the mouth of the Amazon River (Fig. 4), will rise at a rate of 3 mm yr^{-1} by the end of the 21st century (Larour et al., 2017), which is in the range of rates obtained globally from ~2.5 mm yr^{-1} in the 1990s to ~3.4 mm yr^{-1} nowadays (Nerem et al., 2018). The RSL rise has caused coastal erosion and landward sand migration, which accumulates on mudflats and causes mangrove degradation along the lower tidal flats (Cohen & Lara, 2003). In contrast, the RSL rise has caused a landward displacement of boundaries of high pore-water salinity areas (Cohen & Lara, 2003; Lara & Cohen, 2006; Cohen et al., 2018), expanding the zones where coastal wetlands can grow to higher surfaces (Cohen et al., 2018; Cardenas et al., 2022). The interaction between RSL rise and fluvial discharge has affected the estuarine salinity gradients, the sediment supply and the coastal morphology, thus influencing the coastal vegetation distribution (Cohen & Lara, 2003; Cohen et al., 2009; Lara & Cohen, 2006). The current sea-level trends in the ARD are rather scanty (due to the lack of tidal gauge records), compounded by large observation gaps and high measurement noise due to monthly averaging (Anthony et al., 2021).

3.4 *Floods and Droughts*

A conceptual model by Mansur et al. (2016) showed that more than 1.2 million people representing 41% of the total urban population are exposed to risk of flooding during extreme water level events in the ARD. Some communities in the delta have adapted to the regular seasonal flooding by building houses on stilts or using floating structures (Almudi & Sinclair, 2022a). There is a marked increase in severe floods, due to the strengthened Walker circulation, as revealed by the analysis of water levels from the Amazon River at Óbidos from 1970 to 2015, as well as precipitation and atmospheric data (Barichivich et al., 2018). The frequency and intensity of extreme climatic events (especially extreme floods) have new and significantly more severe impacts on local communities (Almudi & Sinclair, 2022b). Deforestation in the Amazon basin can also influence the frequency and intensity of flooding events (Anderson, 1990), such as in 1989, 1999, 2009, 2012, 2017 and 2021 (Espinoza et al., 2022), affecting the human population, flora and fauna (Brando et al., 2014; Davidson et al., 2012; Doughty et al., 2015). Usually, floods occur during La Niña years, while droughts (e.g., 1987, 1998, 2010, 2016, 2010 and 2015/16) occur during El Niño years (Marengo et al., 2013). However, the drought in 2005 happened in a neutral year. The warmer Tropical North

Atlantic Ocean may also have contributed to the dry years (2005, 2010 and 2016) (Shukla et al., 1990). An intense heat wave during a drought in 1926 caused forest fires for several months and killed thousands of rubber gatherers (Meggers, 1994; Williams et al., 2005). The 2005 and 2010 droughts killed large trees, increasing the number of forest fires (Marengo & Camargo, 2008; Marengo et al., 2011; dos Santos et al., 2017). The 2015–2016 drought, which was more intense and extensive than the 2005 and 2010 drought, impacted up to 13% of the rainforests (Aragão et al., 2018; Erfanian et al., 2017). In addition to the direct consequences of these climatic extremes for people, fauna and flora, the intensification of hydrological regime contrasts in Amazonia may be affecting the freshwater discharge into the Atlantic Ocean, where the salinity along the water export pathway of the Amazon River plume decreased by ~3.5% between 2002 and 2016. This process may be part of the ocean and atmosphere interaction influencing the climate (Gouveia et al., 2019). In general, extreme historical droughts have been linked to intense ENSO events combined with sea surface temperatures (SST) anomalies in the tropical Atlantic (Jimenez-Muños et al., 2016).

3.5 Fishing and Pollution

Estuaries are among the most biologically productive waters in the world's ecosystems. The ARD has one of the most extensive mangrove ecosystems on the planet, largely dominated by red mangrove (*Rhizophora mangle*), black mangrove (*Avicennia germinans*) and white mangrove (*Laguncularia racemosa*). The mangrove ecosystem has links with salt marshes, coastal restinga vegetation and the Great Amazon Reef System offshore of the ARD (Mahiques et al., 2019). In these environments, gradients of freshwater and saltwater habitats occur from the coast of Amapá, Pará and Maranhão, forming a mosaic of unique environmental conditions. These are critical to the life cycles of several species that are important for feeding communities, dimensioning the artisanal and high-impact fishing activities, as well as the intense mobility of fishing fleets between sectors. Some species that stand out in terms of socio-biodiversity in the ARD are fish (e.g., *Cynoscion acoupa*), oysters (e.g., *Crassostrea gasar*), estuarine mussels (*Mytella falcata*), crabs (*Ucides cordatus*) and shrimps (Penaeidae), among other aquatic animals, and other organisms that they feed (Marceniuk et al., 2019, 2023; Santos et al., 2010; Sampaio et al., 2020).

The Brazilian North coast around the ARD is one of the world's most important shrimp fishery grounds, with a total area of approximately 223,000 km^2 (Marceniuk et al., 2019). An inventory of the teleost fishes captured by the industrial outrigger trawling operations off the North coast of Brazil recorded 201 fish species belonging to 64 families and 20 orders, and revealed a unique fauna, characterized by 17 endemic species, and a mixture of estuarine-dependent and marine species, mainly associated with coral reefs.

The catfish piramutaba (*Brachyplatystoma vaillantii*) is the most abundant fishery resource in the ARD, therefore the information on the bycatch of piramutaba fishery in the Amazon estuary is important to support regional fisheries policies and the management of protected areas. According to Marceniuk et al. (2023), the conservation status of the bycatch species is classified as vulnerable (VU), and of seven as critically endangered (CR) in the region of the Amazon River estuary (Marceniuk et al., 2023). Additionally, using landings data from a portion of catches in the Amazon, it was found a consistent decline was found in large-bodied fish species at various trophic levels (Heilpern et al., 2022). These authors further suggested that declining evenness, in combination with a declining average body size of fish species in the Amazon, may signal that fisheries are losing resilience and possibly progress towards collapse (ICMBio, 2018). The observed decline in the populations of specific fish species within the region of the Amazon River estuary, especially of the Sciaenidae and Ariidae families, has also been noticed by fishermen and has affected their way of life (Jimenez et al., 2021). This issue is further compounded by the international trade of Amazonian fish byproducts, such as the swim bladder (maw).

Industrial, mining and agricultural activities in the Amazon River basin are causing water pollution, which is affecting the health of both humans and wildlife in the ARD (Castello et al., 2013; Guimarães et al., 2022). Moderate to severe sewage contamination in sediments of streams within the Amazon River watershed near Manaus (upstream Óbidos) reveals the need for immediate implementation of effective sewage treatment strategies (de Melo et al., 2019). Concentrations of most metals in waters and sediment of Amazon rivers cannot be clearly associated with human activities by lack of regional systematic studies to appraise background concentrations before intensification of human activities (Ribeiro et al., 2017). However, human activities are possibly responsible for increased concentration of metals at the local scale, like in the Pará River (Dos Santos et al., 2023). In addition, ~182,000 metric tons of plastic are being discarded into Brazil's Amazonian environment each year (Giarrizzo et al., 2019). This plastic degrades into microplastics that can be incorporated into the soil and/or carried back into the water. Studies in the Xingu River and the Amazon River estuary revealed microplastic particles in the digestive tracts of fish, including species commonly consumed by humans (Andrade et al., 2019; Pegado et al., 2018). However, the impact of microplastics pollution on fishery resources for the densely populated coastal regions of the ARD (Almeida et al., 2003; Santos et al., 2023) is still poorly constrained.

Shifting from fishing in wild environments to fish farming can also increase disease transmission. Favorable conditions for mosquito breeding in fish farm ponds combined with concentration of human populations have promoted the dissemination of malaria (Reis et al., 2015). Fertilizer use and deforestation fueling agriculture across the Amazon edge could also contribute to the increase in nutrients delivery to the ARD and ocean, causing enhanced primary production (Aquino et al., 2022) in regions far away from the river mouth through ocean circulation. This is possibly associated with recent pelagic Sargassum blooms in the tropical North Atlantic Ocean (Lapointe et al., 2021). However, there is no consensus so far about the link between human derived nutrients and primary production in the ARD and the ocean (Johns et al., 2020; Jouanno et al., 2021; Marsh et al., 2023).

4 Current Measures to Solve the Problem

The environmental issues highlighted above are driven by climate change combined with regional anthropogenic activities, mainly represented by deforestation induced by land use change, illegal gold mining and hydropower expansion in the Amazon River watershed. Global efforts are needed to solve the climate crisis, but regional threats and adaptative measures at the national level will likely be needed to mitigate the direct impacts of climate change to the population and ecosystems of the ARD. Overall deforestation rates in the Amazon basin have cascade effects downstream, including the ARD. Data from the PRODES (Monitoramento do Desmatamento na Amazônia Legal por Satélite) program, operated by the Brazilian Institute for Space Research (INPE, 2015), showed a marked decline in deforestation in the Brazilian Amazon, where it was reported to have fallen by ~84%, from a record 27,772 km^2 in 2004 (the historical peak) to 4,571 km^2 in 2012 (INPE, 2015). The declining deforestation trends mainly result from government policy, particularly the *Action Plan for the Prevention and Control of Deforestation in the Legal Amazon* (PPCDAm) launched in 2004. This encompasses, e.g., improved satellite-based monitoring, a "blacklist" of municipalities, expansion of inland and coastal protected areas, land tenure regularization, credit access restrictions, market restrictions and market demand regulation (the "Soy Moratorium", Rudorff et al., 2011). Furthermore, deforestation declined as a result of economic factors of the reported period (2004–2012) such as commodity prices and the currency exchange rates that affect the profitability of agricultural exports (West & Fearnside, 2021) and public policy and interventions (Nepstad et al., 2014). Yet, these successes are being compromised by development pressures and shifts in legislation (Ferreira et al., 2014) and weakening of the PPCDAm policy from 2016, leading to another trend of increasing deforestation rate until 2021. Measures to inhibit illegal gold mining were also weakened in the period of 2016–2022 and robust diagnostics about the downstream impact of hydropower plants are still lacking as a national policy.

5 Future Challenges

Urbanization and a forest-based economy in the ARD area have existed side-by-side as a product of economic and demographic forces since the mid-1970s, and the coupling process of urbanization and forest revitalization, including mangrove areas, in this region is a prominent feature of the region's transformation in recent decades (Hecht et al., 2013). An assessment of urban vulnerability indicates that 60–90% of the urban population lives in conditions of moderate to high degree of vulnerability, caused by inadequate planning coupled with rapid urbanization (Mansur et al., 2016).

Most municipalities of the ARD have various resource economies, such as agroforestry, fishing, mining, agriculture, and ranching, but lack aggregates industries (Mansur et al., 2016). The overfishing of certain species of fish in the Amazon River and its tributaries is disrupting the delicate balance of the delta's ecosystem (Tregidgo et al., 2021). In the near future, crop-livestock-forest integration systems, including silvopastoral ones, which involve animal production integrated with silviculture, are expected to gain more and more relevance, as they are based on the productive incorporation of areas altered by inappropriate previous use (da Silva et al., 2021).

Hydropower dams are considered for fulfilling the energy needs, accounting for up to 71% of global renewable energy in 2016 (Moran et al., 2018). However, economic benefits are often overestimated, and long-term ripple effects on ecosystem structure and diversity have been hitherto underestimated. Yet, these potentially have far-reaching impacts on biodiversity, especially in parts of fisheries, agriculture, people and property (Grill et al., 2015; Winemiller et al., 2016; Moran et al., 2018). In addition to severe local socioecological losses, long-distance downstream effects of dams built upstream in Amazon rivers tributaries will likely influence the ARD by altering the transport of sediment and nutrients (Almeida et al., 2015; Timpe & Kaplan, 2017). These issues have questioned the sustainability of hydropower expansion across the Amazon, since the local and regional impacts surpass the presumed benefits of carbon emission reduction by far, especially when considering the potential high CH_4 emissions from tropical reservoirs (de Faria et al., 2015).

Balancing the need for development with the need to protect the ARD natural resources is a major challenge, and will require transparent governance, careful planning and management, and accurate predicting to ensure that the delta and its inland watershed remains a healthy and sustainable ecosystem.

References

Agência Nacional de Mineração (ANM). (2023). Anuário Mineral Brasileiro: Principais substâncias metálicas. Coordenação técnica de Karina Andrade Medeiros. ANM.

Agência Nacional do Petróleo, Gás Natural e Biocombustíveis (ANP). (2022). Anuário estatístico brasileiro do petróleo, gás natural e biocombustíveis: 2022. ANP.

Aquino, R., Noriega, C., Mascarenhas, A., Costa, M., Monteiro, S., Santana, L., Silva, I., Prestes, Y., Araujo, M., & Rollnic, M. (2022). Possible Amazonian contribution to Sargassum enhancement on the Amazon continental shelf. *Science of the Total Environment, 853*, 158432.

Almeida, C. T., Oliveira-Júnior, J. F., Delgado, R. C., Cubo, P., & Ramos, M. C. (2017). Spatiotemporal rainfall and temperature trends throughout the Brazilian Legal Amazon, 1973–2013. *International Journal of Climatology, 37*, 2013–2026.

Almeida, O. T., Lorenzen, K., & Mcgrath, D. G. (2003). Commercial fishing in the Brazilian Amazon: Regional differentiation in fleet characteristics and efficiency. *Fisheries Management and Ecology, 10*, 109–115.

Almeida, O. T., McGrath, D. G., & Ruffino, M. L. (2001). The commercial fisheries of the lower Amazon: An economic analysis. *Fisheries Management and Ecology, 8*, 253–269.

Almeida, R. M., Shi, Q., Gomes-Selman, J. M., Wu, X., Xue, Y., Angarita, H., Barros, N., Forsberg, B. R., García-Villacorta, R., Hamilton, S. K., Melack, J. M., Montoya, M., Perez, G., Sethi, S. A., Gomes, C. P., & Flecker, A. S. (2019). Reducing greenhouse gas emissions of Amazon hydropower with strategic dam planning. *Nature Communications, 10*, 4281.

Almeida, R. M., Tranvik, L., Huszar, V. L. M., Sobek, S., Mendonça, R., Barros, N., Boemer, G., Durval Arantes Jr., J., & Roland, F. (2015). Phosphorus transport by the largest Amazon tributary (Madeira River, Brazil) and its sensitivity to precipitation and damming. *Inland Waters, 5*, 275–282.

Almudi, T., & Sinclair, A. J. (2022a). Community-level responses to extreme flooding: The case of Brazilian Amazon *ribeirinhos*. *Climate and Development, 14*, 781–793.

Almudi, T., & Sinclair, A. J. (2022b). Extreme hydroclimatic events in rural communities of the Brazilian Amazon: Local perceptions of change, impacts, and adaptation. *Regional Environmental Change, 22*, 27.

Agência Nacional de Águas (ANA). (2023). Hidroweb: Sistemas de Informações Hidrológicas. Retrieved September 28, 2023, from http://www.snirh.gov.br/hidroweb/

Anderson, A. B. (1990). Alternatives to deforestation: Steps toward sustainable use of the Amazon rain forest. Columbia University Press.

Andrade, M. C., Winemiller, K. O., Barbosa, P. S., Fortunati, A., Chelazzi, D., Cincinelli, A., & Giarrizzo, T. (2019). First account of plastic pollution impacting freshwater fishes in the Amazon: Ingestion of plastic debris by piranhas and other serrasalmids with diverse feeding habits. *Environmental Pollution, 244*, 766–773.

Anthony, E. J., Brondizio, E. S., dos Santos, V. F., Gardel, A., & Besset, M. (2021). Sustainable management, conservation, and restoration of the Amazon River Delta and Amazon-influenced Guianas coast: A review. *Water, 13*, 1371.

Aragão, L. E. O. C., Anderson, L. O., Fonseca, M. G., Rosan, T. M., Vedovato, L. B., Wagner, F. H., Silva, C. V. J., Silva Junior, C. H. L., Arai, E., Aguiar, A. P., Barlow, J., Berenguer, E., Deeter, M. N., Domingues, L. G., Gatti, L., Gloor, M., Malhi, Y., Marengo, J. A., Miller, J. B., Phillips, O. L., & Saatchi, S. (2018). 21st Century drought-related fires counteract the decline of Amazon deforestation carbon emissions. *Nature Communications, 9*, 536.

Araujo, L. S., Magdalena, U. R., Louzada, T. S., Salomon, P. S., Moraes, F. C., Ferreira, B. P., Paes, E. T. C., Bastos, A. C., Pereira, R. C., Salgado, L. T., Lorini, M. L., Yager, P., & Moura, R. L. (2021). Growing industrialization and poor conservation planning challenge natural resources' management in the Amazon Shelf off Brazil. *Marine Policy, 128*, 104465.

Armijos, E., Crave, A., Espinoza, J. C., Filizola, N., Espinoza-Villar, R., Ayes, I., Fonseca, P., Fraizy, P., Gutierrez, O., Vauchel, P., Camenen, B., Martıınez, J. M., Dos Santos, A., Santini, W., Cochonneau, G., & Guyot, J. L. (2020). Rainfall control on Amazon sediment flux: Synthesis from 20 years of monitoring. *Environmental Research Communications, 2*, 051008.

Arraut, J. M., Nobre, C., Barbosa, H. M. J., Obregon, G., & Marengo, J. (2012). Aerial rivers and lakes: Looking at large-scale moisture transport and its relation to Amazonia and to subtropical rainfall in South America. *Journal of Climate, 25*, 543–556.

Azevedo-Ramos, C., Moutinho, P., Arruda, V. L. D., Stabile, M. C. C., Alencar, A., Castro, I., & Ribeiro, J. P. (2020). Lawless land in no man's land: The undesignated public forests in the Brazilian Amazon. *Land Use Policy, 99*, 104863.

Baker, P. A., Fritz, S. C., Battisti, D. S., Dick, C. W., Vargas, O. M., Asner, G. P., Martin, R. E., Wheatley, A., & Prates, I. (2020). Beyond Refugia: New insights on quaternary climate variation and the evolution of biotic diversity in tropical South America. In V. Rull & A. C. Carnaval (Eds.), *Neotropical diversification: patterns and processes* (pp. 51–70). Springer International Publishing.

Baird, I. G., Silvano, R. A. M., Parlee, B., Poesch, M., Maclean, B., Napoleon, A., Lepine, M., & Hallwass, G. (2021). The downstream impacts of hydropower dams and indigenous and local knowledge: Examples from the Peace-Athabasca, Mekong, and Amazon. *Environmental Management, 67*, 682–696.

Barichivich, J., Gloor, E., Peylin, P., Brienen, R. J. W., Schöngart, J., Espinoza, J. C., & Pattnayak, K. C. (2018). Recent intensification of Amazon flooding extremes driven by strengthened Walker circulation. *Science Advances, 4*, eaat8785.

Behling, H., Cohen, M. C. L., & Lara, R. J. (2001). Studies on Holocene mangrove ecosystem dynamics of the Bragança Peninsula in north-eastern Pará, Brazil. *Palaeogeography, Palaeoclimatology, Palaeoecology, 167*, 225–242.

Bernardino, A. F., Mazzuco, A. C. A., Souza, F. M., Santos, T. M. T., Sanders, C. J., Massone, C. G., Costa, R. F., Silva, A. E. B., Ferreira, T. O., Nóbrega, G. N., Silva, T. S. F., & Kauffman, J. B. (2022). The novel mangrove environment and composition of the Amazon Delta. *Current Biology, 32*, 3636–3640.

Bertassoli, Jr., D. J., Sawakuchi, H. O., Araújo, K .R. d., Camargo, M. G. P. d., Alem, V. A. T., Pereira, T. S., Krusche, A. V., Bastviken, D., Richey, J. E., & Sawakuchi, A. O. (2021). How green can Amazon hydropower be? Net carbon emission from the largest hydropower plant in Amazonia. *Science Advances, 7*, eabe1470.

Besset, M., Anthony, E. J., & Bouchette, F. (2019). Multi-decadal variations in delta shorelines and their relationship to river sediment supply: An assessment and review. *Earth-Science Reviews, 193*, 199–219.

Best, R. C. (1984). The aquatic mammals and reptiles of the Amazon. In H. Sioli (Ed.), *The Amazon: Limnology and landscape ecology of a mighty tropical river and its basin* (pp. 371–412). Springer.

Boers, N., Marwan, N., Barbosa, H. M. J., & Kurths, J. (2017). A deforestation-induced tipping point for the South American monsoon system. *Scientific Reports, 7*, 41489.

Boulton, C. A., Lenton, T. M., & Boers, N. (2022). Pronounced loss of Amazon rainforest resilience since the early 2000s. *Nature Climate Change, 12*, 271–278.

Brando, P. M., Balch, J. K., Nepstad, D. C., Morton, D. C., Putz, F. E., Coe, M. T., Silvério, D., Macedo, M. N., Davidson, E. A., Nóbrega, C. C., Alencar, A., & Soares-Filho, B. S. (2014). Abrupt increases in Amazonian tree mortality due to drought-fire interactions. *Proceedings of the National Academy of Sciences of the United States of America, 111*, 6347–6352.

Cardenas, S. M. M., Cohen, M. C. L., Ruiz, D. P. C., Souza, A. V., Gomez-Neita, J. S., Pessenda, L. C. R., & Culligan, N. (2022). Death and regeneration of an Amazonian Mangrove Forest by Anthropic and natural forces. *Remote Sensing, 14*, 6197.

Castello, L., & Macedo, M. N. (2016). Large-scale degradation of Amazonian freshwater ecosystems. *Global Change Biology, 22*, 990–1007.

Castello, L., McGrath, D. G., Hess, L. L., Coe, M. T., Lefebvre, P. A., Petry, P., Macedo, M. N., Renó, V. F., & Arantes, C. C. (2013). The vulnerability of Amazon freshwater ecosystems. *Conservation Letters, 6*, 217–229.

Cheng, H., Sinha, A., Cruz, F. W., Wang, X., Edwards, R. L., d'Horta, F. M., Ribas, C. C., Vuille, M., Stott, L. D., & Auler, A. S. (2013). Climate change patterns in Amazonia and biodiversity. *Nature Communications, 4*, 1411.

Chiessi, C. M., Mulitza, S., Pätzold, J., Wefer, G., & Marengo, J. A. (2009). Possible impact of the Atlantic Multidecadal Oscillation on the South American summer monsoon. *Geophysical Research Letters, 36*, L21707.

Cohen, M. C. L., & Lara, R. J. (2003). Temporal changes of mangrove vegetation boundaries in Amazonia: Application of GIS and remote sensing techniques. *Wetlands Ecology and Management, 11*, 223–231.

Cohen, M. C. L., Filho, P. W. M. S., Lara, R. J., Behling, H., & Angulo, R. J. (2005). A model of Holocene mangrove development and relative sea-level changes on the Bragança Peninsula (Northern Brazil). *Wetlands Ecology and Management, 13*, 433–443.

Cohen, M. C. L., Pessenda, L. C. R., Behling, H., de Fátima Rossetti, D., França, M. C., Guimarães, J. T. F., Friaes, Y., & Smith, C. B. (2012). Holocene palaeoenvironmental history of the Amazonian mangrove belt. *Quaternary Science Reviews, 55*, 50–58.

Cohen, M. C. L., Behling, H., Lara, R. J., Smith, C. B., Matos, H. R. S., & Vedel, V. (2009). Impact of sea-level and climatic changes on the Amazon coastal wetlands during the late Holocene. *Vegetation History and Archaeobotany, 18*, 425–439.

Cohen, M. C. L., de Souza, A. V., Rossetti, D. F., Pessenda, L. C. R., & França, M. C. (2018). Decadal-scale dynamics of an Amazonian mangrove caused by climate and sea level changes: Inferences from spatial-temporal analysis and digital elevation models. *Earth Surface Processes and Landforms, 43*, 2876–2888.

Cohen, M. C. L., Camargo, P. M. P., Pessenda, L. C. R., Lorente, F. L., De Souza, A. V., Corrêa, J. A. M., Bendassolli, J., & Dietz, M. (2021). Effects of the middle Holocene high sea-level stand and climate on Amazonian mangroves. *Journal of Quaternary Science, 36*, 1013–1027.

Cruz, C. A., Ribeiro, H. J. P. S., & Silva, E. B. (2021). Exploratory plays of the Foz do Amazonas Basin, NW portion, in deep and ultra-deep waters, Brazilian Equatorial Margin. *Journal of South American Earth Sciences, 111*, 103475.

da Silva, J. A. R., Garcia, A. R., de Almeida, A. M., Bezerra, A. S., & de Brito Lourenço Junior, J. (2021). Water buffalo production in the Brazilian Amazon Basin: A review. *Tropical Animal Health and Production, 53*, 343.

Davidson, E. A., De Araüjo, A. C., Artaxo, P., Balch, J. K., Brown, I. F., Mercedes, M. M., Coe, M. T., Defries, R. S., Keller, M., Longo, M., Munger, J. W., Schroeder, W., Soares-Filho, B. S., Souza, C. M., & Wofsy, S. C. (2012). The Amazon basin in transition. *Nature, 481*, 321–328.

de Area Leão Pereira, E. J., de Santana Ribeiro, L. C., da Silva Freitas, L. F., & de Barros Pereira, H. B. (2020). Brazilian policy and agribusiness damage the Amazon rainforest. *Land Use Policy, 92*, 104491.

de Faria, F. A. M., Jaramillo, P., Sawakuchi, H. O., Richey, J. E., & Barros, N. (2015). Estimating greenhouse gas emissions from future Amazonian hydroelectric reservoirs. *Environmental Research Letters, 10*, 124019.

de Melo, M. G., da Silva, B. A., Costa, G. D. S., da Silva Neto, J. C. A., Soares, P. K., Val, A. L., Chaar, J. D. S., Koolen, H. H. F., & Bataglion, G. A. (2019). Sewage contamination of Amazon streams crossing Manaus (Brazil) by sterol biomarkers. *Environmental Pollution, 244*, 818–826.

Diniz, C., Cortinhas, L., Nerino, G., Rodrigues, J., Sadeck, L., Adami, M., & Souza-Filho, P. W. M. (2019). Brazilian mangrove status: Three decades of satellite data analysis. *Remote Sensing, 11*, 808.

dos Santos, S. R. Q., Braga, C. C., Sansigolo, C. A., de Araujo Tiburtino Neves, T. T., & dos Santos, A. P. P. (2017). Droughts in the Amazon: Identification, characterization and dynamical mechanisms associated. *American Journal of Climate Change, 6*, 425–442.

Dos Santos, C. C. M., Rodrigues Nauar, A., Azevedo Ferreira, J., da Silva Montes, C., Rovasi Adolfo, F., Leal, G., Moraes Reis, G., Lapinsky, J., Machado de Carvalho, L., & Amado, L. L. (2023). Multiple anthropogenic influences in the Pará River (Amazonia, Brazil): A spatial-temporal ecotoxicological monitoring in abiotic and biotic compartments. *Chemosphere, 323*, 138090.

Doughty, C. E., Metcalfe, D. B., Girardin, C. A. J., Amézquita, F. F., Cabrera, D. G., Huasco, W. H., Silva-Espejo, J. E., Araujo-Murakami, A., Da Costa, M. C., Rocha, W., Feldpausch, T. R., Mendoza, A. L. M., Da Costa, A. C. L., Meir, P., Phillips, O. L., & Malhi, Y. (2015). Drought impact on forest carbon dynamics and fluxes in Amazonia. *Nature, 519*, 78–82.

Edmonds, D. A., Caldwell, R. L., Brondizio, E. S., & Siani, S. M. O. (2020). Coastal flooding will disproportionately impact people on river deltas. *Nature Communications, 11*, 4741.

Erfanian, A., Wang, G., & Fomenko, L. (2017). Unprecedented drought over tropical South America in 2016: Significantly under-predicted by tropical SST. *Scientific Reports, 7*, 5811.

Espinoza, J. C., Marengo, J. A., Schongart, J., & Jimenez, J. C. (2022). The new historical flood of 2021 in the Amazon River compared to major floods of the 21st century: Atmospheric features in the context of the intensification of floods. *Weather and Climate Extremes, 35*, 100406.

Espinoza, J. C., Ronchail, J., Marengo, J. A., & Segura, H. (2019). Contrasting North-South changes in Amazon wetday and dry-day frequency and related atmospheric features (1981–2017). *Climate Dynamics, 52*, 5413–5430.

Fearnside, P. M. (1985). Agriculture in Amazonia. In G. T. Prance, & T. E. Lovejoy (Eds.), *Key environments: Amazonia* (pp. 393–418). Pergamon Press.

Fearnside, P. M. (1995). Hydroelectric Dams in the Brazilian Amazon as sources of 'Greenhouse' Gases. *Environmental Conservation, 22*, 7–19.

Fearnside, P. M. (2015). Hidrelétricas na Amazônia: impactos ambientais e sociais na tomada de decisões sobre grandes obras, 2nd ed. Instituto Nacional de Pesquisas da Amazônia.

Ferreira, J., Aragão, L. E. O. C., Barlow, J., Barreto, P., Berenguer, E., Bustamante, M., Gardner, T. A., Lees, A. C., Lima, A., Louzada, J., Pardini, R., Parry, L., Peres, C. A., Pompeu, P. S., Tabarelli, M., & Zuanon, J. (2014). Brazil's environmental leadership at risk. *Science, 346*, 706–707.

Filizola, N., & Guyot, J. L. (2009). Suspended sediment yields in the Amazon basin: An assessment using Brazilian national data set. *Hydrological Processes, 23*, 3207–3215.

Flecker, A. S., Shi, Q., Almeida, R. M., Angarita, H., Gomes-Selman, J. M., García-Villacorta, R., Sethi, S. A., Thomas, S. A., LeRoy Poff, N., Forsberg, B. R., Heilpern, S. A., Hamilton, S. K., Abad, J. D., Anderson, E. P., Barros, N., Bernal, I. C., Bernstein, R., Cañas, C. M., Dangles, O., et al. (2022). Reducing adverse impacts of Amazon hydropower expansion. *Science, 375*, 753–760.

Forsberg, B. R., Melack, J. M., Dunne, T., Barthem, R. B., Goulding, M., Paiva, R. C. D., Sorribas, M. V., Silva, U. L., & Weisser, S. (2017). The potential impact of new Andean dams on Amazon fluvial ecosystems. *Plos One, 12*, e0182254.

Franco, V. dos S., Souza, De Barreiros, E., Lima, & Meiguins, A.M. (2018). Floods and social vulnerability: Study on the Xingu river in Altamira/PA. *Ambiente and Sociedade, 21*, e01573.

Futemma, C., De Castro, F., & Brondizio, E. S. (2020). Farmers and social innovations in rural development: Collaborative arrangements in eastern Brazilian Amazon. *Land Use Policy, 99*, 104999.

Gatti, L. V., Basso, L. S., Miller, J. B., Gloor, M., Domingues, L. G., Cassol, H. L. G., Tejada, G., Aragão, L. E. O., Nobre, C., Peters, W., Marani, L., Arai, E., Sanches, A. H., Corrêa, S. M., Anderson, L., Randow, C. V., Correia, C. S. C., Crispim, S. P., & Neves, R. A. L. (2021). Amazonia as a carbon source linked to deforestation and climate change. *Nature, 595*, 388–393.

Gensac, E., Martinez, J.-M., Vantrepotte, V., & Anthony, E. J. (2016). Seasonal and inter-annual dynamics of suspended sediment at the mouth of the Amazon river: The role of continental and oceanic forcing, and implications for coastal geomorphology and mud bank formation. *Continental Shelf Research, 118*, 49–62.

Giam, X. (2017). Global biodiversity loss from tropical deforestation. *Proceedings of the National Academy of Sciences of the United States of America, 114*, 5775–5777.

Giarrizzo, T., Andrade, M. C., Schmid, K., Winemiller, K. O., Ferreira, M., Pegado, T., Chelazzi, D., Cincinelli, A., & Fearnside, P. M. (2019). Amazonia: The new frontier for plastic pollution. *Frontiers in Ecology and the Environment, 17*, 309–310.

Gibbs, R. J. (1967). The geochemistry of the Amazon River system: Part I. The factors that control the salinity and the composition and concentration of the suspended solids. *Geological Society of America Bulletin, 78*, 1203–1232.

Gouveia, N. A., Gherardi, D. F. M., & Aragão, L. E. O. C. (2019). The role of the Amazon River plume on the intensification of the hydrological cycle. *Geophysical Research Letters, 46*, 12221–12229.

Giosan, L., Syvitski, J., Constantinescu, S., & Day, J. (2014). Protect the world's deltas. *Nature, 516*, 31–33.

Gloor, M., Brienen, R. J. W., Galbraith, D., Feldpausch, T. R., Schöngart, J., Guyot, J. L., Espinoza, J. C., Lloyd, J., & Phillips, O. L. (2013). Intensification of the Amazon hydrological cycle over the last two decades. *Geophysical Research Letters, 40*, 1729–1733.

Goldberg, L., Lagomasino, D., Thomas, N., & Fatoyinbo, T. (2020). Global declines in human-driven mangrove loss. *Global Change Biology, 26*, 5844–5855.

Goulding, M., Barthem, R., & Ferreira, E. J. G. (2003). *The Smithsonian atlas of the Amazon*. Smithsonian Institution Press.

Grill, G., Lehner, B., Lumsdon, A. E., MacDonald, G. K., Zarfl, C., & Reidy Liermann, C. (2015). An index-based framework for assessing patterns and trends in river fragmentation and flow regulation by global dams at multiple scales. *Environmental Research Letters, 10*, 015001.

Guimarães, R. H. E., Wallner-Kersanach, M., & Correa, J. A. M. (2022). Assessment of anthropogenic metals in shipyard sediment in the Amazon delta estuary in northern Brazil. *Environmental Science and Pollution Research, 29*, 77007–77025.

Häggi, C., Chiessi, C. M., Merkel, U., Mulitza, S., Prange, M., Schulz, M., & Schefuß, E. (2017). Response of the Amazon rainforest to late Pleistocene climate variability. *Earth and Planetary Science Letters, 479*, 50–59.

Haghtalab, N., Moore, N., Heerspink, B. P., & Hyndman, D. W. (2020). Evaluating spatial patterns in precipitation trends across the Amazon basin driven by land cover and global scale forcings. *Theoretical and Applied Climatology, 140*, 411–427.

Hecht, S. B., Morrison, K. D., & Padoch, C. (2013). *The social lives of forests: Past, present, and future of woodland resurgence*. University of Chicago Press.

Heilpern, S. A., Sethi, S. A., Brathem, R. B., Batista, V. S., Doria, C. R., Duponchelle, F., Vasquez, A. g., Goulding, M., Isaac, V., Naeem, S., & Flecher, A. S. (2022) Biodiversity underpins fisheries resilience to exploitation in the Amazon river basin. *Proceedings of the Royal Society B, 289*, 20220726.

Höppner, N., Lucassen, F., Chiessi, C. M., Sawakuchi, A. O., & Kasemann, S. A. (2018). Holocene provenance shift of suspended particulate matter in the Amazon River basin. *Quaternary Science Reviews, 190*, 66–80.

IBGE. (2020). Population Estimates. Retrieved June 1, 2023, from https://www.ibge.gov.br/estatisticas/sociais/populacao/9103-estimativas-de-populacao.html?=&t=o-que-e

Instituto Chico Mendes de Conservação da Biodiversidade (ICMBio). (2018). Livro Vermelho da Fauna Brasileira Ameaçada de Extinção: Volume VI - Peixes. ICMBio.

INPE. (2015). Projeto PRODES: Monitoramento da floresta Amazônica Brasileira por satélite. www.obt.inpe.br/prodes

Jimenez, E. A., Barboza, R. S. L., Garcia, J. S., Correa, E. C. S., Amaral, M. T., & Frédou, F. L. (2021). International trade of Amazon fish byproducts: Threats and opportunities for coastal livelihoods. *Ocean & Coastal Management, 212*, 105812.

Jiménez-Muñoz, J. C., Mattar, C., Barichivich, J., Santamaría-Artigas, A., Takahashi, K., Malhi, Y., Sobrino, J. A., Schrier, G. Van Der. (2016). Record-breaking warming and extreme drought in the Amazon rainforest during the course of El Niño 2015–2016. *Scientific Reports, 6*, 33130.

Johns, J. S., Barreto, P., & Uhl, C. (1996). Logging damage during planned and unplanned logging operations in the eastern Amazon. *Forest Ecology and Management, 89*, 59–77.

Johns, E. M., Lumpkin, R., Putman, N. F., Smith, R. H., Muller-Karger, F. E., Rueda-Roa, D. T., Hu, C., Wang, M., Brooks, M. T., Gramer, L. J., & Werner, F. E. (2020). The establishment of a pelagic Sargassum population in the tropical Atlantic: Biological consequences of a basin-scale long distance dispersal event. *Progress in Oceanography, 182*, 102269.

Jouanno, J., Moquet, J. S., Berline, L., Radenac, M. H., Santini, W., Changeux, T., Thibaut, T., Podlejski, W., Ménard, F., Martinez, J. M., Aumont, O., Sheinbaum, J., Filizola, N., & N'Kaya, G. D. M. (2021). Evolution of the riverine nutrient export to the Tropical Atlantic over the last 15 years: Is there a link with Sargassum proliferation? *Environmental Research Letters, 16*, 034042.

Kalacska, M., Lucanus, O., Sousa, L., & Arroyo-Mora, J. P. (2020). High-resolution surface water classifications of the Xingu River, Brazil, pre and post operationalization of the Belo Monte Hydropower Complex. *Data, 5*, 75.

Kato, M. d. S., Kato, O. R., Denich, M., & Vlek, P. L. (1999). Fire-free alternatives to slash-and-burn for shifting cultivation in the eastern Amazon region: The role of fertilizers. *Field Crops Research, 62*, 225–237.

Kauffman, J. B., Bernardino, A. F., Ferreira, T. O., Giovannoni, L. R., Gomes, L. E. O., Romero, D. J., Jimenez, L. C. Z., & Ruiz, F. (2018). Carbon stocks of mangroves and salt marshes of the Amazon region, Brazil. *Biology Letters, 14*, 20180208.

Kemenes, A., Forsberg, B. R., & Melack, J. M. (2011). CO_2 emissions from a tropical hydroelectric reservoir (Balbina, Brazil). *Journal of Geophysical Research: Biogeosciences, 116*, 3004.

Lapointe, B. E., Brewton, R. A., Herren, L. W., Wang, M., Hu, C., McGillicuddy, D. J., Lindell, S., Hernandez, F. J., & Morton, P. L. (2021). Nutrient content and stoichiometry of pelagic Sargassum reflects increasing nitrogen availability in the Atlantic Basin. *Nature Communications, 12*, 3060.

Lara, R. J., & Cohen, M. C. L. (2006). Sediment porewater salinity, inundation frequency and mangrove vegetation height in Bragança, North Brazil: An ecohydrology-based empirical model. *Wetlands Ecology and Management, 14*, 349–358.

Larour, E., Ivins, E. R., & Adhikari, S. (2017). Should coastal planners have concern over where land ice is melting? *Science Advances, 3*, e1700537.

Latrubesse, E. M., Arima, E. Y., Dunne, T., Park, E., Baker, V. R., d'Horta, F. M., Wight, C., Wittmann, F., Zuanon, J., Baker, P. A., Ribas, C. C., Norgaard, R. B., Filizola, N., Ansar, A., Flyvbjerg, B., & Stevaux, J. C. (2017). Damming the rivers of the Amazon basin. *Nature, 546*, 363–369.

Laurance, W. F., Albernaz, A. K. M., Schroth, G., Fearnside, P. M., Bergen, S., Venticinque, E. M., & Da Costa, C. (2002). Predictors of deforestation in the Brazilian Amazon. *Journal of Biogeography, 29*, 737–748.

Li, T., Wang, S., Liu, Y., Fu, B., & Gao, D. (2020). Reversal of the sediment load increase in the Amazon basin influenced by divergent trends of sediment transport from the Solimões and Madeira Rivers. *CATENA, 195*, 104804.

Machado, L. A. T., Calheiros, A. J. P., Biscaro, T., Giangrande, S., Dias, M. A. F. S., Cecchini, M. A., Albrecht, R., Andreae, M. O., Araujo, W. F., Artaxo, P., Borrmann, S., Braga, R., Burleyson, C., Eichholz, C. W., Fan, J., Feng, Z., Fisch, G. F., Jensen, M. P., Martin, S. T., … Wendisch, M. (2018). Overview: Precipitation characteristics and sensitivities to environmental conditions during GoAmazon2014/5 and ACRIDICON-CHUVA. *Atmospheric Chemistry and Physics, 18*, 6461–6482.

Magris, R. A., & Barreto, R. (2010). Mapping and assessment of protection of mangrove habitats in Brazil. *Pan-American Journal of Aquatic Sciences, 5*, 546–556.

Mahiques, M. M., Siegle, E., Francini-Filho, R. B., Thompson, F. L., Rezende, C. E., Gomes, J. D., & Asp, N. E. (2019). Insights on the evolution of the living great Amazon Reef system, equatorial West Atlantic. *Scientific Reports, 9*, 13699.

Mansur, A. V., Brondízio, E. S., Roy, S., Hetrick, S., Vogt, N. D., & Newton, A. (2016). An assessment of urban vulnerability in the Amazon Delta and Estuary: A multi-criterion index of flood exposure, socio-economic conditions and infrastructure. *Sustainability Science, 11*, 625–643.

Marceniuk, A. P., Rotundo, M. M., Caires, R. A., Cordeiro, A. P. B., Wosiacki, W. B., Oliveira, C., Souza-Serra, R. R. M. D., Romão-Júnior, J. G., Santos, W. C. R. D., & Reis, T. D. S. (2019). The bony fishes (Teleostei) caught by industrial trawlers off the Brazilian north coast, with insights into its conservation. *Neotropical Ichthyology, 17*, e180038.

Marceniuk, A. P., Soares, B. E., Rotundo, M. M., Caires, R. A., Rosa, R. D. S., Santos, W. C. R. D., Cordeiro, A. P. B., Romão Junior, J. G., Akama, A., & Wosiacki, W. B. (2023). The bycatch of piramutaba, Brachyplatystoma vaillantii industrial fishing in a salinity and depth gradient in the Amazon estuary, Brazil. *Acta Amazonica, 53*, 93–106.

Marengo, J. A. (2004). Interdecadal variability and trends of rainfall across the Amazon basin. *Theoretical and Applied Climatology, 78*, 79–96.

Marengo, J. A., Alves, L. M., Soares, W. R., Rodriguez, D. A., Camargo, H., Riveros, M. P., & Pabló, A. D. (2013). Two contrasting severe seasonal extremes in Tropical South America in 2012: Flood in Amazonia and Drought in Northeast Brazil. *Journal of Climate, 26*, 9137–9154.

Marengo, J. A., & Camargo, C. C. (2008). Surface air temperature trends in Southern Brazil for 1960–2002. *International Journal of Climatology, 28*, 893–904.

Marengo, J. A., & Espinoza, J. C. (2016). Extreme seasonal droughts and floods in Amazonia: Causes, trends and impacts. *International Journal of Climatology, 36*, 1033–1050.

Marengo, J. A., Souza, C. M., Thonicke, K., Burton, C., Halladay, K., Betts, R. A., Alves, L. M., & Soares, W. R. (2018). Changes in climate and land Use over the Amazon Region: Current and future variability and trends. *Frontiers in Earth Science, 6*, 228.

Marengo, J. A., Tomasella, J., Alves, L. M., Soares, W. R., & Rodriguez, D. A. (2011). The drought of 2010 in the context of historical droughts in the Amazon region. *Geophysical Research Letters, 38*, L12703.

Marques, T. S., Gutierrez, G., & Pina, B. P. (2023). A step back from oil exploration in Brazil. *Science, 381*, 1164.

Marsh, R., Skliris, N., Tompkins, E. L., Dash, J., Almela, V. D., Tonon, T., Oxenford, H. A., & Webber, M. (2023). Climate-sargassum interactions across scales in the tropical Atlantic. *Plos Climate, 2*, e0000253.

Martinelli, L. A., Victoria, R. L., Devol, A. H., Richey, J. E., & Forsberg, B. R. (1989). Suspended sediment load in the Amazon basin: An overview. *GeoJournal, 19*, 381–389.

Martinez, J. M., Guyot, J. L., Filizola, N., & Sondag, F. (2009). Increase in suspended sediment discharge of the Amazon River assessed by monitoring network and satellite data. *CATENA, 79*, 257–264.

Meggers, B. J. (1994). Archeological evidence for the impact of mega-Niño events on Amazonia during the past two millennia. *Climatic Change, 28*, 321–338.

Milliman, J. D., & Meade, R. H. (1983). World-wide delivery of river sediment to the oceans. *The Journal of Geology, 91*, 1–21.

Montanher, O. C., Novo, E. M. L. D. M., & Souza Filho, E. E. D. (2018). Temporal trend of the suspended sediment transport of the Amazon River (1984–2016). *Hydrological Sciences Journal, 63*, 1901–1912.

Moran, E. F., Lopez, M. C., Moore, N., Müller, N., & Hyndman, D. W. (2018). Sustainable hydropower in the 21st century. *Proceedings of the National Academy of Sciences of the United States of America, 115*, 11891–11898.

Murray, N. J., Worthington, T. A., Bunting, P., Duce, S., Hagger, V., Lovelock, C. E., Lucas, R., Saunders, M. I., Sheaves, M., Spalding, M., Waltham, N. J., & Lyons, M. B. (2022). High-resolution mapping of losses and gains of Earth's tidal wetlands. *Science, 376*, 744–749.

Nace, T. E., Baker, P. A., Dwyer, G. S., Silva, C. G., Rigsby, C. A., Burns, S. J., Giosan, L., Otto-Bliesner, B., Liu, Z., & Zhu, J. (2014). The role of North Brazil current transport in the paleoclimate of the Brazilian Nordeste margin and paleoceanography of the western tropical Atlantic during the late Quaternary. *Palaeogeography, Palaeoclimatology, Palaeoecology, 415*, 3–13.

Nepstad, D., McGrath, D., Stickler, C., Alencar, A., Azevedo, A., Swette, B., Bezerra, T., DiGiano, M., Shimada, J., Seroa da Motta, R., Armijo, E., Castello, L., Brando, P., Hansen, M. C., McGrathHorn, M., Carvalho, O., & Hess, L. (2014). Slowing Amazon deforestation through public policy and interventions in beef and soy supply chains. *Science, 344*, 1118–1123.

Nerem, R. S., Beckley, B. D., Fasullo, J. T., Hamlington, B. D., Masters, D., & Mitchum, G. T. (2018). Climate-change-driven accelerated sea-level rise detected in the altimeter era. *Proceedings of the National Academy of Sciences of the United States of America, 115*, 2022–2025.

Nittrouer, C. A., DeMaster, D. J., Kuehl, S. A., Jr., A. G. F., Sternberg, R. W., Faria, L. E. C., Silveira, O. M., Allison, M. A., Kineke, G. C., Ogston, A. S., Filho, P. W. M. S., Asp, N. E., Nowacki, D. J., & Fricke, A. T. (2021). Amazon sediment transport and accumulation along the continuum of mixed fluvial and marine processes. *Annual Review of Marine Science, 13*, 501–536.

NOAA National Centers for Environmental Information. (2022). ETOPO 2022 15 arc-second global relief model. Retrieved September 28, 2023, from https://doi.org/10.25921/fd45-gt74

Nobre, C. A., Sampaio, G., Borma, L. S., Castilla-Rubio, J. C., Silva, J. S., & Cardoso, M. (2016). Land-use and climate change risks in the Amazon and the need of a novel sustainable development paradigm. *Proceedings of the National Academy of Sciences of the United States of America, 113*, 10759–10768.

Ohl-Schacherer, J., Mannigel, E., Kirkby, C., Shepard, G. H., & Yu, D. W. (2008). Indigenous ecotourism in the Amazon: A case study of 'Casa Matsiguenka' in Manu National Park, Peru. *Environmental Conservation, 35*, 14–25.

Ometto, J. P., Aguiar, A. P. D., & Martinelli, L. A. (2011). Amazon deforestation in Brazil: Effects, drivers and challenges. *Carbon Management, 2*, 575–585.

Pegado, T. D. S. E. S., Schmid, K., Winemiller, K. O., Chelazzi, D., Cincinelli, A., Dei, L., & Giarrizzo, T. (2018). First evidence of microplastic ingestion by fishes from the Amazon River estuary. *Marine Pollution Bulletin, 133*, 814–821.

Pereira, L. C. C., de Sousa-Felix, R. C., da Costa, R. M., & Jiménez Quintana, J. A. (2018). Challenges of the recreational use of Amazon beaches. *Ocean Coastal Management, 165*, 52–62.

Pezzuti, J. C. B., Zuanon, J., Ribas, C. C., Wittmann, F., d'Horta, F., Sawakuchi, A. O., Lopes, P. F. M., Carneiro, C. C., Akama, A., Garzón, B. R., Mantovanelli, T., Fearnside, P. M., & Stringer, L. C. (2022). Belo Monte through the food-water-energy nexus: The disruption of a unique socioecological system on the Xingu River. In F. A. Moreira, M. D. Fontana, T. F. Malheiros, & G. M. D. Giulio (Eds.), *The Water-energy-food nexus: what the Brazilian research has to say* (pp. 22–40). School of Public Health, University of São Paulo.

Reis, I. C. D., Honório, N. A., Barros, F. S. M. D., Barcellos, C., Kitron, U., Camara, D. C. P., Pereira, G. R., Keppeler, E. C., da Silva-Nunes, M., & Codeço, C. T. J. P. O. (2015). Epidemic and endemic malaria transmission related to fish farming ponds in the Amazon frontier. *Plos One, 10*, e0137521.

Remsen, J. V., & Parker, T. A. (1983). Contribution of river-created habitats to bird species richness in Amazonia. *Biotropica, 15*, 223–231.

Resende, A. F., Schöngart, J., Streher, A. S., Ferreira-Ferreira, J., Piedade, M. T. F., & Silva, T. S. F. (2019). Massive tree mortality from flood pulse disturbances in Amazonian floodplain forests: The collateral effects of hydropower production. *Science of the Total Environment, 659*, 587–598.

Ribeiro, D. R. G., Faccin, H., Molin, T. R. D., de Carvalho, L. M., & Amado, L. L. (2017). Metal and metalloid distribution in different environmental compartments of the middle Xingu River in the Amazon, Brazil. *Science of the Total Environment, 605–606*, 66–74.

Ronchail, J., Cochonneau, G., Molinier, M., Guyot, J. L., De Miranda Chaves, A. G., Guimarães, V., & de Oliveira, E. (2002). Interannual rainfall variability in the Amazon basin and sea-surface temperatures in the equatorial Pacific and the tropical Atlantic Oceans. *International Journal of Climatology, 22*, 1663–1686.

Roosevelt, A. C. (2013). The Amazon and the Anthropocene: 13,000 years of human influence in a tropical rainforest. *Anthropocene, 4*, 69–87.

Röpke, C. P., Amadio, S., Zuanon, J., Ferreira, E. J., Deus, C. P. D., Pires, T. H., & Winemiller, K. O. (2017). Simultaneous abrupt shifts in hydrology and fish assemblage structure in a floodplain lake in the central Amazon. *Scientific Reports, 7*, 1–10.

Rudorff, B. F. T., Adami, M., Aguiar, D. A., Moreira, M. A., Mello, M. P., Fabiani, L., Amaral, D. F., & Pires, B. M. (2011). The soy moratorium in the Amazon biome monitored by remote sensing images. *Remote Sensing, 3*, 185–202.

Sampaio, D. S., Santos, M. L., Tagliaro, C. H., & Beasley, C. R. (2020). Variation in environmental characteristics of waters among Amazon coast oyster culture units. *Acta Amazonica, 50*, 295–304.

Santos, H. S. S. D., Beasley, C. R., & Tagliaro, C. H. (2010). Changes in population characteristics of beds of an exploited tropical estuarine mussel Mytella falcata (d'ORBIGNY, 1846). *Boletim Do Instituto De Pesca (são Paulo), 36*, 85–97.

Santos, L. F. O., Neu, V., Monteiro, R. C. P., Kütter, V. T., Morais, L. M. S., Soares-Gomes, A., Chelazzi, D., Giarrizzo, T., & Martinelli-Filho, J. E. (2023). Microplastics and microfibers in the Guajará Bay, Amazon delta: Potential sources and variability. *Marine Pollution Bulletin, 195*, 115525.

Scott, D. A., & Carbonell, M. (1986). A directory of Neotropical wetlands. IUCN Conservation Monitoring Centre.

Shukla, J., Nobre, C., & Sellers, P. (1990). Amazon deforestation and climate change. *Science, 247*, 1322–1325.

Silva Junior, C. H. L., Pessôa, A. C. M., Carvalho, N. S., Reis, J. B. C., Anderson, L. O., & Aragão, L. E. O. C. (2021). The Brazilian Amazon deforestation rate in 2020 is the greatest of the decade. *Nature Ecology and Evolution, 5*, 144–145.

Silveira, M. I., Oliveira, E. R., Kern, D. C., Costa, M. L., & Rodrigues, S. F. S. (2011). Site Jabuti, in Bragança, State of Pará, in the archaeological scenario of the Amazonian coastal landscape. *Boletim do Museu Paraense Emílio Goeldi. Ciências Humanas*, 6, 335–345.

Skole, D., & Tucker, C. (1993). Tropical deforestation and habitat fragmentation in the Amazon: Satellite data from 1978 to 1988. *Science, 260*, 1905–1910.

Soares, W. R., & Marengo, J. A. (2009). Assessments of moisture fluxes east of the Andes in South America in a global warming scenario. *International Journal of Climatology, 29*, 1395–1414.

Spracklen, D. V., & Garcia-Carreras, L. (2015). The impact of Amazonian deforestation on Amazon basin rainfall. *Geophysical Research Letters, 42*, 9546–9552.

Sternberg, H. O'R. (1987). Aggravation of floods in the Amazon River as a consequence of deforestation? *Geografiska Annaler: Series A, Physical Geography, 69,* 201–219.

Syvitski, J.P.M., Kettner, A.J., Overeem, I., Hutton, E.W.H., Hannon, M.T., Brakenridge, G.R., Day, J., Vörösmarty, C., Saito, Y., Giosan, L., & Nicholls, R.J. (2009). Sinking deltas due to human activities. *Nature Geoscience, 2*, 681–686.

Szabo, S., Brondizio, E., Renaud, F. G., Hetrick, S., Nicholls, R. J., Matthews, Z., Tessler, Z., Tejedor, A., Sebesvari, Z., Foufoula-Georgiou, E., da Costa, S., & Dearing, J. A. (2016). Population dynamics, delta vulnerability and environmental change: Comparison of the Mekong, Ganges-Brahmaputra and Amazon delta regions. *Sustainability Science, 11*, 539–554.

Tenório, G. S., Souza-Filho, P. W. M., & Ramos Alves, P. J. O. (2015). Mangrove shrimp farm mapping and productivity on the Brazilian Amazon coast: Environmental and economic reason for coastal conservation. *Ocean and Coastal Management, 104*, 65–77.

Timpe, K., & Kaplan, D. (2017). The changing hydrology of a dammed Amazon. *Science Advances, 3*, e1700611.

Tregidgo, D., Parry, L., Barlow, J., & Pompeu, P. S. (2021). Urban market amplifies strong species selectivity in Amazonian artisanal fisheries. *Neotropical Ichthyology, 19*, e210097.

van der Sleen, P., & Albert, J. S. (2017). *Field guide to the fishes of the Amazon, Orinoco, and Guianas*. Princeton University Press.

Veiga, J., Tourrand, J., Poccard-Chapuis, R., & Piketty, M. (2002). Cattle ranching in the amazon rainforest. *Animal Production, 24*, 253–256.

Venticinque, E., Forsberg, B., Barthem, R., Petry, P., Hess, L., Mercado, A., Cañas, C., Montoya, M., Durigan, C., & Goulding, M. (2016). An explicit GIS-based river basin framework for aquatic ecosystem conservation in the Amazon. *Earth System Science Data, 8*, 651–661.

Vogt, N., Pinedo-Vasquez, M., Brondízio, E. S., Rabelo, F. G., Fernandes, K., Almeida, O., Riveiro, S., Deadman, P. J., & Dou, Y. (2016). Local ecological knowledge and incremental adaptation to changing flood patterns in the Amazon delta. *Sustainability Science, 11*, 611–623.

Vuille, M., Burns, S. J., Taylor, B. L., Cruz, F. W., Bird, B. W., Abbott, M. B., Kanner, L. C., Cheng, H., & Novello, V. F. (2012). A review of the South American monsoon history as recorded in stable isotopic proxies over the past two millennia. *Climate of the past, 8*, 1309–1321.

Wang, X., Edwards, R. L., Auler, A. S., Cheng, H., Kong, X., Wang, Y., Cruz, F. W., Dorale, J. A., & Chiang, H. W. (2017). Hydroclimate changes across the Amazon lowlands over the past 45,000 years. *Nature, 541*, 204–207.

Ward, N. D., Keil, R. G., Medeiros, P. M., Brito, D. C., Cunha, A. C., Dittmar, T., Yager, P. L., Krusche, A. V., & Richey, J. E. (2013). Degradation of terrestrially derived macromolecules in the Amazon River. *Nature Geosciences, 6*, 530–533.

Ward, N. D., Krusche, A. V., Sawakuchi, H. O., Brito, D. C., Cunha, A. C., Moura, J. M. S., da Silva, R., Yager, P. L., Keil, R. G., & Richey, J. E. (2015). The compositional evolution of dissolved and particulate organic matter along the lower Amazon River—Óbidos to the ocean. *Marine Chemistry, 177*, 244–256.

West, T. A. P., & Fearnside, P. M. (2021). Brazil's conservation reform and the reduction of deforestation in Amazonia. *Land Use Policy, 100*, 105072.

Williams, E., Dall' Antonia, A., Dall' Antonia, V., Almeida, J. M. de, Suarez, F., Liebmann, B., & Malhado, A.C.M. (2005). The drought of the century in the Amazon Basin: An analysis of the regional variation of rainfall in South America in 1926. *Acta Amazonica, 35*, 231–238.

Winemiller, K. O., McIntyre, P. B., Castello, L., Fluet-Chouinard, E., Giarrizzo, T., Nam, S., Baird, I. G., Darwall, W., Lujan, N. K., Harrison, I., Stiassny, M. L. J., Silvano, R. A. M., Fitzgerald, D. B., Pelicice, F. M., Agostinho, A. A., Gomes, L. C., Albert, J. S., Baran, E., Petrere, M., Zarfl, C., et al. (2016). Balancing hydropower and biodiversity in the Amazon, Congo, and Mekong. *Science, 351*, 128–129.

WinklerPrins, A. M. (2006). Jute cultivation in the lower Amazon, 1940–1990: An ethnographic account from Santarém, Pará, Brazil. *Journal of Historical Geography, 32*, 818–838.

Worthington, T., & Spalding, M. (2018). Mangrove Restoration Potential: A global map highlighting a critical opportunity. Apollo-University of Cambridge Repository.

Zalasiewicz, J., Waters, C. N., Williams, M., & Summerhayes, C. (2018). The anthropocene as a geological time unit: a guide to the scientific evidence and current debate. Cambridge University Press.

Zemp, D. C., Schleussner, C. F., Barbosa, H. M. J., & Rammig, A. (2017). Deforestation effects on Amazon forest resilience. *Advancing Earth and Space Sciences, 44*, 6182–6190.

Changjiang River Delta

Leicheng Guo, Linjing Ren, Haifei Yang, Weiguo Zhang, Lei Gao, Xiuzhen Li, Ben Li, and Wenzhen Zhao

1 Physical Setting

1.1 The Changjiang River

The Changjiang River (also known as the Yangtze River) is the longest river in China and the third longest in the world. Its main stem is approximately 6,300 km long, with a drainage area over 1.8 million km^2 (Fig. 1). The Changjiang River basin is an important economic zone in China, home to a population of 460 million and an annual GDP output of approximately 45 trillion RMB (as to 2020). The main river downstream of Yibin, approximately 2,840 km inland from the sea, is navigable, and is known as the Golden Waterway. It is thus of particular socio-economic importance to understand the water-land-human nexus throughout the river-delta-ocean continuum.

The Changjiang River basin has a humid subtropical climate influenced by Asian monsoon. The basin receives an average annual precipitation of approximately 1,100 mm, and the annual streamflow is between 667 and 1,359 billion m^3 at the Datong gauge. Precipitation and streamflow are concentrated in the summer wet seasons (May–October), which account for 71% of the annual flow. This concentrated rainfall makes the low-lying middle-lower basin and the delta plain vulnerable to both floods and droughts. Historical catastrophic floods have occurred in 1931, 1954, 1998, and 2020, while extreme drought events occurred in 1978, 2011, and 2022–2023. To cope with extreme flooding and drought hazards, many reservoirs and dams were constructed in the upper basin, including the Three Gorges Dam (TGD) and the cascade dams along the Jinsha River. In addition to flood control, these dams

L. Guo (✉) · L. Ren · H. Yang · W. Zhang · L. Gao · X. Li · B. Li · W. Zhao
State Key Lab of Estuarine and Coastal Research, East China Normal University, Shanghai 200241, China
e-mail: lcguo@sklec.ecnu.edu.cn

W. Zhang and H. de Vriend (eds.), *Delta Sustainability*,
https://doi.org/10.1007/978-981-97-7259-9_3

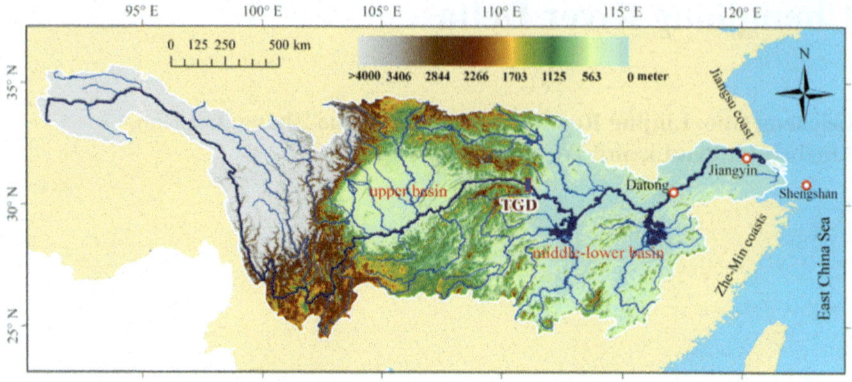

Fig. 1 The Changjiang River basin, with the three Gorges Dam (TGD) separating the upper and middle-lower sub-basins. The hydrological station Datong marks the tidal wave limit

function in water supply, power generation, and navigation; although their impacts on the aquatic ecosystem and river connectivity remains debated.

1.2 The Changjiang River Delta

The Changjiang River Delta (CRD, also called Yangtze River Delta), is a low-lying delta plain seaward of Datong, that faces the marginal East China Sea in the west Pacific (Fig. 2). It comprises a subaerial land area of over 20,000 km^2 seaward of Datong, and a large tongue-shaped subaqueous delta protruding to the southeast. In physical geography, the CRD refers to the low-lying flood plain downstream of Datong (Fig. 1a). Meanwhile, in social science, the Yangtze River Delta Economic Zone covers a larger area, including four state-level provinces: Anhui, Jiangsu, Zhejiang, and Shanghai. The Yangtze River mouth is widely called the Changjiang River Estuary (CRE, or Yangtze Estuary) in literatures. In this chapter, CRD is used when talking about the whole land-water system, and CRE is used when referring the water regions between dikes, including the tidal flats.

The CRD is predominantly shaped by high river discharge, strong tides, and moderate waves. River discharge at the tidal wave limit, Datong, varies in the range of 4,600–92,600 m^3/s, with a multi-year mean value of 28,500 m^3/s. The river also supplies a sediment load of 470 Mt yr^{-1} (1950–1985) in building up the delta. The mean and maximal tidal ranges are 2.7 m and 5.9 m, respectively, indicating a meso-tidal regime. At Sheshan near the mouth, the average wave height is 0.9 m, but under stormy weather, the maximum wave height may reach 6 m. Waves from the south prevail in winter and waves from the north in summer. The East China Sea outer the CRD is a wide and shallow epi-continental marginal sea (Liu et al., 2006). The seaward river plume in the inner shelf zone is affected by strong coastal currents,

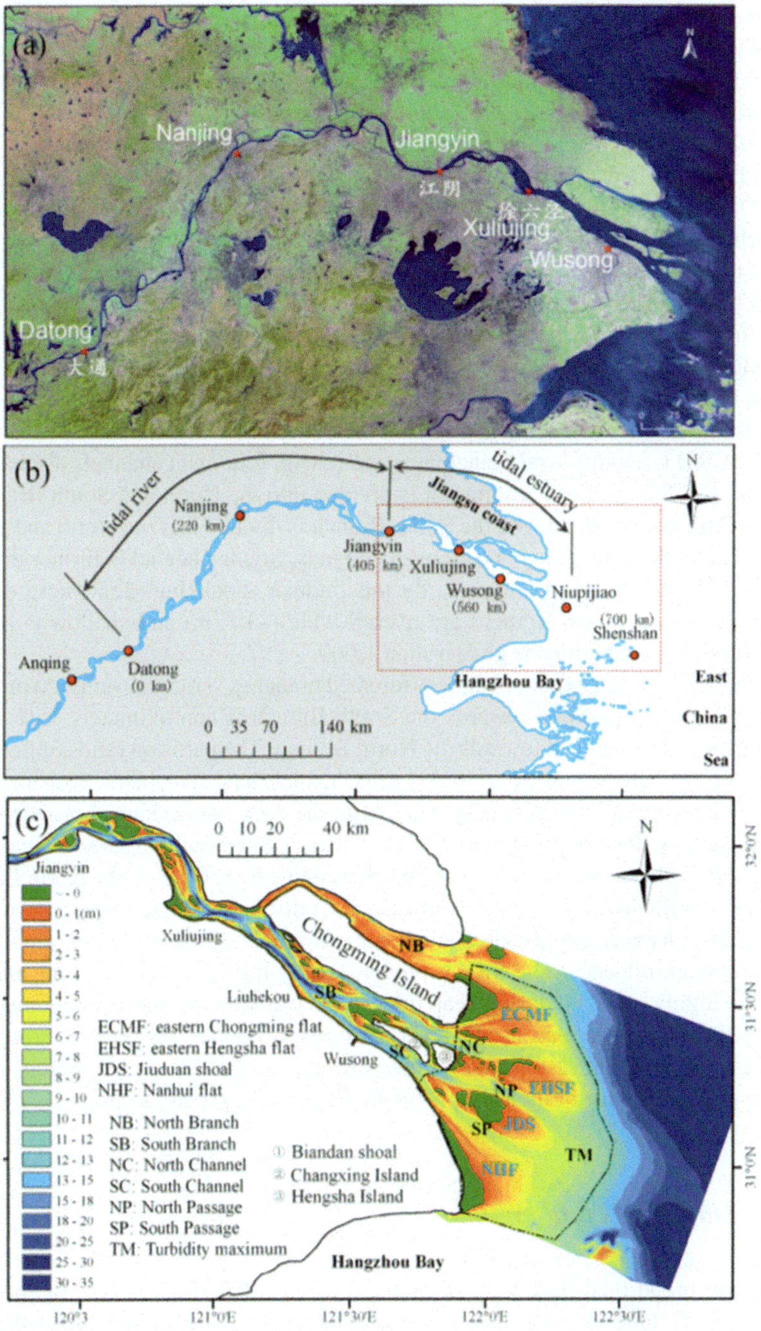

Fig. 2 **a** Satellite image of the CRD, **b** a top view of the river course, and **c** the underwater bathymetry in 1997 seaward Jiangyin. The water depth in **c** is referenced to the datum of lowest low tide. Subplots (**a**) and (**b**) are. Modified from Guo et al. (2021)

including the northward Taiwan Warm Current and southward Zhe-Min Current. Coastal and shelf currents have an impact on the dispersion of sediment carried by rivers and can affect the development of deltas and long-term geomorphological evolution. Tidal waves propagate approximately 650 km inland in the CRD. This ranks the CRD among the longest tidal delta systems globally, only after the Amazon River Estuary where tidal waves propagate inland ~1,100 km (Gallo & Vinzon, 2005). The Datong-Jiangyin (tidal current limit in wet seasons) reach is identified as the tidal river segment in space, with the reach between Jiangyin and the mouth bar zone being the tidal estuary segment, while the seaward regions are the subaqueous delta (see Fig. 2c). The area between Wusong and the 10 m isobath in the subaqueous delta front is also known as the turbidity maximum zone due to its higher turbidity and high-suspended sediment concentration. The water depth in the main channels of the turbidity maximum zone is smaller than in the more landward channels and seaward nearshore, whence it is also called the mouth bar region.

The CRD is a multi-level branching system, with four main channels discharging into the sea. Seaward of Xuliujing, it is divided into the North and South Branches by the Chongming Island, and the South Branch is divided into the North and South Channels by the Changxing and Hengsha Islands. South Channel is further divided into the North and South Passages by the Jiuduan shoal (Fig. 2c). These deltaic channels (except North Branch) are overall wide (~10^4 m) and shallow (~10 m), with width-to-depth ratios is greater than 1,000.

Subtidal flow distribution in these bifurcated branches varies unevenly. At present, the subtidal flow partition towards the South Branch is approximately 95% while only about 5% is directed towards the North Branch. The partition ratios of flow and sediment flux between the South and North Channels have remained ~50% for many years, with a variation of less than 10%. After the deep-water channel project in the North Passage, the ebb flow partition ratio towards the South Passage increased to ~60%, while that to the North Passage decreased to ~40%. Overall, the North Branch is a tide-dominated channel; the North Channel discharges a major portion of river flow and has a lower salinity; the funnel-shaped South Passage has larger salinity and stronger tidal influence. The CRE is predominantly mixed most of the time and can become highly stratified during neap tides in the wet seasons, and strongly mixed at spring tide in the dry seasons. The North Branch is overall well-mixed throughout the year. Saltwater intrusion poses a threat to access of freshwater resources in the CRD, particularly for millions of people in Shanghai in the dry seasons.

1.3 Tidal Flats and Wetlands

There are broad tidal flats and salt-marshes in the CRD (Fig. 3). The majority of tidal flats are located in the mouth bar zone, e.g., the eastern Chongming flat, the Hengsha flat, Jiuduan shoal, and the Nanhui flat. Their total area, which includes regions with elevations below 5 m and above ~6 m (referenced to the lowest low tide datum), is approximately 440 km^2 seaward of Xuliujing (Li et al., 2020). The

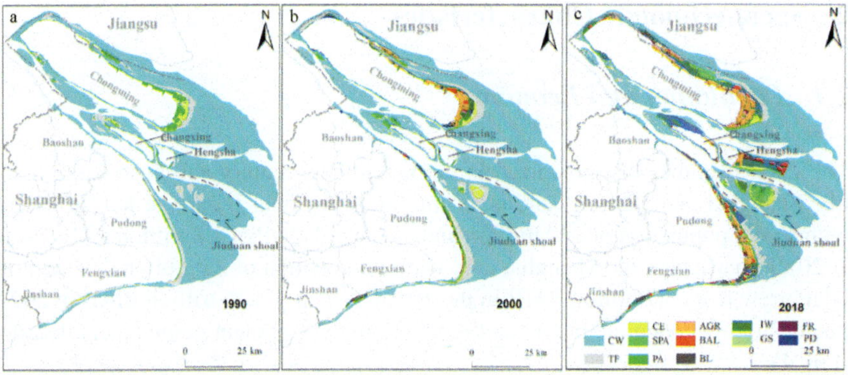

Fig. 3 Distribution of tidal flats and land use changes since 1990 in the CRD. CW: Coastal water, TF: Tidal flat, CE: *Cyperaceae*, PA: *Phragmites australis*, SPA: *Spartina alterniflora*, FR: Forest, BL: Built-up land, BAL: Bare land, AGR: Agricultural land, GS: Grassland, PD: Pond/river, IW: Inland wetland. Modified from Lin et al. (2021)

eastern Chongming flats and the Jiuduan shoal are recognized as very significant ecosystems of regional and global importance. They serve as crucial spawning and feeding grounds for many brackish water fish and 50 million migratory birds along the East Asia–Australian Flyway.

Tidal flats accrete fast in the CRD owing to the large riverine sediment supply. This fast accretion has attracted reclamation and embankment in the aim to create new land for urban development. As a result, a total area of 1,500 km^2 was gained from tidal flat accretion and embankment in the CRD over the 70 years since 1950 (Guo et al., 2021), suggesting intensive human interferences in the delta.

The inter- and supra-tidal flats are largely muddy and vegetated by *Scirpus mariqueter*, *Spartina alterniflora* and *Phragmites australis* in the direction of increasing flat elevation, while the subtidal flats are largely bare. *Spartina alterniflora* is not a native species in the CRD and was artificially introduced in the 1970s, in the aim to stimulate sediment trapping and flat accretion and to enhance coastal safety. *Spartina alterniflora* is highly productive and becomes an aggressive species that covers a major portion of intertidal flats nowadays. Fast expansion of *Spartina* reduces biodiversity and endangers the native species and ecosystem function; therefore, many kinds of measures have been employed to tackle this issue, including physical cutting, burying, chemical killing and extinguisher. Even though, it is expensive to do that and remains a big challenge to minimize *Spartina* and restore the ecosystem, while flat erosion is another concern if Spartina is removed too fast.

2 Socio-economic Characteristics

2.1 Population and Economy

The CRD becomes fast urbanized in the recent half century and hosts large cities like Shanghai and Nanjing, and is an important regional economic hub in China, with a total population of 230 million and an annual GDP of 2.8 trillion RMB (as to 2020). In particular, the Shanghai city, with a population of 25 million, has seen its GDP grew from 3.7 billion RMB in the 1950s to nearly 4.46 trillion RMB in 2022, demonstrating economic bloom and rapid urban development in the recent decades (Fig. 4).

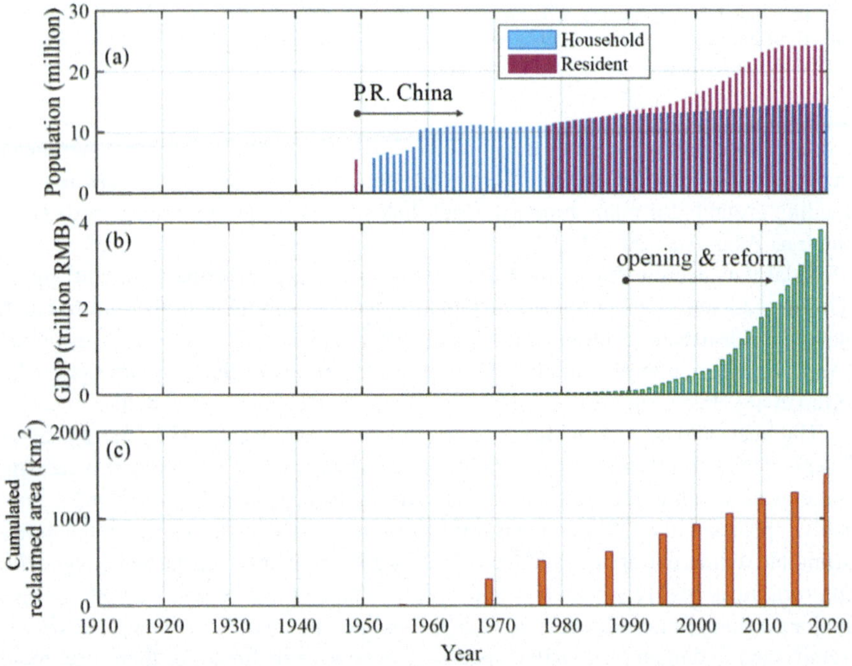

Fig. 4 Historical changes of **a** population, **b** Gross domestic production (GDP) of Shanghai City, and **c** cumulated land area increase in the delta. Data in (**a**) and (**b**) were collected from Shanghai Statistical Yearbook 2020 (http://tjj.sh.gov.cn/tjnj/nj20.htm?d1=2020tjnjen/E0201.htm; http://tjj.sh.gov.cn/tjnj/nj20.htm?d1=2020tjnjen/E0401.htm). Data in (**c**) is from Guo et al. (2021)

2.2 *Major Industries*

The CRD grows into a national and international transportation and trade center, stimulating economic growth and urbanization. Since the mid-19th century, industrialization and urbanization have transformed the delta from an agriculture dominated society to an industrial center in China. According to the economic statics in 2022, the primary, second and third industries contribute 0.2, 25.7 and 74.1% to the GDP of 4.46 trillion RMB in Shanghai.

Transportation has been an important force in the CRD for a long time. The Shanghai Port is the largest container port in China, e.g., a total of 47.3 million twenty-foot container equivalent units (TEU) in 2022. Such an achievement is made possible owing to the deep waterway and harbors located in the delta. Developing and maintaining deep-water navigational channels is essential for economy.

Among the secondary industries, traditional industries such as steel works and refinery show a decreasing contribution to the regional economy, which is being replaced by automotive industry, pharmacy, and other high-tech industries. Pollution discharges to the deltaic water from traditional industries are decreasing, leading to better air and water quality in the delta.

3 Changing Conditions and Management Concerns

The CRD is currently in a critical transition state given continuously changing boundary forcing conditions and conflicting interest between delta development and ecosystem conservation. The CRD is vulnerable and exposed to multiple natural hazards that urgently need integrated management strategies, e.g., coastal erosion, land subsidence, flooding and saltwater intrusion. These threats will be elaborated in the following subsections.

3.1 *Reduced Fluvial Sediment Discharge*

Field monitoring of sediment fluxes began in the early 1950s at Datong, suggesting a mean sediment load of 470 Mt yr^{-1} between 1950 and 1985. A step-wide decrease emerged since the mid-1980s, owing to the combined influences of reforestation, soil conservation and hydropower dams within the catchment (Guo et al., 2019). The TGD accelerated the decrease since 2003 when it started to store water, and the yearly sediment load has approached a temporal low level of 114.2 Mt yr^{-1} in 2006–2023, suggesting a 75% of decline (Fig. 5).

The sediment reaching the delta is mainly fine material composed of clay and silt, with a median grain size of ~27 μm at Datong in the pre-TGD period, while the sediment on the channel bottom has a median grain size of ~166 μm. Along with

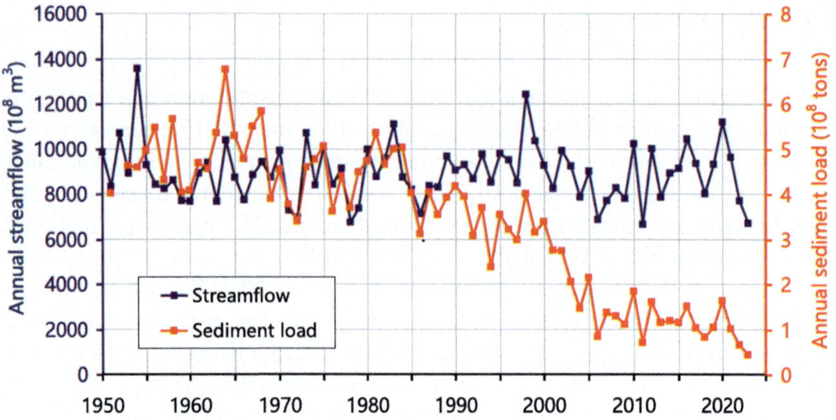

Fig. 5 Annual streamflow and sediment load at Datong in 1950–2023

sediment load reduction, the incoming sediment becomes finer downstream of the TGD, with the median grain size decreased to ~11.7 μm at Datong between 2000 and 2020. Considering the combined effects of hydropower dams in the upper main river and the armoring effect along the downstream river, river-borne sediment to the CRD is likely to persist at a low value in the coming 100–400 years (highly likely smaller than 120 Mt yr^{-1}), and the subsequent impact on the deltaic morphodynamics and wetland safety arises broad management concern.

3.2 *Coastal Erosion*

Coastal erosion and land loss due to sediment shortage is a severe problem in the CRD like in many deltas worldwide (Giosan et al., 2014; Syvitski et al., 2009). The initial decline in riverine sediment supply since the mid-1980s, followed by an accelerated reduction since the early 2000s, has reduced tidal flat accretion since 2016 (Luan et al., 2021; Zhao et al., 2018). This reduction has led to a change from sedimentation to erosion (Fig. 6). Protected by sea dikes and jetties, the existing shorelines in the CRD are relatively stable, and erosion is predominantly detected in the submerged delta front regions. Owing to the fast tidal flat accretion, the total tidal flat area in the CRD seems not to have decreased in the past decades, despite the extensive reclamation. As coastal erosion continues, tidal flat accretion has significantly slowed down and started to shrink, particularly the subtidal flats. This could result in the loss of tidal wetlands and salt marshes, which are of profound ecosystem functions and services.

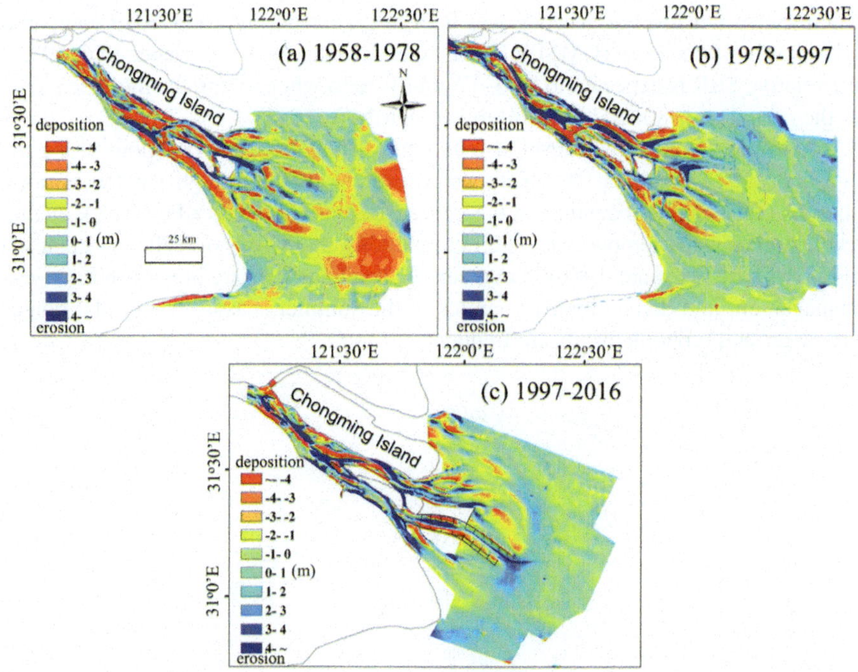

Fig. 6 Erosion and deposition pattern of the Changjiang Estuary (except the North Branch) in three periods, **a** 1958–1978, **b** 1978–1997, **c** 1997–2016. Adopted from Zhang et al. (2022a, 2022b)

3.3 Sea-Level Rise and Flooding Hazards

The CRD is exposed to multiple sources of flooding sources, including high river discharge, high tide, storm surge, and sea-level rise (SLR) (Wang et al., 2012; Yin et al., 2021). The annual precipitation in the CRD is 1,037 mm on average, which is likely to cause waterlogging in the urban areas. Shanghai city, with a mean elevation of 2.2 m (above mean sea level), is at the top of the list of deltas that are at high risk of severe flooding (Balica et al., 2012; Tessler et al., 2015). The TGD has regulated the frequency and magnitude of extremely high river discharge, but a high river discharge may still occur, e.g., the large flood in July 2020 with a peak discharge of 84,500 m^3/s at Datong (Wei et al., 2020). The CRD is hit by an average of 2–4 typhoons each summer and autumn. Typhoons cause high winds, large waves, and storm surge etc., which causes many disastrous damages. For instance, Typhoon Yuni in 1997 caused catastrophic property and life losses in Shanghai due to extreme tides that exceeded historical records, resulting in dike breaches and overbank flow.

Sea-level rise worsens the flooding risk in the CRD. SLR rate was 3–4 mm yr^{-1} between the 1980s and 1990s along Chinese coasts (ESDCAS, 1994). Meanwhile, the mean SLR rate in the CRD was ~1 mm yr^{-1} between 1912 and 1993 (Fig. 7a; Qin & Li, 1997) and increased to 2.4 mm yr^{-1} in 1981–2015 (Wang et al., 2018).

The mean water level at Wusong was 0.14 m higher during 2009–2019 compared with that in 1910–1930 (DMMW, 2021). Considering land subsidence (Xu, 1995), the relative SLR is expected to be 6–17 cm by 2050 when taking the mean sea level in the period of 1980–2020 as a reference (DMMW, 2021).

Moreover, the mean sea level also exhibits strong seasonal variations along the coasts of East China Sea. The mean water level is 0.2–0.5 m higher in the summer seasons (July–September) than in the winter seasons (December-February), due up to the higher water temperature, shelf circulations, and winds over the marginal sea (Fig. 7b). This seasonal variation in mean sea level has a strong impact on flood risk management and deltaic hydrodynamics in the summer season, when tidal ranges are larger and typhoon events are frequent.

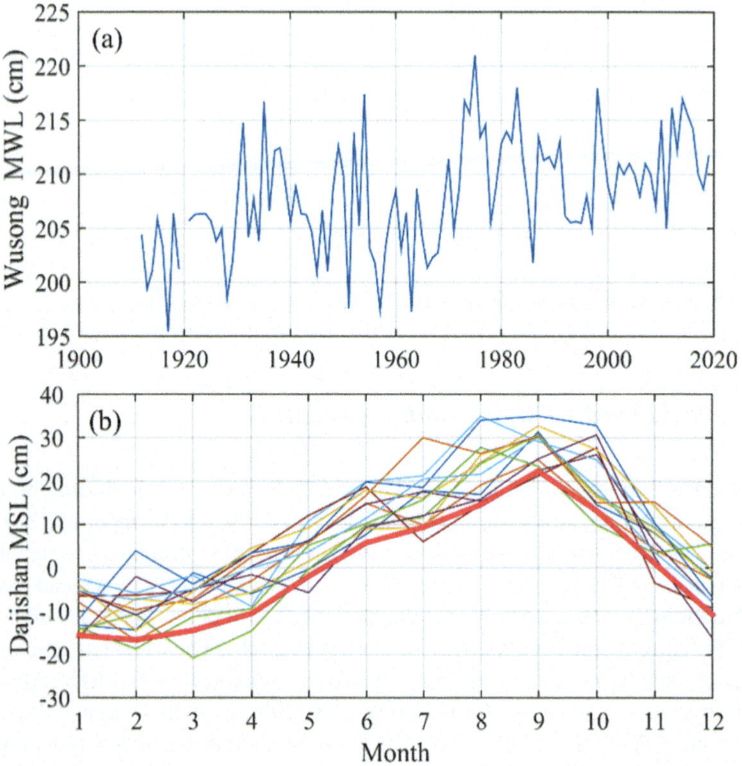

Fig. 7 **a** Inter-annual changes of mean water level (MWL) at Wusong since 1912, and **b** monthly changes of mean sea levels (MSL) at Dajishan in the period of 2007–2019. Data in (**a**) are from Chen (1991), Xu (1995), and Chen et al. (2018), and data in (**b**) are from the annual bulletins of sea-level changes on China's coasts published by the Ministry of Natural Resources of PRC (http://www.mnr.gov.cn/sj/sjfw/hy/gbgg/zghpmgb/; SOA, 2020). The data in (**a**) are referenced to the Wusong Datum, and the records in (**b**) are referenced to the long-term mean sea level, which is 2.09 m above the Wusong Datum. The thick line in (**b**) indicates the mean value between 1975 and 1993. Modified from Guo et al. (2021)

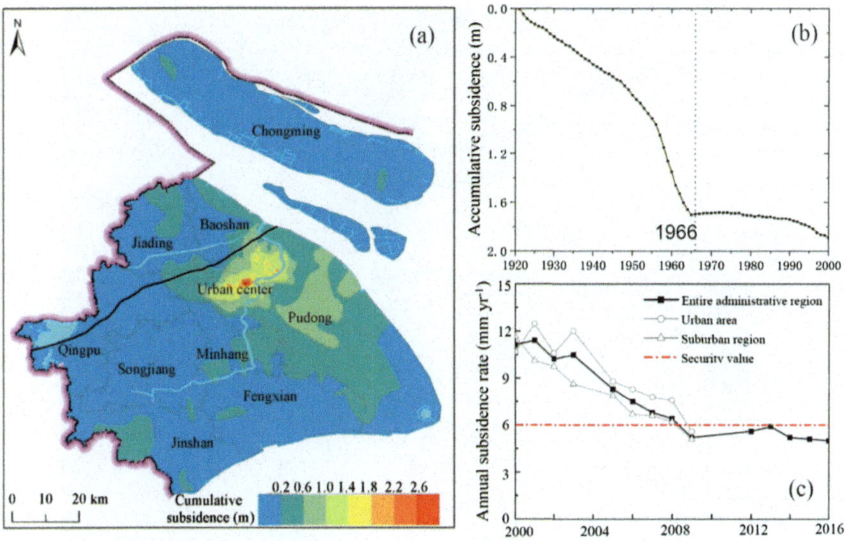

Fig. 8 **a** Distribution of mean land subsidence in Shanghai for the period of 1920–2016, **b** cumulative subsidence changes from 1920 to 2000, and **c** the changes of subsidence rate between 2000 and 2016. Panels (**a**) and (**c**) are from He et al. (2019a, 2019b), and panel (**b**) is redraw based on data from Xue et al. (2005)

Land subsidence may further exacerbate the flood risk and damage. The mean land subsidence rate was approximately 3.9 mm yr^{-1} between 1921 and 1965 in Shanghai owing to groundwater extraction but decreased to 0.9 mm yr^{-1} when groundwater extraction stopped, and recharge was implemented since 1966. Rapid urbanization and intensive construction of high buildings and underground infrastructures resulted in an increased subsidence rate to 10.2 mm yr^{-1} between 1986 and 1997 again. The maximum cumulated subsidence in the downtown regions of Shanghai was 6.04 m between 1980 and 2005, i.e., a rate of 24 mm yr^{-1}. Land subsidence, combined with SLR, is to increase the vulnerability of CRD to both pluvial and coastal flooding (Fig. 8).

3.4 *Saltwater Intrusion*

Freshwater supply is a big issue for tens of millions of people living in the CRD, particularly Shanghai. There are three crucial reservoirs in the CRD supplying drinking water for Shanghai, i.e., the Qingcaosha Reservoir, the Chenhang Reservoir, and the Dongfengxisha Reservoir. However, these reservoirs are threatened by saltwater intrusion, including the intruded saltwater from the North Channel and saltwater overspill from the North Branch (Fig. 9). Saltwater intrusion is particularly severe during the dry seasons (December-February) when river discharge at

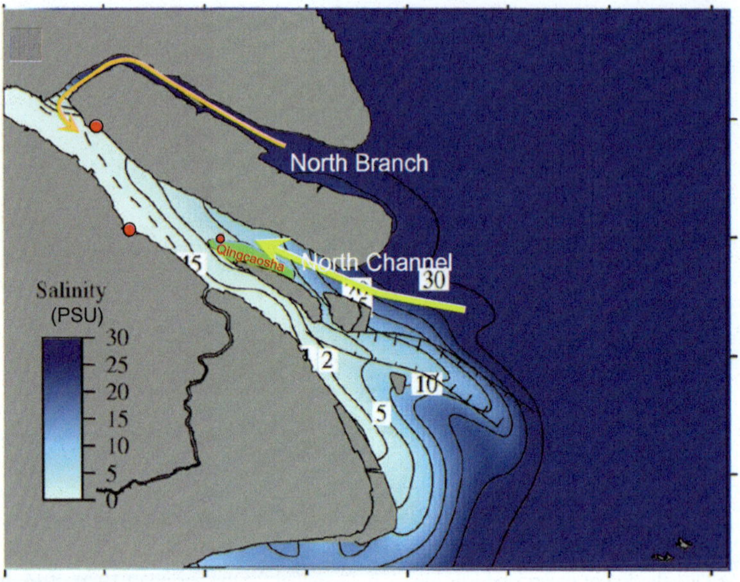

Fig. 9 Distribution of salinity near surface at spring tide in a dry season under northerly wind in the CRD. The red dots indicate the location of the reservoirs, and the yellow arrows indicate saltwater intrusion paths. Modified from Zhu et al. (2020)

Datong is below a threshold of about 12,000 m^3/s. Northerly wind would enhance saltwater intrusion and increase salinity at the intake of the Qingcaosha Reservoir, as occurred in winter 2014 (Zhu et al., 2020). In 2022, river discharge at Datong begins to decrease to <12,000 m^3/s in September, 2–3 months earlier than in the past. Continued low river discharge until the spring of 2023 substantially challenged freshwater supply in Shanghai, although extra river discharge released from the TGD help mitigating saltwater intrusion provisionally. Future SLR may further enhance saltwater intrusion and increase the water shortage risk in Shanghai.

3.5 Nutrient Flux and Aquatic Ecosystem

Since the 1980s, urbanization, industrialization, intensive agricultural activities, and wastewater discharge have led to increased nutrient and contaminant flux to the CRD and water quality degeneration (Fig. 10; Yang et al., 2023). There is a particulate organic carbon (POC) burial sink shift from the coastal shelves to the reservoirs in the upstream catchment. Furthermore, downstream of the TGD in 2003–2011, there is a 77% reduction of total phosphorous and an 84% reduction of particulate phosphorous compared to 1950–1990 (Zhou et al., 2013). The dams have trapped 60% of dissolved silicate load, causing substantial silica flux decreasing in downstream rivers (Ran

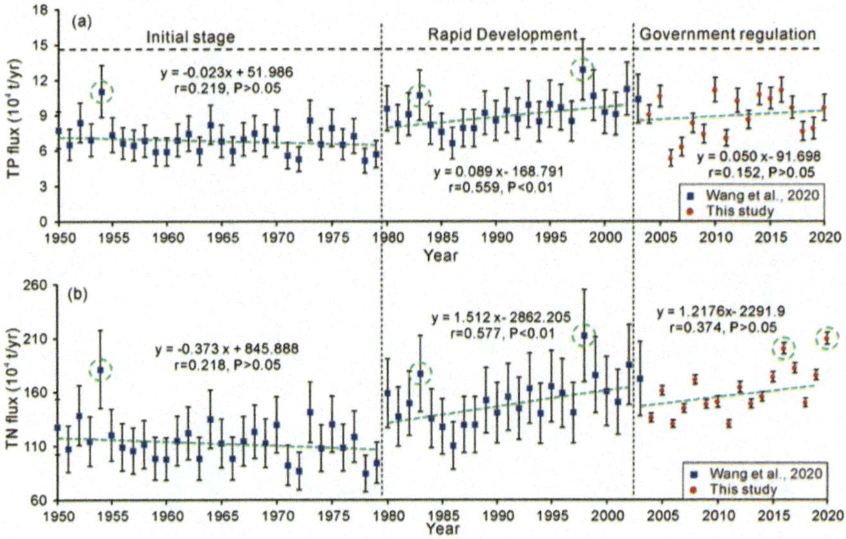

Fig. 10 Long-term variations of the annual fluxes of river-borne, **a** total phosphorus (TP), and **b** total nitrogen (TN) between 1950 and 2020. Extreme high values (green dashed circles) related to extreme summer floods were excluded from trend analysis (Wang et al., 2021a, 2021b; Yang et al., 2023)

et al., 2016). The total nitrogen and dissolved inorganic nitrogen flux, however, have increased by a factor of >2 due to increased fertilizer usage (Zhang et al., 2015; Liu et al., 2019). Soaring fertilizer usage may play a more important role in modulating the flux of total nitrogen and total phosphorous than dam retention (Liu et al., 2019). In response, the Si:N ratio has decreased by a factor of 10, from ~10 in 1960 to ~1 in 2000, while the N:P ratio has increased enormously from 50–100 in 1960s to 300–500 in 2000s in the river (Zhang et al., 2015).

These changes have extensive impacts on the nutrient level, primary production and associated riparian ecosystem in the river-delta-coastal ocean continuum (Zhang et al., 2015) and on the land-to-ocean biogeochemical cycling. For instance, elevated nutrient flux to the sea caused algae bloom, red tides, and hypoxia in the nearshore water of East China Sea (Fig. 11). Red tides occupy an area >100 km^2 in the nearshore water of East China Sea and occur frequently in spring and summer seasons during the past two decades, which is partially ascribed to nutrient flux changes. Fluvial flux of P and N has slowly increased owing to extensive usage of fertilizer in the watershed, while that of Si decreased due to sediment load decline (Wang et al., 2023; Yang et al., 2023). Despite these first signs of improvement, the decline of the fluvial sediment input to the CRD reduces water turbidity, which in turn has strengthened irradiation and photosynthesis. This may shift the zone of high primary production landwards and pose severe challenges to the deltaic ecosystem (Huang et al., 2022; He et al., 2022).

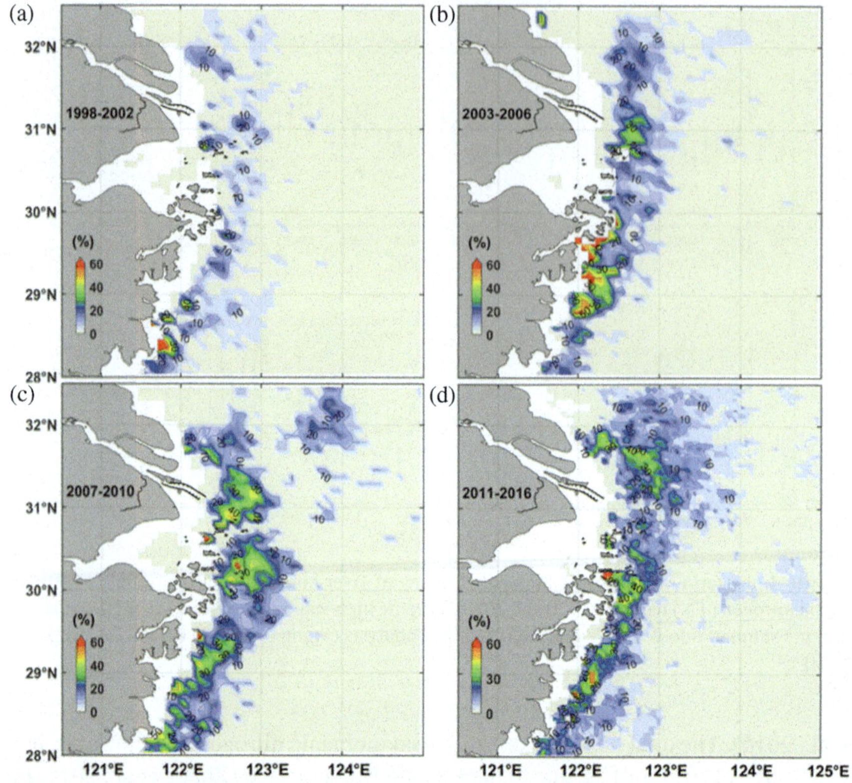

Fig. 11 The probability of algal bloom occurrence based on the threshold of chlorophyll 10 mg/m^3, in the periods of **a** 1998–2002, **b** 2003–2006, **c** 2007–2010, **d** 2011–2016. Data of chlorophyll concentration are retrieved from satellite products. Modified from He et al. (2022)

3.6 Alien Species and Wetland Ecosystem Diversity

Coastal wetlands provide crucial ecosystem functions, supporting biodiversity and bioproductivity, enhancing nearshore wave attenuation and sediment trapping, and stimulating carbon sequestration and pollutant retention. The introduction and expansion of *Spartina alterniflora* can significantly impact native species and reduce biodiversity in the wetlands of the CRD. *Spartina alterniflora* was intentionally imported from North America to China in 1979 due to its beneficial ecological effects in safeguarding embankments and promoting sedimentation. In 1986, the Shanghai Municipal Government introduced *Spartina* on the Nanhui flat. To ensure aviation safety at Pudong International Airport, *S. alterniflora* was also planted in 1997 on the Jiuduan Shoal, with the aim of stimulating saltmarsh expansion and attracting birds.

Over the past 40 years, the coastal wetland area of the CRD and its vegetation communities have changed greatly, with drastic expansion of spartina and shrinkage

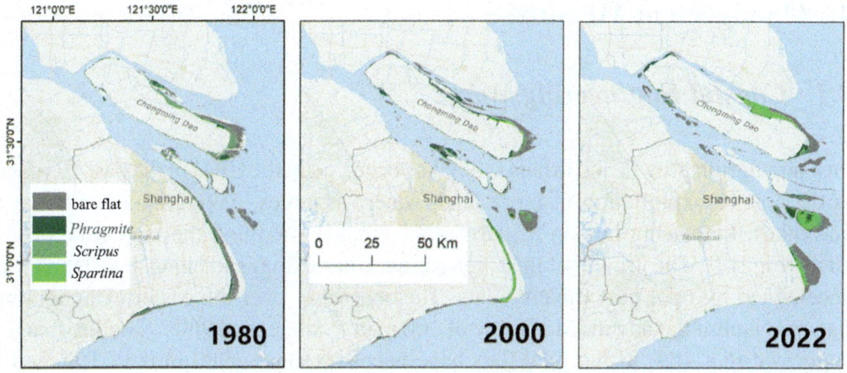

Fig. 12 Decadal changes in the major vegetation species over the wetlands of the CRD from 1980 to 2022. Modified from Lin et al. (2021) with latest data in 2022 updated

of native species (Fig. 12). The areas covered by the native species *Scirpus mariqueter* and *Phragmites australis* decreased from 67 to 22%, and from 33 to 31%, respectively. Meanwhile, the area covered by exotic species expanded from 0 to 47%. The invasion of *S. alterniflora* has enhanced the number and diversity of the number and biodiversity of epiphytic nematodes, while the structure of the soil food web supported by *S. alterniflora* is simpler than that with native vegetation. The total density of benthic animals in the *S. alterniflora* habitat is 36% and 22% lower than that of the adjacent *P. australis* and *S. mariqueter* habitats, respectively. The invasion of *S. alterniflora* also weakens the complexity of arthropod food webs, negatively impacting insect biodiversity.

The invasion of *S. alterniflora* has resulted in a reduction of habitat heterogeneity in the saltmarshes. *Charadriiformes* and *Anseriformes* are the two major waterbird groups in the CRD. The invasion of *Spartina alterniflora* alters the habitat characteristics and food resources of birds, although it provides habitat for some birds (e.g., *Megalurus pryeri*). The diversity of migratory birds in spring, breeding birds in summer and migratory birds in winter in the *S. alterniflora* community is significantly lower than that in native vegetation communities. In total, the rapid expansion of *S. alterniflora* has led to the loss of habitat and a decrease in habitat quality for most birds.

4 Management Measures

4.1 Coastal Engineering Works

Fast population growth and urbanization has occurred in the CRD in the past 70 years, at the cost of exploitation of more land, water resources, navigational waterways, and dikes. Increasing human activities have greatly modified the landscape of the CRD (Fig. 13). The growth of the Chongming, Changxing, and Hengsha Islands was accelerated by tidal flat embankments. The coastlines over the eastern Chongming flat and Nanhui flat advanced at a rate of 100 m yr^{-1} since the 1960s. Satellite images suggested that 40% of the tidal flats have been lost since 1980 in the CRD, and a cumulated area of ~1500 km^2 was embanked between 1950s and 2020 (Guo et al., 2021). Dikes have been continuously built and upgraded along the mainland and islands to protect the low-lying delta plain and property around Shanghai. The dikes have stabilized a major portion of the coastlines and narrowed the channels in the CRD.

Moreover, a giant reservoir, the Qingcaosha Reservoir with a storage capacity of 380 million m^3 to supply Shanghai with freshwater was constructed in 2010 by reclaiming a sand bar to the north of Changxing Island. It reduced the width of the upper North Channel by half and may have affected the flow and sediment partition between the North and South Channels.

To overcome the barrier effect of the mouth bars and facilitate navigation, two long jetties and dozens of groins were constructed between 1998 and 2012 in the selected main waterway, the North Passage. The channel was deepened from 6–7 m to 12.5 m, stimulating waterborne transportation along the whole Changjiang River. The long jetties, with an elevation of 2.5 m above the Wusong Datum, largely blocked inter-channel water and sediment exchange (Zhu et al., 2016). Together with

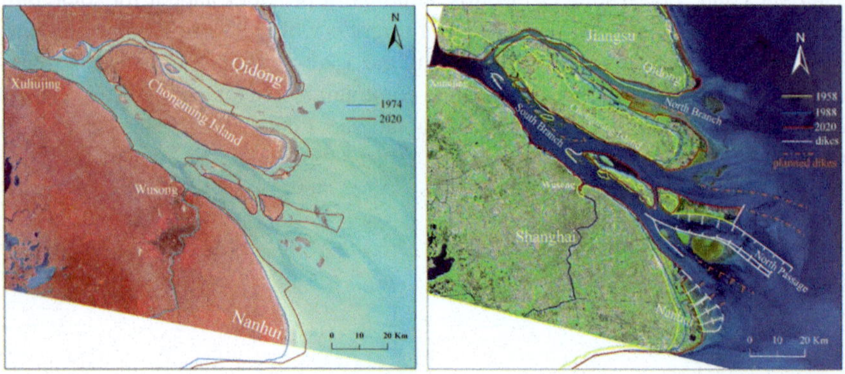

Fig. 13 Coastline changes since 1958 and human activities over the last century based on the satellite images in 1974 and 2020 during a low tide, with regulation dikes possibly imposed in future identified. From Guo et al. (2021)

a huge amount of maintenance dredging, up to 50–100 million m^3 per year along the 350 m wide and 60 km long navigation channel, the North Passage became narrower and deeper, and its subtidal flow partition rate has decreased from 50 to 40% (Zhu et al., 2019). Dumping of the dredged sediment over the landward part of the Hengsha flat and following embankment measures accelerated the sedimentation and flat accretion therein. While the short-term and local impact of human activities has been extensively studied, the long-term and large-scale impact of the human activities combined is still largely unknown and requires further investigation.

4.2 *Environment Protection*

Human activities have inevitably modified the natural condition of the delta environment, causing water pollution and ecological degradation. To address these issues, significant efforts have been made to protect the environment and achieve a balance between development and conservation in national policies. For instance, the Yangtze River Protection Law has been established in 2020. A 3–4 months of fishing ban in the coastal ocean is implemented. National natural reserves for migratory birds, Chinese sturgeon (*Acipenser sinensis*) and salt marsh wetland have been set up (Fig. 14). Since 2017, the central government has strictly controlled coastal reclamation. Sewage discharge has been collected by wastewater pipe networks and treated before discharging. For instance, the Bailonggang Sewage Treatment ranks the biggest in Asia with a treatment capacity of 280 million m^3/day. The second largest treatment is Zhuyuan, which has a treatment capacity of 220 million m^3/day and will be upgraded to treatment ability of 340 million m^3/day.

4.3 *Wetland Restoration*

In order to increase both the area and quality of coastal wetlands and saltmarshes in the CRD and to improve the ecosystem service, the China government have proclaimed some laws and regulations for wetlands protection and restoration. For instance, (1) aggressive land reclamation is banned since 2018 with limited exceptions; (2) it is aimed to remove 90% of *Spartina alterniflora* in the coastal wetlands of Shanghai, Jiangsu and Zhejiang; and (3) there is a target to maintain over 35% of the natural coastline (in length) compared to those that have been hardened with concrete and dikes.

Shanghai government launched the Chongming Dongtan Ecological Restoration Project in September 2013 to control further seaward expansion of Spartina, which provide new habitats for birds, and accelerate siltation and expansion of native species. Indeed, the species richness and abundance of birds have been in the Chongming Dongtan has been steadily increasing from 2017 to 2023 (Fig. 15).

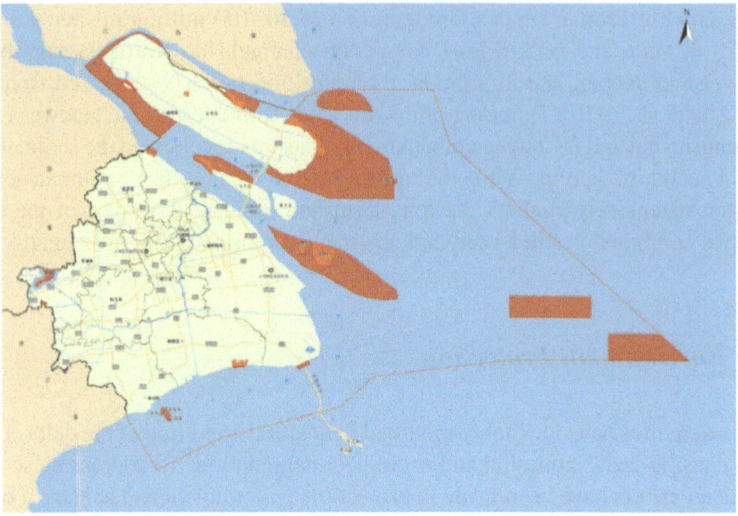

Fig. 14 The identified red zone in the CRD, including the Chongming Dongtan Nature Reserve and Jiuduansha Wetland Reserve etc., where human activities are strictly banned. From planning report of Shanghai Municipality published in 2018

Fig. 15 The effect of wetland restoration by attracting waterbirds in eastern Chongming Island

5 Management Perspectives and Solutions

Sustainable delta development needs to take account of multiple needs and adopt multi-objective management strategy. This includes ensuring the safety of shorelines and infrastructure, defending against coastal flooding, conserving ecosystems, and ensuring freshwater supply. To accomplish this, it is important to promote knowledge transfer, involve more stakeholders, manage conflicts of interest, and understand the tradeoffs involved in decision-making and strategic planning. Sediment deficits associated with SLR and reduced sediment supply from rivers is a point of concern in deltas and coasts. Rather than worrying about irreversible sediment decline that will persist in the coming decades to centuries, it would be more important to focus on properly managing the remaining sediment supply and to promote sedimentation within the delta. Lessons learned from the restoration of the Mississippi Delta, for instance, may be of use to the management of the CRD. For example, land-building sediment diversion by breaching dikes and opening barrier gates have proven effective and beneficial in sustaining natural resilience in the Mississippi Delta. Sediment diversion stimulates sediment deposition in regions close to coasts, rather than letting the river-supplied sediment flush along the channels and be lost to deeper water (Day et al., 2007; Xu et al., 2019). The utilization of dredged sediment to create new marshes and wetlands can help mitigate habitat loss caused by sediment deficiency and land subsidence. In the Rhine-Meuse Delta, flood-control defense measures have traditionally involved constructing dikes and barriers. However, opening the barrier gates to let tides do their constructive job in promoting tidal flat accretion is proved successful in conserving tidal wetlands and habitats (Smit et al., 1997). Instead of fighting against nature, living and building with nature and managed retreat to adapt SLR are options to mitigate coastal erosion and increase resilience to flood hazards, which in the meantime benefits ecosystem restoration (Temmerman et al., 2013). These pioneering experiences in Mississippi Delta and Rhine-Meuse Delta have implication for management and protection of the CRD, although they have different physical processes and social dynamics. For instance, aggressive tidal flat reclamation is no longer allowed banned since 2016, as a national strategy in achieving green and eco-development in China. This change may benefit the development of natural tidal flats and associated ecosystem.

Despite the sediment shortage and the consequent coastal erosion, the CRD is facing high siltation in the Deep-Water Navigation Channel in the North Passage. The average annual dredging volume is presently 50–60 million m^3. This dredged sediment was initially deposited in the landward part of the Hengsha flat but was later brought out to the sea as the Hengsha flat has accreted to an elevation above high tide. Disposing of dredged sediment in the sea, however, can be considered as a waste of resources. One ongoing plan is to embank the seaward part of the Hengsha flat, an area of approximately 300 km^2, as a place for dumping of dredged sediment in the coming 10–20 years. It will render the subtidal flat inter-tidal, using submerged dikes and jetties. Yet, it remains an open question how this intervention would influence the dynamics at the system scale of the CRD.

As more and more dikes and jetties have been built, the morphodynamics of the channels and shoals in the CRD have changed. This change is expected to reduce the sediment trapping efficiency and endanger the delta safety in the long run. Fortunately, tidal flat embankment and reclamation are largely prohibited, and it is time to resort to future solutions such as building with nature and adopt nature-based solutions to address erosion and flooding.

The coastal dikes surrounding Shanghai have a flood protection standard with a recurrence interval of 100–200 years (Wang et al., 2012). Raising dikes in response to SLR and land subsidence has become increasingly unsustainable given the costs and the risk of dike failure. Future flood risk management requires a comprehensive understanding of the combined impact multiple flood drivers and their joint probability of occurrence, improved forecasting of river floods and typhoon paths, as well as strategies for managed retreat. Preserving and restoring waterfront tidal wetlands, instead of reclaiming them, can help to dissipate waves and surges and mitigate flood risk.

Possible solutions to saltwater intrusion include real-time operation of the TGD by releasing more river flow to counteract saltwater intrusion in the dry seasons, and regulation of the North Branch to mitigate overspill etc. It is worth noting that saltwater intrusion in the North Branch has been alleviated in the recent decade due to channel aggradations and reduced channel volume and tidal prism. There was a plan to construct a barrier at the mouth of North Branch, but the impact on the branch and the delta system as whole remains insufficiently understood. Another long-term solution of water supply is to plan a new reservoir to fundamentally increase the freshwater supply capacity and the resilience under SLR.

The previous approach to managing the CRD was focused on constructing dikes, jetties, groins, and embankments to increase its morphodynamic stability and promote infrastructure development, e.g., fixing the division of the branches, reducing channel migration and sand bar movement, and reclaiming tidal flats. Now it is time to reevaluate this stabilization strategy and its benefits for and long-term impacts on the delta system behavior, and to work out more sustainable and resilient measures (Fig. 16). As the delta becomes increasingly constrained, the tidal prism may decrease and the canalized deltaic channels may become more efficient in flushing fluvial sediment to the sea, which would mean a decline in sediment residence time and sediment trapping efficiency, and higher vulnerability to erosion and delta shrink in longer term. Thus, instead of being deposited in the delta to build up shoals and wetlands, a larger portion of the sediment will be lost to the sea due to the pipeline effect, like in the Mississippi Delta (Paola et al., 2011). This trend is exacerbated by the decline in riverine sediment supply and SLR. Tidal wetlands and ecosystems are increasingly valued in the coastal management community, and there is an ongoing mindset change from coastal development to ecosystem conservation throughout the Changjiang river-delta continuum. After all, hazard mitigation and multi-objective management of the CRD require integrated system thinking and long-term perspectives based on improved interdisciplinary scientific understanding of the deltaic hydro-morphodynamics and their coupled interaction with society and the ecosystem, including the responses and feedback mechanisms.

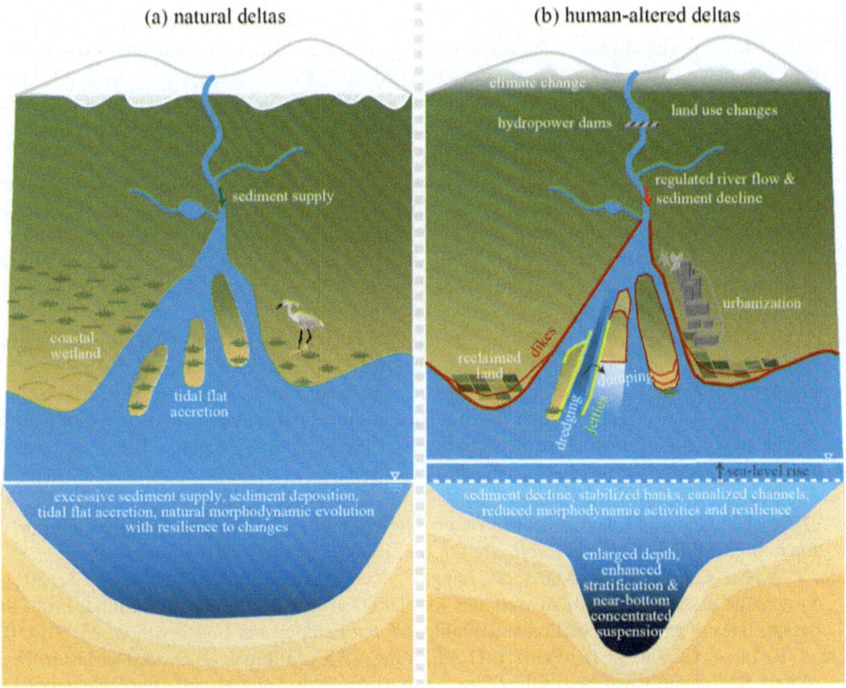

Fig. 16 Sketches of **a** natural deltas and **b** deltas under significant anthropogenic interventions with conditions mimicking the CD. Adopted from Guo et al. (2021)

References

Balica, S. F., Wright, N. G., & van der Meulen, F. (2012). A flood vulnerability index for coastal cities and its use in assessing climate change impacts. *Natural Hazards, 64*, 73–105.

Chen, X. Q. (1991). Sea-level changes since the early 1920's from the long records of two tidal gauges in Shanghai, China. *Journal of Coastal Research, 7*(3), 787–799.

Chen, H. Q., Chen, J. Y., Chen, Z. J., Ruan, R. L., Xu, G. Q., Zeng, G., Zhu, J. R., Dai, Z. J., Chen, X. Y., Gu, S. H., Zhang, X. L., Wang, H. M. (2018). Mapping sea level rise behavior in an estuarine delta system: A case study along the Shanghai coast. *Engineering, 4*(1), 156–163.

Day, J. W., Boesch, D. F., Clairain, E. J., Kemp, G. P., Laska, S. B., Mitsch, W. J., Orth, K., Mashriquie, H., Reed, D. J., Shabman, L., Simenstad, C. A., Streever, B. J., Twilley, R. R., Watson, C. C., Wells, J. T., & Whigham, D. F. (2007). Restoration of the Mississippi Delta: Lessons from Hurricanes Katrina and Rita. *Science, 315*, 1679–1684.

Division of Marine Monitoring and Warning (DMMW), State Oceanic Administration of China. (2021). Bulletin of sea level changes in China 2020.

Earth Science Division of Chinese Academy of Sciences. (1994). Impact of sea level rise on the deltaic regions of China and its mitigation. Science Press.

Gallo, M. N., & Vinzon, S. B. (2005). Generation of overtides and compound tides in the Amazon Estuary. *Ocean Dynamics, 55*, 441–448.

Giosan, L., Syvitski, J., Constantinescu, S., & Day, J. (2014). Climate change: Protect the world's deltas. *Nature, 516*, 31–33.

Guo, L. C., Su, N., Townend, T., Wang, Z. B., Zhu, C. Y., Zhang, Y. N., Wang, X. Y., & He, Q. (2019). From the headwater to the delta: A synthesis of the basin-scale sediment load regime in the Changjiang River. *Earth-Science Reviews, 197*, 102900.

Guo, L. C., Zhu, C. Y., Xie, W. M., Xu, F., Wu, H., Wan, Y. Y., Wang, Z. H., Zhang, W. G., Shen, J., Wang, Z. B., & He, Q. (2021). Changjiang Delta in the Anthropocene: Multi-scale hydro-morphodynamics and management challenges. *Earth-Science Review, 223*, 103850.

He, X. C., Yang, T. L., Shen, S. L., Xu, Y. S., & Arulrajah, A. (2019a). Land subsidence control zone and policy for the environmental protection of Shanghai. *International Journal of Environmental Research and Public Health, 16*, 2729.

He, Z. F., Li, F. L., Dominech, S., Wen, X. H., & Yang, S. Y. (2019b). Heavy metals of surface sediments in the Changjiang (Yangtze River) Estuary: Distribution, speciation and environmental risks. *Journal of Geochemical Exploration, 198*, 18–28.

He, Y. R., Wang, Y. H., & Wu, H. (2022). Regulation of algal bloom hotspots under mega estuarine constructions in the Changjiang River Estuary. *Frontiers in Marine Science, 8*, 791956.

Huang, Y. G., Yang, H. F., Jia, J. J., Li, P., Zhang, W. X., Wang, Y. P., Ding, Y. F., Dai, Z. J., Shi, B. W., & Yang, S. L. (2022). Declines in suspended sediment concentration and their geomorphological and biological impacts in the Yangtze River Estuary and adjacent sea. *Estuarine Coastal Shelf Science, 265*, 107708.

Li, X., Zhang, X., Qiu, C. Y., Duan, Y. Q., Liu, S. A., Chen, D., Zhang, L. P., & Zhu, C. M. (2020). Rapid loss of tidal flats in the Yangtze River Delta since 1974. *International Journal of Environmental Research and Public Health, 17*, 1636.

Lin, S. W., Li, X. Z., Yang, B., Ma, Y. X., Jiang, C., Xue, L. M., Wang, J. J., & Yan, Z. Z. (2021). Systematic assessments of tidal wetlands loss and degradation in Shanghai, China: From the perspectives of area, composition and quality. *Global Ecology and Conservation, 25*, e01450.

Liu, J. P., Li, A. C., Xu, K. H., Velozzi, D. M., Yang, Z. S., Milliman, J. D., & DeMaster, D. J. (2006). Sedimentary features of the Changjiang River-derived along-shelf clinoform deposit in the East China Sea. *Continental Shelf Research, 26*, 2141–2156.

Liu, Y. X., Deng, B., Du, J. Z., Zhang, G. S., & Hou, L. J. (2019). Nutrient burial and environmental changes in the Yangtze Delta in response to recent river basin human activities. *Environmental Pollution, 249*, 225–235.

Luan, H. L., Ding, P. X., Yang, S. L., & Wang, Z. B. (2021). Accretion-erosion conversion in the subaqueous Yangtze Delta in response to fluvial sediment decline. *Geomorphology, 382*, 107680.

Paola, C., Twilley, R. R., Edmonds, D. A., Kim, W., Mohrig, D., Parker, G., Viparelli, E., & Voller, V. R. (2011). Natural processes in delta restoration: Application to the Mississippi Delta. *Annual Reviews of Marine Science, 3*, 67–91.

Qin, Z. H., & Li, Y. P. (1997). Discussion of the sea-level variations and its long-term projection methods in Shanghai. *Acta Oceanologica Sinica, 19*, 1–7.

Ran, X. B., Liu, S., Liu, J., Zhang, J. Y., Che, H., Ma, Y. X., Wang, Y. B. (2016). Composition and variability in the export of biogenic silica in the Changjiang River and the effect of the Three Gorges Reservoir. *Science of the Total Environment, 571*, 1191–1199.

Smit, H., Smits, R., van der Velder, G., & Coops, H. (1997). Ecosystem responses in the Rhine-Meuse Delta during two decades after enclosure and step toward estuary restoration. *Estuaries, 20*, 504–520.

State Oceanic Administration People's Republic of China (SOA). (2020). China Sea level rise bulletin 2019. China Ocean Press.

Syvitski, J. P. M., Kettner, A. J., Overeem, I., Hutton, E. W. H., Hannon, M. T., & Brakenridge, G. R. (2009). Sinking deltas due to human activities. *Nature Geoscience, 2*(10), 681–686.

Temmerman, S., Meire, P., Bouma, T. J., Herman, P. M. J., Ysebaert, T., & de Vriend, H. J. (2013). Ecosystem-based coastal defense in the face of global change. *Nature, 504*, 79–83.

Tessler, Z. D., Vorosmarty, C. J., Grossberg, M., Gladkova, I., Aizenman, H., Syvitski, J. P. M., & Foufoula-Georgiou, E. (2015). Profiling risk and sustainability in coastal deltas of the world. *Science, 349*, 638–643.

Wang, J., Gao, W., Xu, S. Y., & Yu, L. Z. (2012). Evaluation of the combined risk of sea level rise, land subsidence, and storm surges on the coastal areas of Shanghai, China. *Climatic Change, 115*, 537–558.

Wang, Q. S., Pan, C. H., & Zhang, G. Z. (2018). Impact of and adaptation strategies for sea-level rise on Yangtze River Delta. *Advances in Climate Change Research, 9*, 154–160.

Wang, Y. N., Hu, H., & Li, M. T. (2021a). Long-term changes in phytoplankton communities in China's Yangtze Estuary driven by altered riverine fluxes and rising sea surface temperature. *Geomorphology, 376*, 107566.

Wang, Y. J., Liu, D. Y., Xiao, W. P., Zhou, P., Tian, C. G., Zhang, C. S., Du, J. Z., Guo, H., & Wang, B. D. (2021b). Coastal eutrophication in China: Trend, sources, and ecological effects. *Harmful Algae, 107*, 102058.

Wang, Y., Gao, L., Ming, Y., & Zhao, L. B. (2023). Recent declines in nutrient concentrations and fluxes in the lower Changjiang River. *Estuaries and Coasts, 46*, 1475–1493.

Wei, K., Ouyang, C. J., Duan, H. T., Li, Y. L., Chen, M. X., Ma, J., An, H. C., & Zhou, S. (2020). Reflections on the catastrophic 2020 Yangtze River basin flooding in southern China. *The Innovation, 1*(2), 100038.

Xu, K. H., Bentley, S. J., Day, J. W., & Freeman, A. M. (2019). A review of sediment diversion in the Mississippi River Deltaic plain. *Estuarine, Coastal and Shelf Science, 225*, 106241.

Xu, G. (1995). Evolution and development of the Changjiang River estuarine delta and its impact on the adjacent sea water. Ph.D. Dissertation of East China Normal University

Xue, Y. Q., Zhang, Y., Ye, S. J., Wu, J. C., & Li, Q. F. (2005). Land subsidence in China. *Environmental Geology, 48*, 716–720.

Yang, H. F., Zhu, Q. Y., Liu, J. A., Zhang, Z. L., Yang, S. L., Shi, B. W., Zhang, W. X., & Wang, Y. P. (2023). Historic changes in nutrient fluxes from the Yangtze River to the sea: Recent response to catchment regulation and potential linkage to maritime red tides. *Journal of Hydrology, 617*, 129024.

Yin, J., Lin, N., Wang, Y. H., Pringle, W. J., Tan, J. K., Westerink, J. J., & Yu, D. P. (2021). Hazard assessment for typhoon-induced coastal flooding and inundation in Shanghai, China. *Journal of Geophysical Research: Oceans, 126*, e2021JC017319.

Zhang, J., Wu, Y., Zhang, Y. Y. (2015). Plant nutrients and trace elements from the Changjiang watershed to the East China Sea. In J. Zhang (Ed.), *Ecological continuum from the Changjiang (Yangtze River) watersheds to the East China Sea continental margin*. Springer.

Zhang, W. G., Xu, Y.-J., Guo, L. C., Lam, S.-N., Xu, K. H., Yang, S. L., Yao, Q., & Liu, K.-B. (2022a). Comparing the Yangtze and Mississippi River Deltas in the light of coupled natural-human dynamics: Lessons learned and implications for management. *Geomorphology, 399*, 108075.

Zhang, J., Zhang, G., Du, Y., Zhang, A. Y., Chang, Y., Zhou, Y. C., Zhu, Z., Wu, Y., Zhang, Z. F., & Liu, S. M. (2022b). From the water sources of the Tibetan Plateau to the ocean: State of nutrients in the Changjiang linked to land use changes and climate variability. *Science China: Earth Sciences, 65*(11), 2127–2174.

Zhao, J., Guo, L. C., He, Q., Wang, Z. B., van Maren, D., & Wang, X. Y. (2018). An analysis on half century morphological changes in the Changjiang Estuary: Spatial variability under natural processes and human intervention. *Journal of Marine Systems, 181*, 25–36.

Zhou, J. J., Zhang, M., Lu, P. Y. (2013). The effect of dams on phosphorus in the middle and lower Yangtze River. *Water Resources Research, 49*(6), 3659–3669.

Zhu, L., He, Q., Shen, J., & Wang, Y. (2016). The influence of human activities on morphodynamics and alteration of sediment source and sink in the Changjiang Estuary. *Geomorphology, 273*, 52–62.

Zhu, C. Y., Guo, L. C., van Maren, D. S., Tian, B., Wang, X. Y., He, Q., & Wang, Z. B. (2019). Decadal morphological evolution of the mouth zone of the Yangtze Estuary in response to human interventions. *Earth Surface Processes and Landforms, 44*, 2319–2332.

Zhu, J. R., Chen, X. Y., Li, L. J., Wu, H., Gu, J. H., & Lyu, H. H. (2020). Dynamics mechanism of an extremely severe saltwater intrusion in the Changjiang Estuary in February 2014. *Hydrology Earth System Science, 24*, 1–14.

Danube River Delta

Liliana Ene, Gabriel Lupu, Adrian Burada, Marian Mierlă, Mihai Doroftei, Silviu Covaliov, Iuliana-Mihaela Tudor, Orhan Ibram, Aurel Năstase, Alexandru Doroşencu, Lucian-Eugen Bolboacă, Mihai Marinov, Vasile Alexe, and Marian Tudor

1 Physical setting

1.1 The Danube River

The Danube River is the second longest European river. It rises in Germany's Black Forest and flows about 2,850 km to the Black Sea. The Danube has been an important international waterway for centuries. The river basin includes 19 countries: Albania, Austria, Bosnia and Herzegovina, Bulgaria, Croatia, Czech Republic, Germany, Hungary, Italy, Republic of Moldova, Montenegro, North Macedonia, Poland, Romania, Serbia, Slovakia, Slovenia, Switzerland and Ukraine (Fig. 1). The average water flow rate is 6,500 m^3/s and the average annual sediment transport is about 67.6 million tons.

1.2 The Danube Delta

The Danube Delta is the second largest river delta in Europe, after the Volga Delta (Fig. 2), and is the best preserved on the continent. The greater part of it lies in Romania (Tulcea County), with a small part in Ukraine (Odesa Oblast). Its approximate surface area is 4152 km^2, of which 3446 km^2 in Romania (Fig. 2). The Danube Delta is a very low flat plain, lying 0.52 m above Mean Black Sea Level (MBSL) with a general gradient of 0.006 m/km. Being close to the Black Sea level, the hypsometry is limited to very narrow range of values. The maximum difference in altitude is 15 m between the highest point (+12.4 m) of the Letea dunes and the lowest lake

L. Ene · G. Lupu (✉) · A. Burada · M. Mierlă · M. Doroftei · S. Covaliov · I.-M. Tudor · O. Ibram · A. Năstase · A. Doroşencu · L.-E. Bolboacă · M. Marinov · V. Alexe · M. Tudor
Danube Delta National Institute for Research and Development, Tulcea 820112, Romania
e-mail: gabriel.lupu@ddni.ro

W. Zhang and H. de Vriend (eds.), *Delta Sustainability*,
https://doi.org/10.1007/978-981-97-7259-9_4

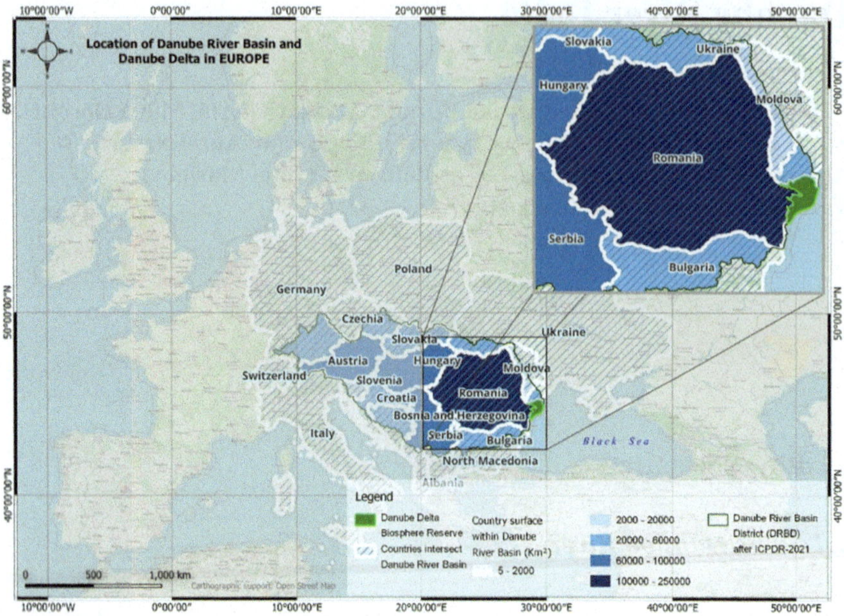

Fig. 1 The location of the Danube River in Europe

bottom (−3 m) from the marine part in the delta. Compared to the Black Sea level, only 20.5% of the delta area is below 0 m, and 54.6% is in the range of 0–1 m above MBSL. More than 93% of the delta area is within the height range of 3 m hypsometry (Bondar et al., 1999; Gâştescu & Driga, 1981, Gâştescu et al.,1985; Vespremeanu, 1984; Vespremeanu et al., 2004).

The main morpho-hydrographic units are: the predeltaic territories, fluvial banks and maritime banks, the hydrographic network (main branches, tributaries and canals), lakes and swampy lands (Gâştescu et al., 1985). Predeltaic territories are located in the Bugeac area to the north of the Chilia branch of the river. The loess deposits were eroded by water and redeposited to form the basis of the Câmpul Chiliei and Stipoc sand bank. They account for 2.4% of the delta area. River sand-banks are located along the margins of the main branches and near crossways. The rate of accretion and the sandbar height tend to decrease closer to the sea. Coastal currents are primarily responsible for the way in which marine sandbanks are formed and they tend to lie parallel to the coastline. Many of these banks (or Grindul) were then 'stranded' as additional marine deposits formed beyond them. Maps of the delta show clearly that the larger sand bars were deposited as series of high dunes with slacks between them. The initial banks were formed in the Letea, Caraorman and Crasnicol areas. The network of rivers, side channels and canals determine the way in which water is able to move through and around the delta. Dredging of the main channels, including some of the smaller ones that are used by commercial fishermen and tourists, means that water flows freely through them. Some of the side channels

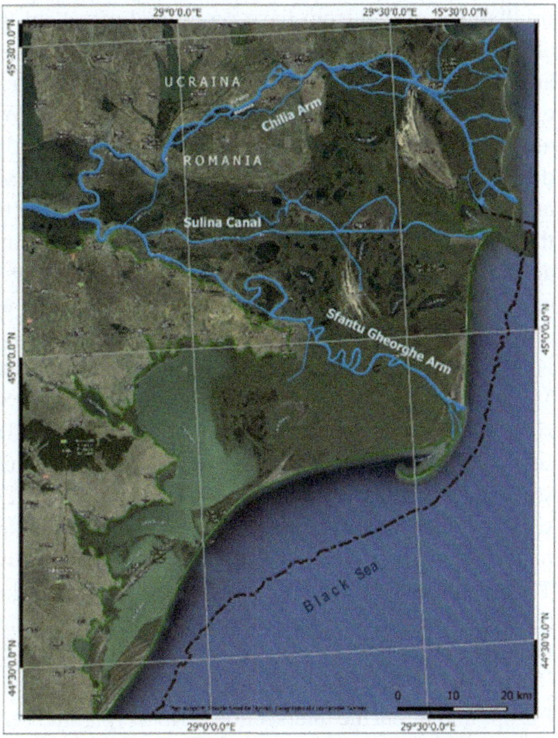

Fig. 2 The Danube Delta (on Google Satellite Hybrid)

have been blocked to reduce pollution and siltation within some of the lakes. The cutting of canals, like Dranov, Sireasa and Litcov, has improved the circulation of water. Other canals like Mila 35, Uzlina, Crişan-Caraorman and Filat have tended to clog ecologically important areas.

The lakes in the delta are arguably the most important morphohydrographic elements, some of which are designated as strictly protected areas. Drainage of some of the lakes in the Pardina and Sireasa areas since 1980, reduced the total lake area in the delta from some 31 262 ha (9.28% of the area of the delta) to 25,794 ha (7.28% of the delta's area). The larger lakes include Dranov (2,170 ha), Gorgova (1,377 ha), Rosu (1,445 ha) and Lumina (1,367 ha). Whereas there were 668 lakes in the Romanian sector of the delta before 1980, drainage has reduced this number to 479. Swamps lie between 0.5 and 1 m above water level and they surround lakes in the depression zones within the delta. They tend to be inundated during early summer, when meltwater from the mountains increase the discharge of the Danube as it runs into the delta. The marshy vegetation that flourishes in these areas includes the economically and ecologically important reed (*Phragmites australis*).

The Romanian seacoast is like a submerged plain, with very few morphological irregularities. The continental flat decreases in width from 170 km in the North to 130 km in the South. Since the Black Sea is almost totally landlocked, tidal oscillations of 7–11 cm have a minimal effect on the shoreline of the delta. The effects of wave action are highly significant, but these vary according to exposure and the distance of wind fetch across the Black Sea. Wave action and the seiching effect vary according to wind direction and speed. Waves heights of up to 2.5–4 m are normal, but waves of 7 m were measured. Steady winds also cause drift up or down the coast and the currents generated are capable of eroding existing coasts and of depositing materials. It has been calculated that 60 km of the shoreline of the Bioshpere Reserve is in retreat (about 50 ha/annum), whereas the coastline is advancing into the sea along a 50 km stretch (adding some 10 ha/annum). The level of the Black Sea is rising as a response to global warming. The most prominent changes in shoreline morphology can be observed in the way where the Chilia arm is accumulating new land at a rate of around 2 ha per annum. Between 1870 and 1992, the length of this arm increased by 7 km. The Sulina arm has been developed since 1862, to provide a major navigation route to Tulcea, Galati and Braila. New sections of channel were cut between Tulcea and Sulina to reduce the distance by some 28 km. The natural formation of Sacalin Island on off shore sandbanks, to the south of St George dates back to 1897. A catastrophic flood deposited plenty of the material and the island had grown to 10 km in length by 1924 and 17 km by 1992. The island has now virtually joined up with the mainland, the Gârla de Mijloc mouth has been blocked and the channel is becoming narrower and narrower (Gâştescu & Driga, 1981).

Danube Delta Biosphere Reserve (D.D.B.R.) is located in the Eastern part of Europe and lies at the intersection of 45°N (parallel of latitude) with 29°E (longitude) (Fig. 3). The total area of D.D.B.R. is about 5800 km^2, more than half of which (3510 km^2) belong to what is commonly called the "Danube Delta". The remaining area is shared between the upstream Danube flood plain (Isaccea-Tulcea sector 102 km^2), the Razim-Sinoie lagoon complex (1145 km^2), the neighboring strip from the Black Sea (1030 km^2) up to the 20 m isobath, and the Danube River between Cotul Pisicii and Isaccea (13 km^2).

The Romanian D.D.B.R has a special universal value, with a triple conservation status: (1) as an entirely biosphere reserve, through inclusion in the world network of biosphere reserves within the UNESCO "Man and Biosphere" Program in 1990; (2) as an wetland area of international importance, especially as a habitat for waterfowl, according to the Ramsar Convention, to which Romania joined in 1991; and (3) as an area of universal natural heritage, included in 1990 on the List of UNESCO World Cultural and Natural Heritage. By its size, the Danube Delta is the most important wetland area in the South-Eastern Europe, having an important contribution to the regional and global water cycle.

The climate around the Danube Delta is continental, with hot dry summers and very cold winters. The climate supports three external influences, as a result of the "buffer" position of the D.D.B.R between the bordering continental land that surrounds it to the north, west and south and the Black Sea to the East: continental, pontic and advection air influences.

Under the influence of the bordering land and the Black Sea, precipitation gradually decreases from west to east. A significant contribution to the annual amount

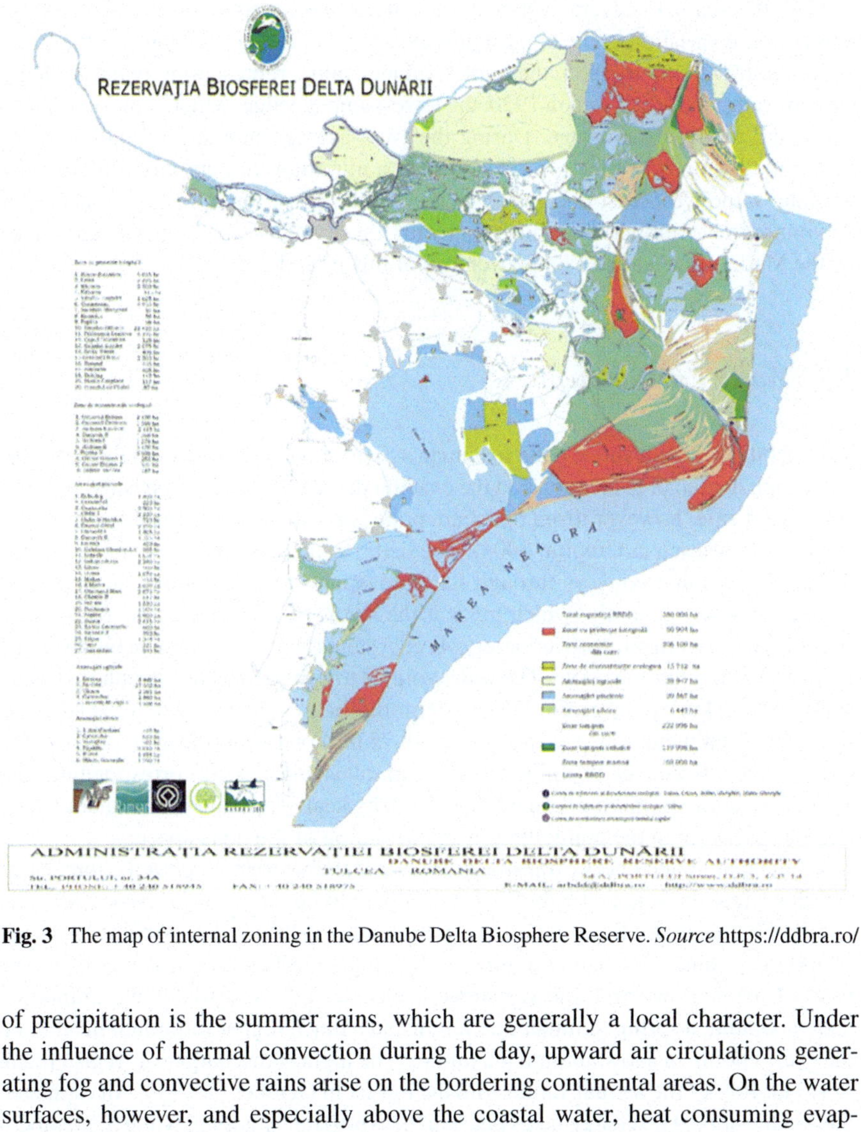

Fig. 3 The map of internal zoning in the Danube Delta Biosphere Reserve. *Source* https://ddbra.ro/

of precipitation is the summer rains, which are generally a local character. Under the influence of thermal convection during the day, upward air circulations generating fog and convective rains arise on the bordering continental areas. On the water surfaces, however, and especially above the coastal water, heat consuming evaporation processes lead to temperature inversions, characterized by downward air currents. These cause the breakdown of cloud systems and a decrease or absence of precipitation. In this context, the average annual rainfall quantities are:

- in the deltaic space: Tulcea 438.4 mm, Gorgova 406.9 mm, Sfântu Gheorghe 403.6 mm, Suli- na-dig 330.5 mm;
- in the Razim-Sinoie Lagoon Complex: Jurilovca 386.6 mm, Dranov 356.5 mm and Gura Portiţei 327.2 mm.

Over the years, the average annual amounts of precipitation have exhibited large non-periodic variations. The biggest pluviometric contrasts occur right on the coast, at Sulina: 690.5 mm in 1935, the highest annual value, representing twice the long-term average, and 132.7 mm in 1920, the lowest annual value, which represents about 1/3 of the multi-year average. During the year, average monthly rainfall amounts record a maximum in June (45–55 mm) and a minimum in February (18–35 mm), with the same west–east decreasing trend. On the coast, there is also a secondary maximum in November–December, but with lower values (30–38 mm), determined by the Mediterranean and Pontic cyclones of this period.

1.3 Water Environment

Vulnerability of the Danube delta to human activities is to a large extent reduced, due to the high degree of protection and the ensuing decrease of economic activities. The delta is not safe, however, from upstream pollution and land loss resulting from sea level rise (Gâştescu & Ciupitu, 2004). The delta, as the terminal area of the Danube River, receives and conveys through its arteries and lake complexes an important volume of water annually, which gives it the status of a wetland. One of the vital conditions of a wetland, with its variety of ecosystems and representative biodiversity, is the water circulation system. The water volume transported by the Danube to Ceatal Chilia is 205 km^3/year, given the long-term discharge of 6515 m^3/s (1921–2000). The annual sediment transport amounts to 58.75 million t/year (the averaged over the same period). Moreover, 90 million ton/year of salt is transported, corresponding with a mineralization of 350 mg/L and 2,576.1 * 10^{12} kcal. This subsystem undoubtedly plays a basic role in the configuration and evolution of the delta space.

Of the matter and energy transported by the Danube, 95% reaches the three arms and flows into the sea, and only 5% is taken over by the network of backwaters and canals. Characteristic flow values: mean discharge 6515 m^3/s; maximum discharge 15500 m^3/s (June, 1970), and minimum discharge 1350 m^3/s (October, 1921). The yearly Danube flow variations registered at Ceatal Chilia record a winter minimum and maximum, a spring-summer maximum and a much higher minimum in autumn, the cycle ending in a low autumn maximum. The highest discharge was registered in May (11.46% of the annual value) with the lowest in October (5.51%). An estimated 33% of the annual discharge takes place in the April-June interval, while September–November is left with only 17–18%. The distribution of discharge over the main Danube arms (Chilia and Tulcea, the latter branching out into Sulina and St George) is uneven, with more or less significant variations over the last 150 years. Until 1890, values on the Chilia Arm seemed to increase (from 63 to 70%), to remain constant (7%) on the Sulina Arm, and to decrease on the Tulcea and the St George arms (from 30 to 23%). Subsequently, the Chilia Arm exhibited steady decreases, from 72% in 1910 to 54.3% in 1991–2000. Correction works, permanent dredging and consolidation of the banks (required by maritime navigation) made the Sulina Arm increases, from 7% in 1893 to 13.8% in 1928/1929, 15.4% in 1970 and 19.5%

in 1978, stabilising at these value (20.1% in 1991–2000). No significant evolution was observed in the St. George Arm, with discharges fluctuating between 23.0% in 1893 and 25.4% in 1991–2000. Most of the Danube discharge on the three Danube arms flows directly into the Black Sea and only a small quantity passes through the network of backwaters, canals and the lacustrine complexes (Bondar, 1991).

The Danube at the apex of the Delta (Ceatalul/River Fork Chiliei) drains a catchment area of 805300 km^2, with 120 major tributaries and an average flow of (6515 m^3/s), distributing its water and sediment over the three main arms (Fig. 2):

- Chilia, with its many ramifications and islets, is the youngest and longest (120 km) arm; its strength lies in the fact that it carries the largest part of the water and sediment volumes (58%). Throughout its evolution, it has formed several secondary deltas the first one in the Pardina depression, the second one downstream of the village of Chilia Veche and the third one beginning downstream of Periprava. River navigation takes place on this arm, the most important ports being Izmail and Vîlcov (Ukraine);
- Sulina, preferred for maritime navigation according to the studies of the Danube Commission (1856), which led to meander corrections and riverbed deepening between 1862 and 1902. As a result, the length was reduced from 92 to 63.7 km, while the water and sediment volumes increased from 7–8 to 18.8% (currently), at the expense of Chilia arm, which used to carry 70%. Sulina Town, as entrance port for marine vessels, saw changes in its economic development. In order to maintain navigability despite the siltation in the arm's mouth, a canal was built that extends progressively into the sea (currently 10 km).
- St. Gheorghe, the Southernmost and oldest arm with a length of 109 km, in 1980–1990 underwent changes by six meander cut-offs, shortening it to 69.7 km. The volume of water and the sediment flowing through this arm is about 23.2% of the Danube's total. Before its end into the Black Sea, another arm branches out on the right side, which in turn bifurcates into the Gârla de Mijloc and the Gârla Turcească. Both reach a bay called Meleaua Sfântu Gheorghe, which resulted from the emergence (1897) and further development of the islands Sacalin (currently connected to the mainland, forming a 19 km long peninsula). On the left bank of the Gârla Turcească, (the Turkish Stream) the wooden Turkish Lighthouse building was there from 1862 until 1978 when it burned down.

The chemical composition of waters within the delta is relatively homogenous. The mean value fordissolved minerals vary between 300 and 500 mg/L, according to seasonal changes and dilution factors. The Danube drains many mountainous areas where limestone is the main geological component. Not surprising, the waters of the river are high in dissolved bicarbonates, calcium, magnesium. Sodium, potassium, sulphates and chlorides are also present. Bearing in mind the influx of calcium-rich floodwaters into the delta in early summer, the fate of calcium and bicarbonates in open waters in the delta is interesting. Both show high values in spring and autumn when the pH of the water is lower. Warm weather brings about a partial transformation, as a result of intense photosynthetic activity. The pH of the water entering the delta tends to range between 7.7 and 8.5. Values in some of the lakes can be lower, due to the decomposition of organic material in vegetation and detritus. Values of pH 6.4 are not unusual and these often occur in deeper and colder water. Eutrophication can become a problem in summer, when nitrogen and phosphorous are readily available (some of these having entered the delta as pollutants in the Danube). Algal blooms reduce light penetration and this can have a damaging effect on macrophyte plants. The death and decomposition of these causes deoxygenation of the water and may lead to fish deaths. Some of the blue/green algae produce toxins and these can be particularly damaging in confined bodies of water.

1.4 Wetland Ecosystem and Biodiversity

The Danube Delta keeps its enormous biodiversity in a better state than most other deltas in Europe, even in the world. It contains a greater range of habitat types, lower and higher plants, invertebrates and vertebrates than all other deltas in Europe. Many of the species that live within the delta are unique to it, these include plants and animals.

The impressive value of biodiversity in R.B.D.D. is represented by a number of taxa (species, subspecies, form, variety) evaluated at the level of 2017 at the figure of 9581: Bacteria (209); Chromista (210); Fungi (145), Plantae (2391); Protozoa (429); Animalia (6197) (Tudor et al., 2017).

In the D.D.B.R. there are delimited areas with a differentiated regime of ecological protection, conservation and resource use. They are meant to insure the protection and conservation of their natural habitat and specific biological diversity. Moreover, their designation should support the valorization of the natural resources available, according to the needs of the local population and within the limits of the natural biological carrying capacity. The different protection regimes are:

- strictly protected areas, which have the conservation regime of scientific reserves;
- buffer zones, with the role of protection of strictly protected areas; here limited activities for capitalizing on available resources are allowed, in accordance with the approved management plans;
- areas of sustainable development, economically exploitable through traditional practices or new, non-invasive, ecologically acceptable technologies;
- ecological reconstruction areas, in which only measures are taken to restore the damaged environment, with the objective to make them sustainable.

2 Socioeconomic Characteristics

2.1 *Population and Settlements*

The first archaeological remains recorded within the delta are burial mounds at Chilia Veche and Murighiol. These date back to the Bronze Age (3,000–500 BC). Both the Caraorman and Letea sandbanks were inhabited by the end of first Iron Age, in the 6th and 7th centuries BC. These settlements continued to be inhabited until the second Iron Age. (Sarmatian tombs dating from the 2nd to the 3rd Centuries BC have also been found in hillocks at Chilia Veche). Scythian and Dacian tribes moved into the north Dobrogea in the 8th to the 6th Century BC. A Greek transit point for merchandise has been excavated near Caraorman and this was used between the 5th to the 3rd Centuries BC. The Greeks founded Histria, as an important fortified trade centre by the 3rd Century BC. There is some evidence that the Greeks may have been active in the delta from the7th to the 5th Century BC. Roman legions arrived around the delta in 29 BC and by the next year they had subjugated the Geto-Dacians and the Bastarnae (Rădulescu, 1987). The Byzantine period (a continuation of the Roman period in the East) saw the development of an intense period of navigation and trade within the delta and the surrounding Dobrogea. A series of fortified settlements were constructed between the 9th and the 14th Centuries and Genovese merchants are recorded to have been active in the Chilia Veche area at the end of the 13th and the beginning of the 14th Century (Giurescu, 1966).

The Danube European Commission (1856–1938) created the conditions that enabled Sulina to expand into a major port at the mouth of the Tulcea to Sulina arm. Farming and fishing became important economic activities that enabled this growth. Sulina itself flourished to become a porto-franco and the population grew to 9,000 residents, with another 5 thousand itinerant stevedores and other workers. The story of Sulina is, like so many other towns and villages in the Delta, one of growth and then decline. In the case of Sulina, the decline came after the end of the 1914–1918 war. More than fourteen nationalities live and coexist peacefully within the D.D.B.R and it is a good example of multi-ethnic tolerance.

The number of people living in the delta declined after the 1939–1945 war and it has fluctuated since then. Periods of economic activity, like the "reed", "fish farming" and "polder construction for farming" periods, saw increases. By 1966 the population reached its maximum, with 19,700 inhabitants in the delta. Subsequently, numbers went down to 14,300 in the delta proper in 2002 and 12,000 in 2011, according to the Census of the population and housing (2011). The results from the 2022 Cesus remains to be published yet. The population is gradually aging, as younger people tend to move to the cities, where they can enjoy the benefits of modern society. The 26 settlements in the delta are divided into seven communities and one town (Sulina).

2.2 The Main Economic Activities

We specified that since ancient times, fishing has been the main occupation of the Danube Delta inhabitants and although today the supply of fish has diminished and changed in quality, it continues to be the basic trade. A second major occupation has been (and still is) sheep and cattle breeding. Traditional agriculture has been practiced successfully by the inhabitants of the settlements situated on the relatively high river banks at low risk from flooding. Agricultural activities are under the jurisdiction of the Tulcea, Town Councils and the Central Board of Agriculture, while the Danube Delta Biosphere Authority monitors the effects of farming activities on ecosystems.

Tourism constitutes a very important activity. In the Danube Delta Biosphere Reserve, the implementation of ecotourism has been a priority since its establishment. The Danube Delta Biosphere Reserve is an area recognized for its tourist potential and a unique tourist product for Romania, due to its geographical, biological and ecological characteristics. Also, both on the territory of the D.D.B.R, and especially in the perimeter bordering the reserve, there are numerous vestiges and monuments that attest to the stages of habitation of this region, from antiquity (Aegyssus Fortress, Halmyris Fortress, Orgame/Argamum Fortress, Noviodunum Fortress, Enisala Fortress, Histria Ancient City) to present days (Art Museum, Museum of History and Archaeology, Museum of Ethnography and Folklore, Babadag Oriental Art Museum). Regarding the number of tourists, a research report accomplished by researchers from the Danube Delta National Institute for Research and Development showed that for the 2020–2021 tourist season, due to the restrictions during the pandemic, there was a decrease in the number of tourists and overnight stays compared to previous years. According to the NIS, the largest number of tourists/ per tourist season was registered in 2019, in Sulina ~7396 tourists and respectively 25689 overnight stays, and in the town of Sfântu Gheorghe ~14,185 tourists and respectively 42,158 overnight stays.

According to observations and investigations carried out during the years 2019–2022, tourism in the tourist and recreation areas of D.D.B.R. (like Sulina, Sfântu Gheorghe, Chilia, Crişan Murighiol, etc.) is in continuous development. In recent years, new types of tourism have developed, such as "cultural experiences", and new hostels have appeared that preserve the architectural and traditional elements. The "relaxation and Spa" type tourism was also noticed by the appearance of the resort and luxury tourist complexes (four and five stars) with a large accommodation capacity (e.g., Lebăda Luxury Resort & Spa) in the Crişan area, and boarding houses with swimming pool and related recreational activities (Spa) in the Murighiol area. Also, recreational activities got better organized such as trips on canals or to Letea and Sulina, through the appearance of resorts in the Crisan area and the appearance of fish restaurants with "a la carte" menu, which favors the development of local tourism.

3 Current Issues

3.1 Pollutants

As a result of the diversification of services and the development of new industries, a new series of chemical compounds has appeared, in addition to the usual anthropogenic contaminants. They can be harmful to the natural capital and implicitly also to human health. Among them, pharmaceutical residues have been recognized as one of the new threats today. In Romania, research in this field is in its infancy and only a few studies are presented, based on the results from 1 to 2 sets of measurements. Random determinations and the absence of a well-defined monitoring program with validated analytical methods are the main precise information about the behavior and effects of these compounds and/or metabolites is lacking. Let alone their transformation byproducts in aquatic ecosystems. Measurements with concrete results are presented in 3 specialized studies. Two of these concerned the concentrations of 15 compounds in the Someş River (Moldovan, 2006; Moldovan et al., 2007).

The most comprehensive study in Romania, "Preliminary survey of pharmaceutical residues in some important Romanian rivers", carried out in 2014, was based on samples taken from the Danube, Prut, Siret, Jijila, Timiş, Prahova and the Galati drinking water network, from which 43 residual pharmaceutical compounds were determined. Of course, the number of studies that only mention or simply refer briefly to the effects of pharmaceutical residue pollution on aquatic ecosystems is somewhat larger.

Another category of contaminants is formed by micro/macro plastics. The term microplastics generally refers to plastic particles with a diameter <5 mm, this being the definition used by most authors. Because they have only been investigated shortly, the long-term effects of exposure to microplastics are not known. Freshwater species are part of a complex food web with a great diversity of food types and different feeding strategies. Research on marine organisms has signalled the phenomenon of malnutrition associated with intensive feeding with microplastics that replace part of the natural food.

A study carried out in 2014 showed that the Danube at the exit from Austria transports an average amount of micro/macro plastics of 48.2 g/s (Lechner et. al., 2014). Moreover, the Austrian researchers warn us that these quantities probably underestimate the total quantity transported into the Black Sea. Another study carried out by ICPDR (2014), mentions that the Danube transports an amount between 6 and 66 kg/day of microplastics and between 7 and 161 kg/day of plastics (micro + macroplastics). These calculations were made based on the average annual hydrographic data from 2009 to 2014.

The agricultural sector has direct and indirect effects on the environment, agriculture being one of the main sources of nitrate pollution of surface and underground waters. Very high concentrations are registered even now, especially in areas with intensively practiced agriculture. Also, more than 90% of ammonia emissions come from agriculture, especially from activities such as the use of nitrogen-based inorganic fertilizers, the storage of organic agricultural fertilizers and the spreading of manure. Although the use of chemical fertilizers in the Danube Delta Biosphere Reserve is prohibited, the Danube transports significant quantities of polluting agents from the excessive use of fertilizers and pesticides upstream. Also, the use of irrigation water of inadequate quality and quantity has visible environmental consequences, ultimately leading to degradation of the deltaic ecosystems.

In 82% of the European watersheds, intensive agriculture and horticulture are the main cause of excess phosphorus. Untreated or insufficiently treated waste water discharges also contribute significantly. Combinations of chemicals and chemical mixtures, according to some studies, can have harmful effects on human health, even when the individual chemicals are below the "safe level".

3.2 Climate Change

Climate change research takes place in the context of increasingly frequent manifestations of extreme weather phenomena (maximum temperatures, minimum temperatures, intense precipitation, extreme droughts, etc.) (Mierlă, 2017). According to a research based on the European Climate Assessment & Dataset (ECAD) for the period 1950–2020 (Klein et al., 2002), the minimum temperatures continuously increased throughout the studied period (1950–2020) and in the entire study area.

The same increasing trend was also recorded for the multiannual average temperatures in the same period (1950–2020) and the same territory studied. The water surface of the Black Sea, through its thermal inertia, leads to a lower thermal amplitude and a higher average temperature. The increase in temperature over the years is a general trend. The increase in maximum temperature has intensified in the last 40 years (since 1980).

4 Management Measures

4.1 National and EU Strategy

The Danube Delta Biosphere Reserve Authority is the institution to manage the natural heritage of the national interest in the Reserve, as well as the restoration and protection of the physical and geographical units within this area. The Authority is responsible for managing public lands, having the authority and duties of environmental protection. Its management measures mainly include:

- hydrotechnical works for improved circulation of water;
- design for the improvement of environmental conditions in areas of natural reproduction of fish;

- ecological reconstructions in fishery arrangements;
- afforestation for consolidation of banks and funding of forest restoration;
- works for sediment removal and to prevent clogging of the main canals and streams.

Danube Strategy (http://www.mae.ro/strategia-duna-rii) is a European Union project developed in 2010 by the collaboration of 14 countries bordering the Danube. The four priority issues are:

- connectivity (in transport, culture, tourism, energy networks);
- environmental protection (water management, biodiversity protection, risk management);
- increasing prosperity of Danube region (through education, research, competitiveness);
- improvement of governance (i.e., institutional capacity and internal security).

This strategy resulted in a plan of action adopted in 2011, which includes a detailed analysis of the current situation and proposes the kind of projects to be implemented in the short, medium or long term. The objectives of the Plan are regularly monitored by the European Commission in the light of changing economic and social trends in the region.

4.2 The Danube Delta's Strategy in Last Decade

The purpose of the Strategy is the economic and social development of the Danube Delta area, through the increase of living standards, valorization of economic and cultural potential, as well as environmental protection and sustainable exploitation of natural heritage. To achieve these objectives the following interventions are proposed:

- increasing the quality of life through training, medical assistance, education;
- conservation of biodiversity;
- development and modernization of shipping infrastructure;
- improving the system of information and management of traffic along the Romanian sector of the Danube;
- prevention of water pollution from ships;
- more efficiency of the management system;
- developing partnerships in support of the Reserve management;
- the protection and preservation of the built heritage already existing in the Danube Delta area, such as ancient and medieval fortresses, churches, etc.;
- the development of cultural tourism, in addition to leisure and sports;
- the development of transport infrastructure in rural areas;
- preserving traditional occupations known in the Danube Delta.

Specific legislation supports the management measures. It regulates fisheries (e.g., prohibition periods and areas of fishing, zones for protection of living aquatic resources, effort and quotas allocated annually, tools and fishing methods permitted commercially, etc.). Rules on access and movement for ships and boats along the canals and inland lakes, tourism-related services, promoting ecotourism, transportation, agriculture and the system of penalties and fines are also among regulatory objectives.

5 Perspectives on Management and Solutions

The Danube Delta, which is one of the spectacular deltas of the world and the second largest in Europe, became a UNESCO World Heritage Site in 1991. This delta is classified as a biosphere reserve, the conservation and protection of biodiversity allowing sustainable economic and social development.

The problems in the delta are multiple and complex, raising even basic research questions. To solve them, everyone involved in the management of Danube Delta needs to comply with the rules and show initiative. It is absolutely necessary to preserve the heritage of the Danube Delta, in parallel with economic and social development.

References

Bondar, C. (1991). Water flow and sediment transport of the Danube at its outlet into the Black Sea. *Meteorology and Hydrology, 21*, 21–25.

Bondar, C., State, I., Cernea, D., & Blendea, V. (1999). Bilanţul morfologic al liniei de coastă a Deltei Dunării între 1962–1997, The Delta's state-of-the-art protection and management (in Romanian).

Gâştescu, P., & Ciupitu, D. (2004). Interconections between a delta and its drainage basin. State-of-the-art (The case of Danube Delta). In V. Sovocovschi (Ed.), Riscuri şi Catastrofe. Casa Cărţii de Ştiinţă. (in Romanian).

Gâştescu, P., & Driga, B. (1981). Évolution du débit liquid á l'embouchure du Danube dans la Mer Noire pendant la période 1850–1980, RRGGG - Géographie (in Romanian).

Gâştescu, P., Driga, B., & Anghel, C. (1985). Caracteristicile morfohidrografice ale Deltei Dunării ca rezultat al modificărilor naturale şi antropice actuale, Hidrologia (in Romanian).

Giurescu, C. C. (1966). Ştiri despre populaţia românească în Dobrogea în hărţi medievale şi modern (in Romanian).

ICPDR. (2014). Report-Plastics and microplastics in the Danube River. https://www.icpdr.org/main/publications/plastics-and-microplastics-danuberiver

Klein, T. A. M. G., Wijngaard, J. B., & Konnen, G. P. (2022). Daily dataset of 20th-century surface air temperature and precipitation series for the European climate assessment. *International Journal of Climatology, 22*, 1441–1453.

Lechner, A., Keckeis, H., Lumesberger-Loisl, F., Bernhard, Z., Krusch, R., Tritthart, M., Glas, M., & Schludermann, E. (2014). The Danube so colourful: A potpourri of plastic litter outnumbers fish larvae in Europe's second largest river. *Environmental Pollution, 188*, 177–181.

Mierlă, M. (2017). Some climate parameters evolution within Danube Delta Biosphere Reserve territory for 1961–2013 period, în Scientific Annals of the Danube Delta Institute.

Moldovan, Z. (2006). Occurrences of pharmaceutical and personal care products as micropollutants in rivers from Romania. *Chemosphere, 64*, 1808–1817.

Moldovan, Z., Schmutzer, G., Tusa, F., Calin, R., & Alder. A. C. (2007). An overview of pharmaceuticals and personal care products contamination along the river Somes watershed, Romania. *Journal of Environmental Monitoring, 9*, 986–993.

Rădulescu, A. (1987). Bazilici şi monumente creştine în contextul etnogenezei româneşti din secolele III – VII în Dobrogea. In *Monumente istorice şi izvoare creştine* (pp. 7-77). Arhiepiscopiei Tomisului şi Dunării de Jos, Galaţi (in Romanian).

Tudor, M., Rakosy, L., Ruicănescu, A., Niţu, E., Stănescu, M., Anastasiu, P., Lupu, G., Levente, S., Iorgu, I., Manci, O., Tudor, I. M., Oprea, A., Covaliov, S., Schneider, E., Török Zs., Doroşencu, A., Skolka, M., Marinov, M., Spiridon, C., Paraschiv, M., Ibram, O., Vasile, A., Răileanu, Ş., Năstase, A., Pricop, E., & Doroftei, M. (2017). An inventory of species from the Danube Delta Biosphere Reserve.

Vespremeanu, E. (1984). Morphological and morphodynamique aspects of the submarine relief in front of the Danube Delta (in the North-West of the Black Sea). RRGGG - Géographie, 28.

Vespremeanu, E., Vespremeanu-Stroe, A., & Constantinescu, Ş.(2004). Evoluţia ţărmului Deltei Dunării în ultimii 40 de ani. *Studii şi cercetări de oceanografie costieră, 1*, 15-30. (in Romanian).

Fly River Delta

Fan Xu and Qinzi Cheng

1 Physical Setting

1.1 The River Basin

The Fly River system, spanning roughly 75,000 km^2 in western Papua New Guinea (Fig. 1), is fed by three primary tributaries. The Ok Tedi originates from the Hindenburg Ranges, the Upper Fly from the southern sections of the Victor Emanuel Range, and the Strickland sources from both the Victor Emanuel and Central Ranges. At D'Albertis Junction, the Upper Fly and Ok Tedi converge to the Middle Fly. The Middle Fly and the Strickland merge at Everill Junction. After running through a length of 1060 km, the river discharges into the Gulf of Paupa.

The Fly basin is located in a humid tropical climate, and rainfall in the region varies with elevation (Moi et al., 2001). The mountains receive consistent heavy rainfall throughout the year, with a slightly drier spell between September and November. The southern floodplain has pronounced seasonal shifts and generally receives less rain than the mountains. The rainfall is 1,847 mm/yr at Obo near Everill Junction. The area's rainfall patterns are influenced by both El Niño and La Niña events. The average water discharge at the coast is estimated to be about 6000 m^3/s (Wolanski et al., 1997). Sediment discharge is high due to the heavy rainfall and erosion rate. It is estimated at 70–80 Mt/yr (Dietrich et al., 1999). Since 1985, mining activities in the catchment have increased the sediment discharge by 40% (Wolanski et al., 1995).

F. Xu (✉) · Q. Cheng
State Key Laboratory of Estuarine and Coastal Research, East China Normal University, Shanghai 200241, China
e-mail: fxu@sklec.ecnu.edu.cn

W. Zhang and H. de Vriend (eds.), *Delta Sustainability*,
https://doi.org/10.1007/978-981-97-7259-9_5

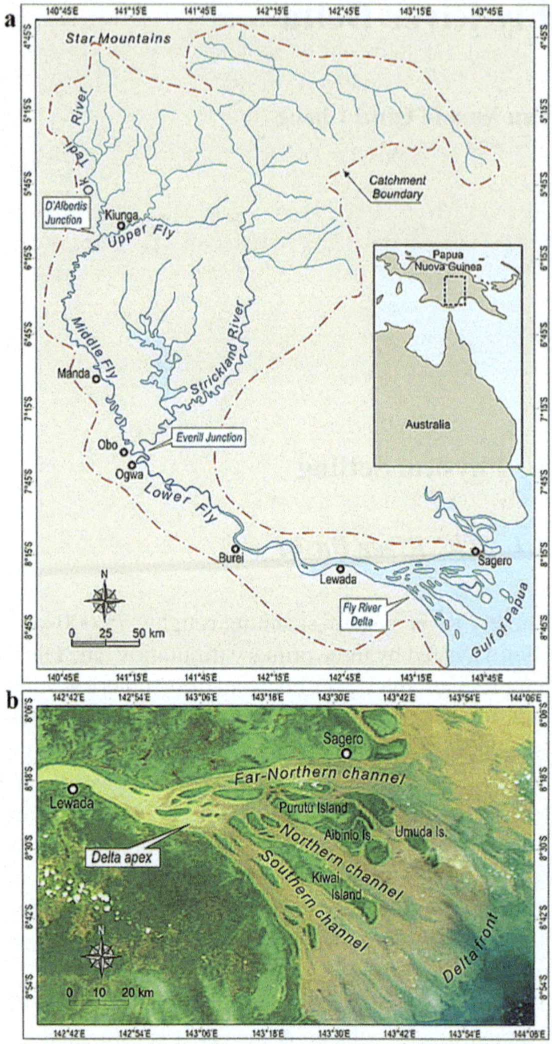

Fig. 1 **a** Catchment and **b** delta of the Fly River, Papua New Guinea (Canestrelli et al., 2010)

1.2 The Fly River Delta

The Fly River delta exemplifies a classic tide-dominated delta morphology with its large, funnel-shaped estuary and three major distributary channels: the Far Northern Entrance, Northern Entrance, and Southern Entrance (Fig. 1; Coleman & Wright, 1975; Galloway, 1975; Wright, 1985). Spanning approximately 100 km at the coast, the delta's three main channels converge at Sumogi Island (delta apex) about 100 km inland. Upstream of the apex, the tidal-fluvial channel region has one primary channel and some islands, exhibiting meandering stream characteristics like point bars, cut banks, and a thalweg along outer bends. Channel width and meander wavelength increase seaward toward the apex (Hughes & Baker, 1996). The major distributaries between the apex and the mouth are typically broad (km-scale), shallow (<10 m, rarely >12 m), relatively straight channels. Long intertidal bars up to several km long with 5–8 m relief subdivide the channels, which are maintained by strong tidal flows. While steep, eroding banks dominate some shorelines, but temporarily prograding shores also occur (Fig. 2). Vegetation colonization of the intertidal bars and prograding shores allows island formation and land expansion. The Fly River delta supports an estimated 80,000 ha of mangroves (Robertson et al., 1991). Mangroves occur on the deltaic islands, forming the main vegetation system of the islands, and in low-lying areas along the main riverbanks (Fig. 3). Seaward of the islands on the delta front are linear distributary mouth bars (Dalrymple et al., 2003). These 1–6 km wide bars are visible up to 15–20 km offshore, reflecting bedload convergence where flood and ebb channels intermingle. The gently sloping sandy delta front comprises the inner shelf of the Gulf of Papua (Harris et al., 1993), part of a large subaqueous delta clinoform feature. Sedimentary facies progresses to prodelta mud with intermixed primary and biogenic structures in the inner Gulf of Papua (Alongi et al., 1992).

Tides dominate currents in the Fly River Delta (Wolanski et al., 1995). Spring tidal ranges are up to 4 m, decreasing to around 1 m during neap tides. The average daily tidal water flux within the estuary is approximately 18 times greater than the river discharge. Tidal currents commonly reach 1 m/s, up to 2 m/s at the apex, and

Fig. 2 Images depicting various sedimentary environments in the Fly River Delta. **a** A bank undergoing erosion, **b** the sediment-accumulating bank of a newly formed island, **c** a bank experiencing progradation, and **d** the interior of an island (Walsh & Ridd, 2008)

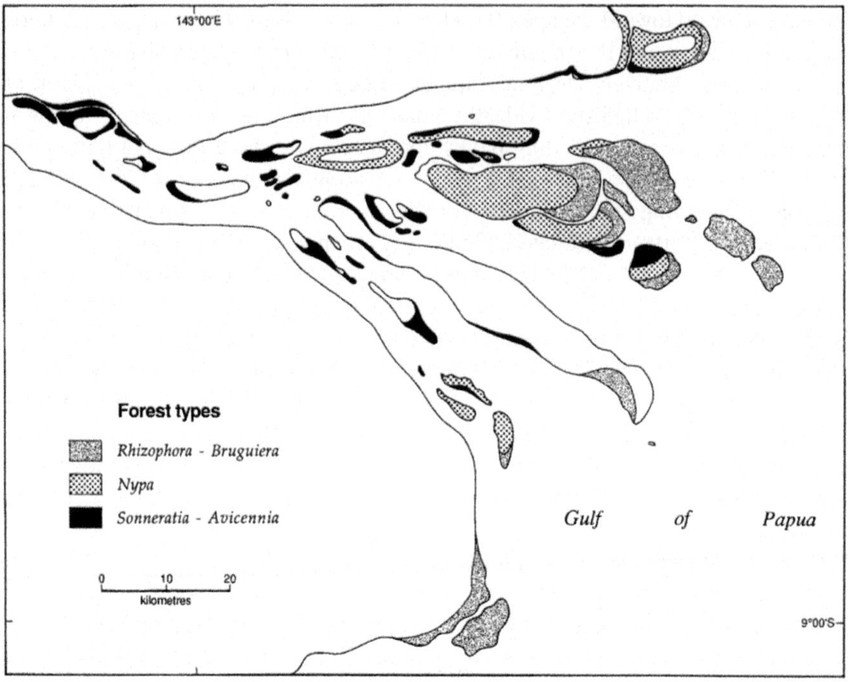

Fig. 3 Map of the Fly River delta illustrating the spatial distribution of major mangrove types (Roberston et al., 1991)

1.5 m/s at the mouth. Tidal influence propagates upstream about 400–570 km from the Gulf, depending on river discharge (Dietrich et al., 1999). Freshwater discharge is highly asymmetric with up to 80% diverted south of Sumogi Island. Seawater also intrudes further upstream in the Northern Entrance. An estimated 55–80% of the freshwater and sediment flux passes through the Southern Entrance channel. This distribution directly influences the mangrove types, with the Southern Channel favoring freshwater varieties and the Northern and Far Northern Channels saltwater ones.

The southeast-facing delta is directly exposed to waves generated by southeast trade winds from April to November with a ~1000 km fetch, allowing long period waves influencing the seabed dynamics. The gently sloping 150-km wide shelf offshore is expected to modify the wave spectrum propagating shoreward. Nearby buoy measurements reveal significant wave heights exceeding 0.25 m for 90%, 0.7 m for 50%, 1.5 m for 10%, and 2.0 m for 1% annually at Kerema (Thom & Wright, 1983). The delta likely experiences more energetic waves, given coastal alignment. Waves generated in the Coral Sea significantly influence the mouth and delta front sediment transport (Harris et al., 2004; Walsh et al., 2004). The median wave height is 1.3 m during December-March trade winds versus 0.3 m in the March–November monsoon (Thom & Wright, 1983), contributing to hypothesized seasonal cross-shelf sediment transport patterns (Harris et al., 1993, 2004; Walsh et al., 2004). Strong along-axis winds generate short-period waves within the delta, unevaluated but expected to resuspend bottom sediments on unstable banks and islands, despite small size (<0.5 m height).

2 Socio-economic Characteristics

Papua New Guinea is an island country in the equatorial Pacific region, with a total area of about 463,000 km^2, a total population of 9.7497 million in 2020, a GDP of 82.5 billion KINA (23.85 billion US dollars), and a per capita GDP of 8609 KINA (2446.1 US dollars). Papua New Guinea has rich mineral resources, such as cobalt, copper, gold, silver, natural gas, nickel and oil. In the past 30 years, mining and oil accounted for about 49% and 26% of Papua New Guinea's total commodity exports, respectively. In 2020, the total output value of mining was 28.3 billion KINA, and the total output value of agriculture, forestry and fisheries fishing and aquaculture was 21.3 billion KINA (Fig. 4). PNG is linguistically and culturally one of the most diverse countries in the world, with over 1,000 tribes and 800 languages (UN in PNG, 2017).

According to the United Nations Sustainable Development Assessment (UNDP, 2023), Papua New Guinea has a high proportion of poor population, more than half (56.6%) of the population are in multidimensional poverty, and 25.8% of the population are in severe multidimensional poverty. The urbanization rate is low, and the rural population accounts for about 80%. On the other hand, the low population density and modernization level also make Papua New Guinea a paradise for many wild animals and plants, with a forest coverage rate of 79.3%, and a biological species of up to 1786 species. It is the most species-rich island in the world, with amphibians, reptiles, birds and mammals accounting for 5–9% of the world's wildlife.

The Fly River Delta belongs to the Western Province, which is the largest province in Papua New Guinea with an area of 97,000 km^2 and a total population of 142,000 people. It is the province with the lowest population density. Due to the low level

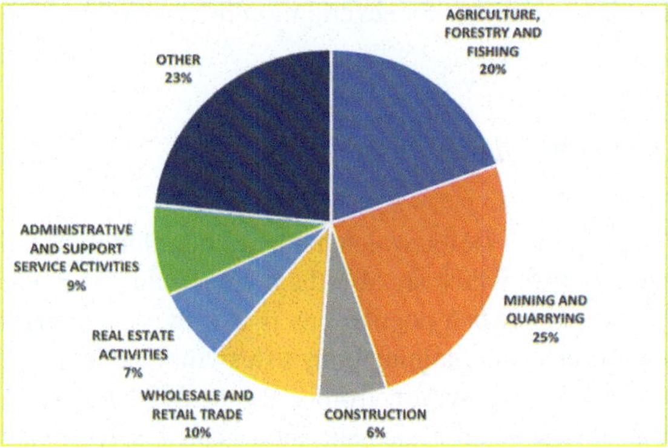

Fig. 4 Industry Share of Papua New Guinea in 2020. *Source* Papua New Guinea National Accounts 2014–2020

of the Fly River Delta, most of it consists of swampy lowlands. The population distribution on the delta is relatively scattered. The main towns are Daru on the south coast and Balimo on the north inland. The rest are scattered small villages. The residents of the Fly River Delta engage in agriculture and hunting. In addition, the Fly River Delta region also has rich mineral resources, including gold, silver, copper.

Daru is the capital of the Western Province, located in the south of the Fly River Delta. It is also the largest town in the Fly River Delta region. The 2020 census results show that Daru has a population of 20,292 people and a total area of 19.59 km^2. The economy is mainly based on fishing, agriculture and services. In terms of agriculture, rice, palm oil, bananas and other crops are planted on a certain scale. Daru is the transportation hub of the Western Province. It used to be a busy trade port for sea cucumbers, pearls and shark skins. It has now developed into a fishing center.

Balimo has a total population of 5,889 people and an area of 2.45 km^2. Planting rice, cassava and bananas are common forms of agriculture. Balimo is surrounded by vast swamps and wetlands. This area is called "Great Wetlands" and is one of the largest wetlands in Papua New Guinea. It has a rich ecosystem and wildlife.

3 Current Issues

3.1 Mining

Ok Tedi is the largest mining company in Papua New Guinea, located in the mining area upstream. It has adverse effects on the basin's up to 30,000 people and destroys farmland and forests. Even the coastal provinces that have the main gold and copper mines face food insecurity problems. The floods during the flood season also cause the spread of mining pollutants. In the Western Province bordering Indonesia, heavy rains in May and June 2012 flooded 15 coastal villages along the Fly River, destroyed food gardens, flooded seagrass meadows, and caused tributaries of the Fly River to overflow. After this, health problems were prevalent, and most marine foods were considered unsafe for several months (Asian Development Bank, 2014).

3.2 Overfishing

The Fly River Delta, which has a low population density, is a "refuge" for river sharks and sawfish, which are also the main species. In recent years, the main threat to river shark and sawfish populations is that they are caught by small-scale fishermen for the purpose of obtaining bony fish swim bladders. The results of a survey conducted during the dry season from 2017 to 2020 in the Fly River Delta region showed that among small-scale fishermen's catches, river sharks accounted for 29.4% in the southern region, and sawfish accounted for 14.8% and 20.3% in the northern and southern regions, respectively.

3.3 *Loss of Mangroves*

Due to the rapid and unsustainable development, mangroves are also shrinking. For generations, mangroves have been cut down for use as building materials and firewood. They are an important source of food and livelihood for the surrounding communities. According to Unua (1992), due to population growth, village establishment, coastal mangroves were cut down. The mangroves near Daru, for instance, have been cleared for slum development. After mangroves have been cut down, their stabilizing effect is lost and the coast often starts to erode, forcing villages to move inland.

3.4 *Drought and Flood Disaster*

The Fly River Delta is located near the equator and has a tropical monsoon climate (monsoon season from November to May). Under the influence of the monsoon, to seasonal heavy rainfall, hurricanes, floods and other event occur. These extreme weather events may cause river overflow, flood disasters and house collapse. In the dry season, prolonged drought can lead to crop reduction, food shortage and water scarcity.

Between January 20 and 26, 2021, the United Nations reported that floods caused by abnormal tides affected thousands of residents in coastal areas of Papua New Guinea. It is estimated that at least 7,000 people have moved as a result. Continuous heavy rain caused flooding along the Fly River and Strickland River in North, Middle and South Fly districts. The most affected communities were Obo Station and Middle Fly, where at least 3,172 people had no access to drinking water (United Nations Office for the Coordination of Humanitarian Affairs, 2021).

The 2015 El Niño Niño caused persistent low rainfall from December 2015 to February 2016 (World Food Programme, 2016) (Fig. 5). The 2022 La Niña caused abnormally low rainfall in most parts of the equatorial Pacific, with annual rainfall in Papua New Guinea (especially in the east) and other regions more than 80% less than the long-term average (World Meteorological Organization, 2022) (Fig. 6).

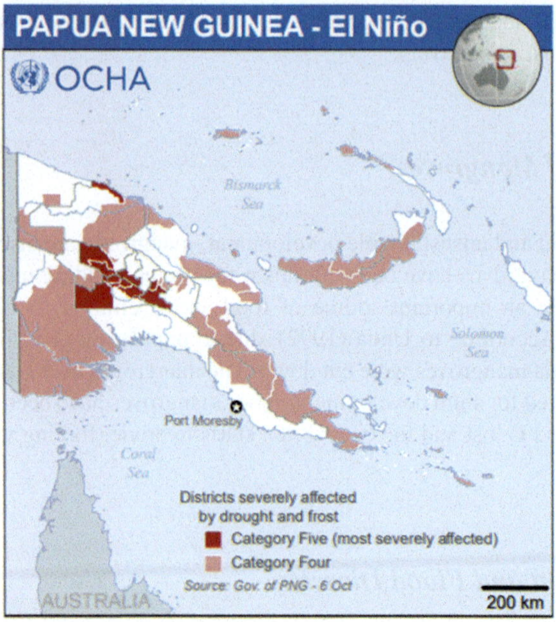

Fig. 5 Drought map for Papua New Guinea, affected by 2015 EI Nino. *Source* Papua New Guinea: Drought-UN Resident Coordinator's Office Situation Report No. 2

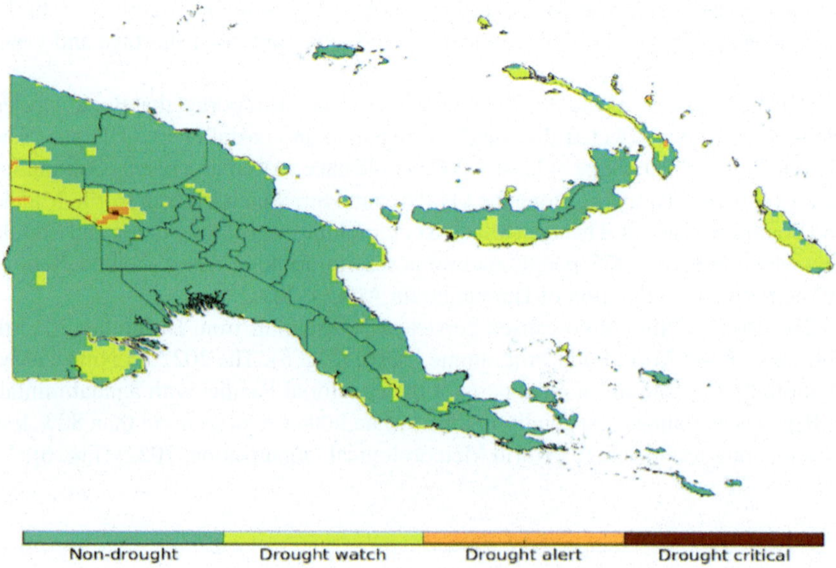

Fig. 6 3-month drought indicator for Mar 2022. *Source* PNG DMT Humanitarian Update No. 4, 20230330

3.5 *Climate Change*

3.5.1 Temperature

According to existing datasets, the average surface temperature in the Southwest Pacific region in 2022 was 0.13 °C [0.05–0.15 °C] (Figs. 7 and 8), higher than the average for 1991–2020 (WMO, 2022), and 0.46 °C [0.27–0.52 °C] higher than the average for 1961–1990. The region was much colder in 2021 and 2022 than in previous years, mainly due to the presence of La Niña conditions for most of these two years.

The "most likely" scenario provided by the Pacific Climate Change Science Program (PCCSP) indicates that by 2030, Papua New Guinea's monthly average temperature will increase by about 0.9 °C [0.4–1.0 °C] (CSIRO, 2010). This increase is higher than the temperature increase recorded in the past 30 years (0.2 °C per decade).

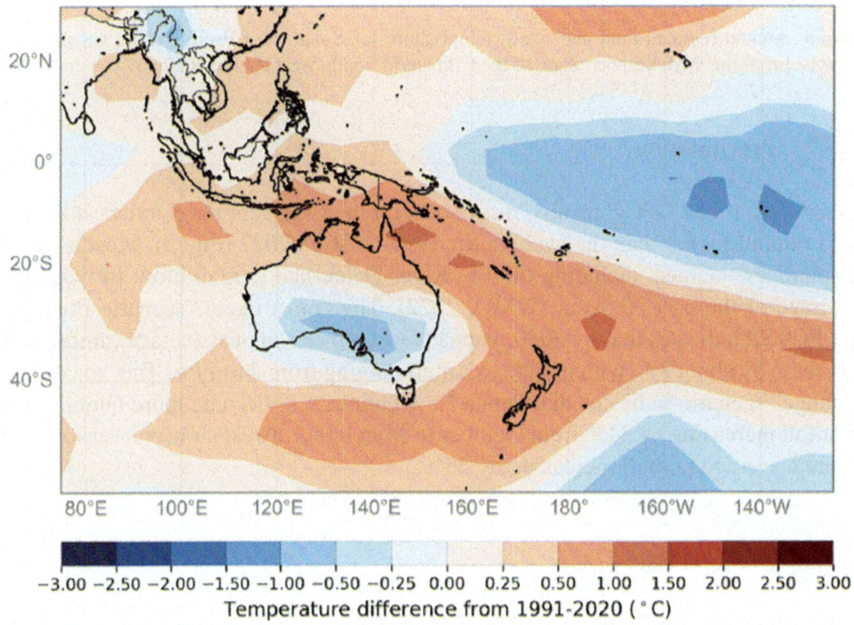

Fig. 7 Annual temperature anomalies for 2022, relative to the 1991–2020 average, using the HadCRUT5 dataset. *Source* From the WMO report, State of the Climate in South-West Pacific 2022

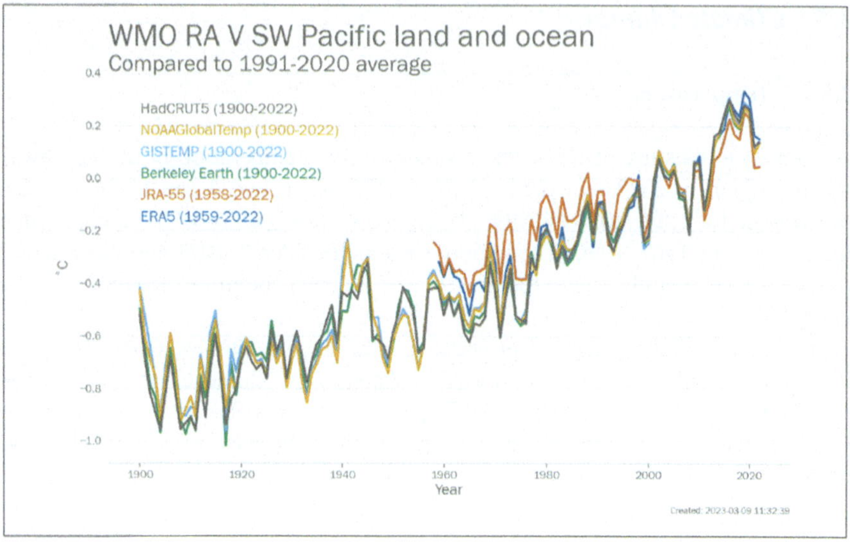

Fig. 8 Annual regional land and ocean temperature for South-West Pacific, using six datasets. *Source* From the WMO report, State of the Climate in South-West Pacific 2022

3.5.2 Precipitation

According to the GPCC results, most areas on the islands of the Southwest Pacific had rainfall higher than the average for 1991–2020 in 2022 (Fig. 9). Most areas of Papua New Guinea, including the Fly River Delta, had rainfall close to twice the average of the past 30 years (WMO, 2022). The "most likely" scenario provided by PCCSP indicates that by 2030, Papua New Guinea's annual average rainfall will increase by about 8%, with monthly rainfall ranging from 2 to 17%. The "maximum change" scenario indicates that Papua New Guinea will become more humid, with rainfall increasing by 34% from October to March (northwest monsoon season) and annual average rainfall increasing by 25%.

3.5.3 Sea level rise

Since the early 1990s, the global average sea level has risen at an average rate of 3.4 ± 0.3 mm per year (Fig. 10) due to ocean warming and land ice melting. The sea level rise in the eastern Indian Ocean, western tropical Pacific and mid-latitude Pacific north and south is higher than the global average (WMO, 2022).

The Australian Bureau of Meteorology and CSIRO predict that by 2030, the sea level in the western Pacific will rise by 40–150 mm. Since the late 19th century, the global average sea level has risen by about 200 mm. Therefore, a further rise of 40–150 mm by 2030 means that the sea level rise is accelerating.

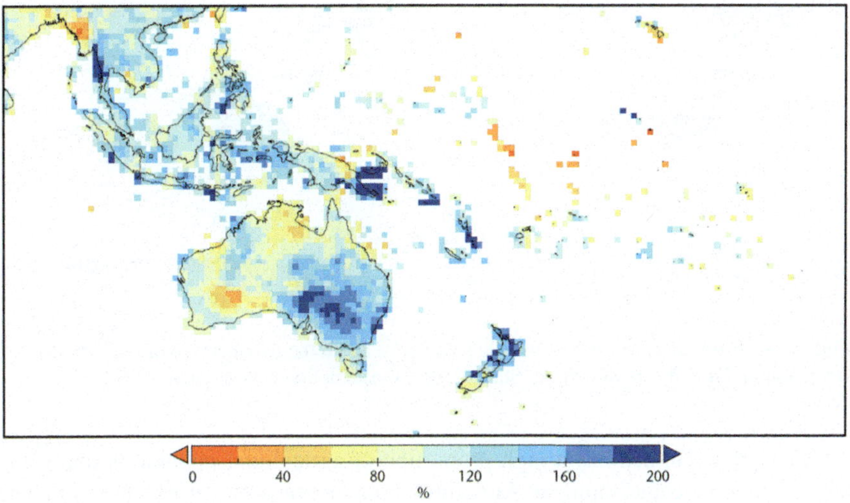

Fig. 9 Precipitation anomalies for 2022, expressed as a percentage of the 1991–2002 average. *Source* From the WMO report, State of the Climate in South-West Pacific 2022

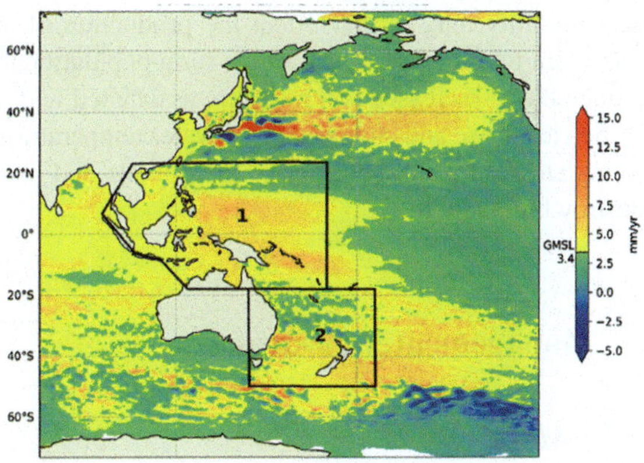

Fig. 10 Annual mean sea level trends from January 1993 to June 2022 for the western Pacific. *Source* From the WMO report, State of the Climate in South-West Pacific 2022

3.5.4 Marine Heatwaves

In 2022, severe marine heatwaves occurred in the Southwest Pacific (Fig. 11), with the most prominent and persistent one in the south of Papua New Guinea, lasting for more than six months. Under the climate warming trend, the frequency and probability of marine heatwaves and their duration may increase. This will threaten the survival of coral reefs and other marine organisms (WMO, 2022).

Climate change will not only directly affect the living conditions of local residents, but also threaten food security, increase the frequency of droughts, floods and other disasters, and increase their vulnerability.

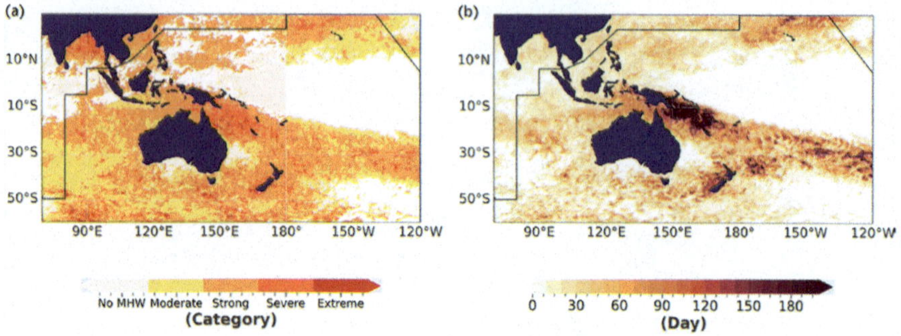

Fig. 11 **a** Maximum categories of marine heatwaves, and **b** the maximum duration over the year 2022. *Source* From the WMO report, State of the Climate in South-West Pacific 2022

Papua New Guinea has a high proportion of agricultural population and a low level of economic development. Agricultural production is easily affected by natural conditions. The population of the Fly River Delta region is sparse and scattered, with weak resistance to risks. Climate change has a great impact on food security.

Swamp taro is the main food for lowland residents. Increased precipitation and sea level rise will change the salinity and humidity of the soil. High temperature and humidity may further affect the production of taro. Sweet potato is the most important food for two-thirds of the rural population in Papua New Guinea. It is also vulnerable to increased soil moisture associated with high rainfall (Bourke, 2018).

Sea level rise and changes in sea water temperature will also threaten the original living environment of coastal aquatic plants and animals such as mangroves, and destroy biodiversity.

4 Management Measures

4.1 Mangrove Protection

An analysis conducted by the Climate Change and Development Authority (CCDA) highlighted that community-based mangrove planting is an economically viable measure for coastal communities, providing income and incentives for local residents to maintain mangroves (Asian Development Bank, 2018).

4.2 Technical Support on Agriculture

To help local residents cope with food production reduction and food security issues caused by climate change, the Food and Agriculture Organization (FAO) and the

World Food Programme (WFP) organized technical experts to provide agricultural technical support to the National Disaster Response Team of the National Agricultural Department of Papua New Guinea. This include providing agricultural tools, distributing planting materials, cultivating agricultural planting technology talents, and enhancing local residents' abilities of water saving, drought adaptation, small-scale irrigation et cetera (Bourke, 2018).

4.3 Disaster Prevention and Mitigation

Papua New Guinea has established a national-provincial disaster response system. Supported by the National Disaster Centre (NDC) as the lead coordinating agency, provincial disaster management is led by their respective Provincial Disaster Committees (PDCs).

The support of international humanitarian partners is coordinated through the Disaster Management Team (DMT), co-led by the Acting Director of the National Development Commission and the UN Resident Coordinator. These patterns support the Papua New Guinea government in responding to disasters in an informal manner (UN Country Team in Papua New Guinea, 2019). For example: In 2018, the International Organization for Migration (IOM) worked closely with the National Disaster Centre (NDC) to support national, provincial and local governments in responding to complex emergencies. This encompassed reducing disaster-induced displacement, and mitigating the negative impacts of displacement on affected populations by strengthening disaster risk reduction, preparedness, response and recovery capacities. IOM is working with local communities to increase awareness of disaster vulnerability and the ability to implement mitigation and response measures through community-based disaster risk reducing planning interventions, reducing exposure to natural disasters, building resilience and promoting adaptation to climate change.

5 Future Challenges

The Fly River Delta region has a sparse and scattered population, a high proportion of agricultural population, and weak resistance to risks. However, the existing social-economic data are mainly at the national level, and the basic statistics in the sub-national level are known to be extremely limited. Understanding the real state of poverty, vulnerability and deprivation in the Fly River Delta can be challenging due to the very low availability of reliable data. A highly decentralized society, unique local cultures and low levels of connectivity amplify the difficulty of eliminating poverty and risk management. Future climate change and sea level rise may threaten the food security and living space of local residents. The national and provincial governments in Papua New Guinea should enhance their monitoring and management systems, as well as improve early warning mechanisms, in order to effectively address the potential consequences of climate change and natural disasters on the well-being and progress of local inhabitants. Additionally, financial and technical support from international organizations and institutions, such as the United Nations, the World Health Organization, the World Bank, and others, is crucial for the development of Papua New Guinea. This assistance will aid in improving the quality of drinking water and sanitation conditions, and will enable local residents to access reliable and affordable environmental sanitation and personal hygiene facilities.

References

Alongi, D. M., Christofferson, P., Tirendi, F., & Robertson, A. I. (1992). The influence of freshwater and material export on sedimentary facies and benthic processes within the Fly Delta and adjacent Gulf of Papua (PNG). *Continental Shelf Research, 12*, 287–326.

Asian Development Bank. (2014). State of the coral triangle. Asian Development Bank.

Asian Development Bank. (2018). Community-based mangrove planting handbook for Papua New Guinea.

Bourke, R. (2018). Impact of climate change on agriculture in Papua New Guinea. In A. R. Quartermain (Ed.), *Climate change: Our environment, livelihoods and sustainability* (pp. 35–50). University of Goroka.

Canestrelli, A., Fagherazzi, S., Defina, A., & Lanzoni, S. (2010). Tidal hydrodynamics and erosional power in the Fly River delta, Papua New Guinea. *Journal of Geophysical Research, 115*, F04033.

Coleman, J. M., & Wright, L. D. (1975). Modern river deltas: Variability of processes and sand bodies. In M. L. Broussard (Ed.), *Deltas, models for exploration* (pp. 99–149). Houston Geological Society.

CSIRO. (2010). Sea level rise: Understanding the past: improving projections for the future. Wealth from the Oceans Flagship, Commonwealth Scientific and Industrial Research Organisation.

Dalrymple, R. W., Baker, E. K., Harris, P. T., & Hughes, M. G. (2003). Sedimentology and stratigraphy of a tide-dominated, foreland basin delta Fly River, Papua New Guinea. In H. Sidi, D. Nummedal, P. Imbert, H. Darman, & H. W. Posamentier (Eds.), *Tropical deltas of southeast Asia-sedimentology, stratigraphy and petroleum geology* (pp. 147–173). SEPM Special Publication No. 76.

Dietrich, W. E., Day, G. M., & Parker, G. (1999). The Fly River, Papua New Guinea: Inferences about river dynamics, floodplain sedimentation, and fate of sediment. In A. J. Miller & A. Gupta (Eds.), *Varieties of fluvial form* (pp. 345–376). Wiley.

Galloway, W. E. (1975). Process framework for describing the morphologic and stratigraphic evolution of deltaic depositional systems. In M. L. Broussard (Ed.), *Deltas, models for exploration* (pp. 87–98). Houston Geological Society.

Harris, P. T., Baker, E. K., Cole, A. R., & Short, S. A. (1993). A preliminary study of sedimentation in the tidally dominated Fly River Delta, Gulf of Papua. *Continental Shelf Research, 13*, 441–472.

Harris, P. T., Hughes, M. G., Baker, E. K., Dalrymple, R. W., & Keene, J. B. (2004). Sediment transport in distributory channels and its export to the prodeltaic environment in a tidally dominated delta: Fly River, Papua New Guinea. *Continental Shelf Research, 24*(10), 2431–2454.

Moi, A. S., Dremdap, T., & Simbina, P. (2001). APL compliance and additional monitoring program. 1999/2000 Annual Report. Appendix 1—Hydrology. Report ENV01-02, Environment Department, Ok Tedi Mining Ltd.,

Robertson, AI., Daniel, P. A., & Dixon, P. (1991). Mangrove forest structure and productivity in the Fly River estuary, Papua New Guinea. *Marine Biology, 111*, 147–155.

Thom, B. G., & Wright, L. D. (1983). Geomorphology of the Purari Delta. In T. Petr (Ed.), *The Purari: Tropical environment of a high rainfall River Basin* (pp. 47–65). Dr. W. Junk Publishers.

UN Country Team in Papua New Guinea. (2023). Papua New Guinea: Multi-Hazard Contingency Plan 2019, 2019. UN Country Team in Papua New Guinea. Management Team Humanitarian Update No. 04 (Updated as of 30th March 2023).

UN in PNG. (2017). United Nations Development Assistance Framework for Papua New Guinea 2018–2022.

UN Resident Coordinator's Office. (2016). Papua New Guinea: Drought—UN Resident Coordinator's Office Situation Report No. 2.

UNDP (United Nations Development Programme). (2023). Multidimensional Poverty Index 2023: Papua New Guinea.

United Nations Office for the Coordination of Humanitarian Affairs. (2021). Papua New Guinea: Floods.

Unua, W. B. O. (1992). Country report: Papua New Guinea. In T. Nakamura (Ed.), *Proceedings Seminar and Workshop on Integrated Research on Mangrove Ecosystems in Pacific Islands Region* (pp. 34–45). Japan International Association for Mangroves.

Walsh, J. P., Nittrouer, C. A., Palinkas, C. M., Ogston, A. S., Sternberg, R. W., & Brunskill, G. J. (2004). Clinoform mechanics in the Gulf of Papua, New Guinea. *Continental Shelf Research, 24*, 2487–2510.

Walsh, J. P., & Ridd, P. V. (2008). Processes, sediments, and stratigraphy of the Fly River Delta. In B. Bolton (Ed.), *The Fly River, Papu New Guinea: Environmental studies in an impacted tropical river system* (pp. 153–176). Elsevier.

Wolanski, E., King, B., & Galloway, D. (1995). Dynamics of the turbidity maximum in the Fly River estuary, Papua New Guinea. *Estuarine, Coastal and Shelf Science, 40*, 321–337.

Wolanski, E., King, B., & Galloway, D. (1997). Salinity intrusion in the Fly River estuary, Papua New Guinea. *Journal of Coastal Research, 13*, 983–994.

World Food Programme. (2016). Papua New Guinea: Drought.

WMO (World Meteorological Organization). (2022). Statement on the State of the Global Climate in 2022.

Wright, L. D. (1985). River deltas. In R. A. Davis (Ed.), *Coastal sedimentary environments* (2nd ed., pp. 1–75). Springer.

Ganges-Brahmaputra-Meghna River Delta

Sheikh Aftab Uddin, Lijun He, Md. Jaker Hossain, Nabila Nusrat, and Milki Debi

1 Physical Setting

1.1 The Ganges-Brahmaputra-Meghna Rivers

The Ganges-Brahmaputra-Meghna river (hereafter referred to as the GBM) is a trans-boundary river system draining through India, China, Nepal, Bhutan, and Bangladesh (Fig. 1). The Ganges River originates in the Himalayan range flowing southwest into India and then turning southeast before meeting with Brahmaputra in Bangladesh. The Ganges totals 2700 km in length and drains a catchment area of 106×10^4 km^2. Average annual water discharge amounts to 550 billion m^3, which is strongly influenced by monsoon climate.

The headwater of the Brahmaputra River is located in the Himalayan range, which flows east through the southern area of China, then flows south and southwest into eastern India and Bangladesh. The Brahmaputra River has a length of 2700 km, a catchment area of 62×10^4 km^2 and annual water discharge of 618 billion m^3. After merging with Ganges, the river is called Padama with a length of 100 km before merging with Meghna.

The Meghna River has a length of 264 km, a catchment area of 8×10^4 km^2, but a high annual water discharge of 142 billion m^3. After merging with Padama, the river flows a further length of 241 km before discharging into Bay of Bengal.

The drainage basin of the Ganges River primarily consists of highly weathered sediments and volcanic areas, resulting in a substantial clay load in the channel.

S. A. Uddin (✉) · Md. J. Hossain · N. Nusrat
Institute of Marine Sciences, University of Chittagong, Chattogram 4331, Bangladesh
e-mail: aftabims@cu.ac.bd

L. He · Md. J. Hossain · M. Debi
State Key Laboratory of Estuarine and Coastal Research, East China Normal University, Shanghai 200241, China

W. Zhang and H. de Vriend (eds.), *Delta Sustainability*,
https://doi.org/10.1007/978-981-97-7259-9_6

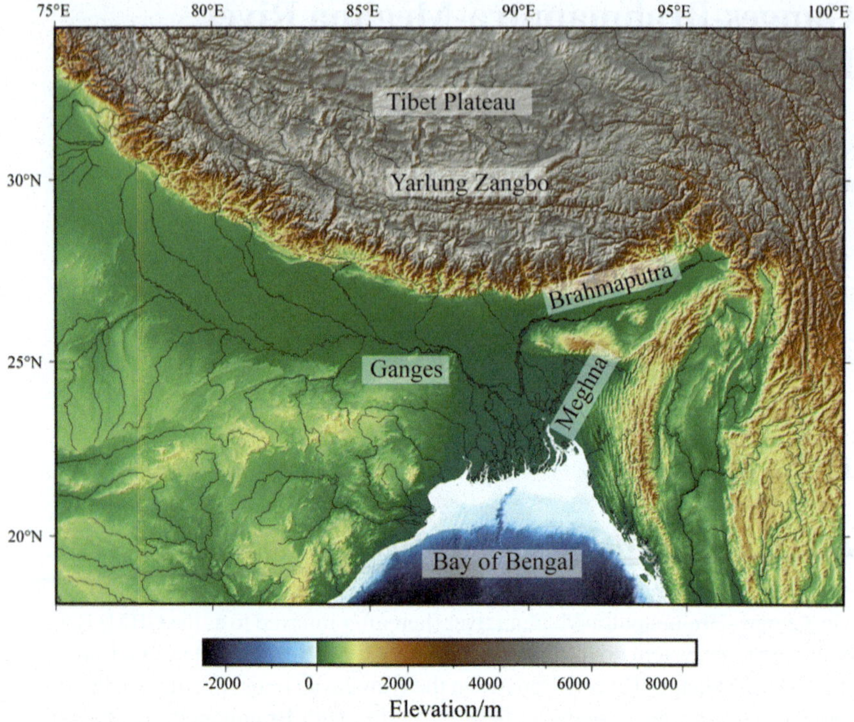

Fig. 1 The Ganges-Brahmaputra-Meghna river system

On the other hand, the Brahmaputra River basin contains young, unweathered sedimentary rocks, with limited clay available for transport. Consequently, the bedload is predominantly composed of silt and fine sand. The Brahmaputra has a slightly higher total sediment concentration compared to the Ganges.

Over the Quaternary period, the GBM rivers have occupied and abandoned numerous distinct courses, leaving a lasting mark on the region's geology. Various factors exert significant control over river migration. These factors encompass recent active tectonic movements, erratic catastrophic floods due to tropical monsoons, and the transportation of exceptionally high sediment loads, among others. The notable alterations in the flow patterns of the main rivers are closely tied to the tectonic evolution of the subcontinent. Historically, both the Ganges and Brahmaputra rivers have undergone multiple course changes. An instance of this occurred around 1787, when the Brahmaputra River avulsed into its current course. This avulsion was triggered abruptly when the Dhiang diverted the Po River in Tibet, China. Subsequently, a major flood event extended this avulsion (Coleman, 1969).

On the whole, the GBM drains an area of 1.76 million km^2 with a water discharge of 1310 billion m^3. The Himalayas, the highest mountain range in the world, have been eroding and releasing massive amounts of sediment into the rivers flowing down from the mountains. Steeper slopes, a surplus of bed load material, a highly variable discharge, and easily erodible banks favor the braided nature of the Ganges and Brahmaputra rivers in their upper reaches. It is estimated that ca. 1 billion tonnes of sediments is discharged into the delta (Coleman, 1969). Situated in the Asian monsoon climate zone, water and sediment discharges show marked seasonal variation, which peaks in June to September.

1.2 The Ganges-Brahmaputra-Meghna Delta

The Ganges-Brahmaputra-Meghna Delta (GBMD), also known as the Bengal Delta (Fig. 2), is one of the largest and most complex river deltas in the world. It is located in the southern part of Bangladesh and extends into the Indian state of West Bengal. The delta covers an area of approximately 115,000 km^2 (Schmuck, 2001; Woodroffe et al., 2006) and is formed by the confluence of three major rivers: the Ganges, Brahmaputra, and Meghna. The delta is the most fertile and vegetated alluvial land with more than 130 million inhabitants (Woodroffe et al., 2006). The delta is home to the world's largest mangrove forest, the Sundarbans, which is a UNESCO World Heritage Site, covering 10,000 km^2 which is shared between Bangladesh (60%) and India (40%) (Rahman et al., 2020).

The main driving force behind the delta formation is the combined effect of the high sediment load and the unique hydrodynamic conditions in the region. When the rivers carrying the sediment-rich water reach the Bay of Bengal, they slow down due to the decrease in slope and the influence of tides and ocean currents. As a result, the river currents lose their energy and deposit the sediment they carry, forming a vast network of distributaries and channels. The deposition of sediment is further facilitated by the large tidal range in the Bay of Bengal, which can reach up to 6 m. The tides push the sediment-laden water farther inland and help in distributing the sediments across the deltaic region. The combination of river discharge, tidal action, and oceanic processes has contributed to the growth and expansion of the delta over time.

According to spatial variation of fluvial-tidal interaction, the subaerial delta can be divided into an inland fan delta and a coastal tidal delta, with the latter being further divided into a fluvial-tidal delta in the east and tidal delta in the west (Wilson & Goodbred, 2015). The fluvial-tidal delta is located at the active river-mouth distributaries, which comprise of $\sim$5,000 km^2 of shoals, islands and channels that define the prograding coastline and subaerial-delta front. In the western part of the coastal plain, fluvial discharge is minor and tidal dominance forms the tidal delta plain, which shows funnel shaped tidal channels and mangrove vegetated islands (Wilson & Goodbred, 2015). This reflects a migration of active river mouth from west to the east in Late Holocene (Allison et al., 2003). About 50% of the GBMD sediment is deposited within a prograding subaqueous delta and delivered into the deep water through the adjacent Swatch of No Ground canyon system (Kuehl et al., 1989).

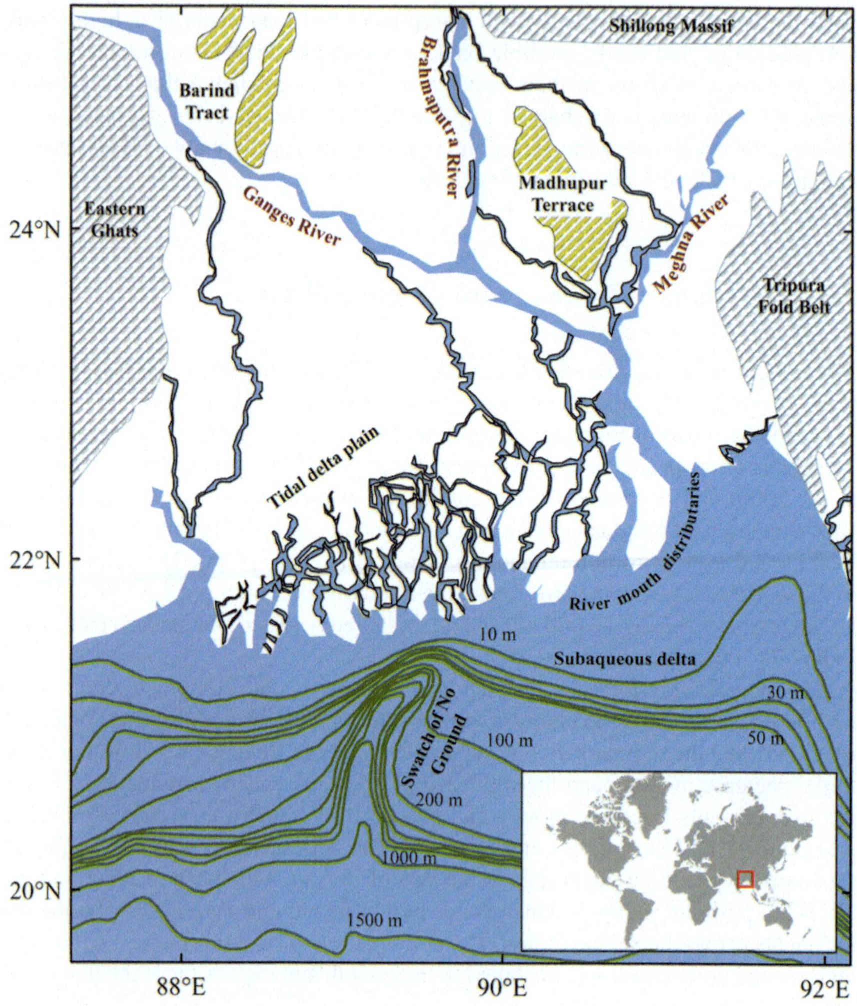

Fig. 2 The Ganges-Brahmaputra-Meghna delta, including subaerial and subaqueous portions of the delta

2 Socio-economic Characteristics of Bangladesh

2.1 Population

Bangladesh is located in South Asia, situated between 20°34′ and 26°38′ north latitude and 88°01′ and 92°42′ east longitude. Bangladesh shares its borders with India on the west, north, and east, while Myanmar is located to the southeast. To the south, it is bound by the Bay of Bengal.

In terms of demography, around 170 million people reside on a broad definition of the delta, with population densities ranging from 688 inhabitants/km^2 to 2156 inhabitants/km^2 (Fig. 3). The majority of the GBM Delta has a population density of approximately 1262 individuals per km^2, which makes it one of the most densely populated regions on the planet. The yearly growth rate of the population is 1.22% (BBS, 2022). The delta region has a population density of 893 people per square kilometer, while the remainder of the country has a density of 1248 people per square kilometer (BBS, 2022). In the delta region, the male to female ratio is 0.98 (BBS, 2022).

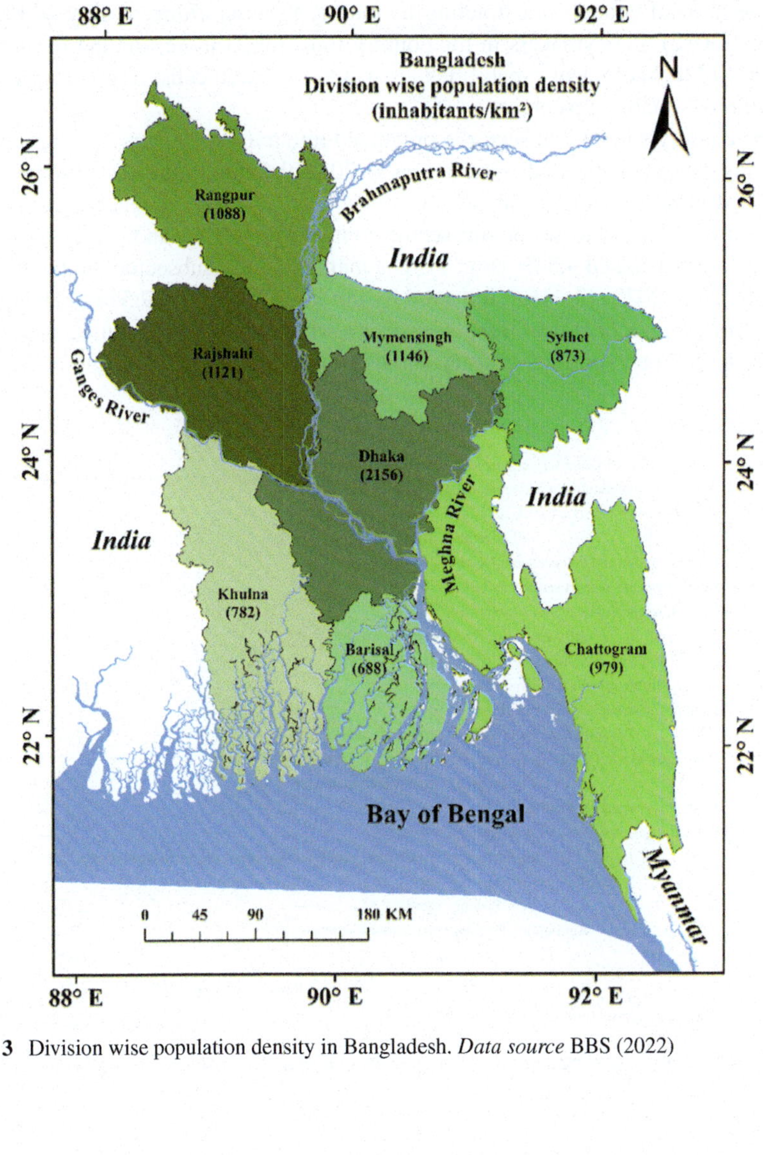

Fig. 3 Division wise population density in Bangladesh. *Data source* BBS (2022)

2.2 Major Economic Sectors

The three main economic sectors of the GBM delta are agriculture, industry, and services, which are further subdivided into 20 sub-sectors. The sub-sector wise contribution to Bangladesh's GDP in 2022 is depicted in Fig. 4. Crops, animal production, forestry, and fisheries are all included in the agricultural sector. The agricultural sector's contribution to the national economy in 2022 was 11.61%, with 9.08% coming from agriculture and forestry and 2.53% from fishing. Agriculture provides the majority of the raw materials used in the service and manufacturing sectors. Although one of the most promising sub-sectors is fishing, the nation has not yet been successful enough in extracting its marine fisheries from the Bay of Bengal (Manik, 2022). Policymakers in the country must invest more funding in building human and technological capabilities, as well as research cells, in order to harvest the country's marine resources.

The country's industrial sector is organized into five sub-sectors: mining, manufacturing, energy, water, and construction. In 2022, the mining sub-sector contributed 1.76%, the manufacturing sub-sector contributed 24.29%, the energy sub-sector contributed 1.23%, the water sub-sector contributed 0.10%, and the construction sub-sector contributed 9.54% (Fig. 7). The manufacturing subsector had the highest growth rate in 2022 (11.41%), followed by the water subsector (9.54%). The industrial sector of the economy of Bangladesh is growing fast than agriculture and service. Bangladesh's top exporting industry is ready-made garment. According to Hossain

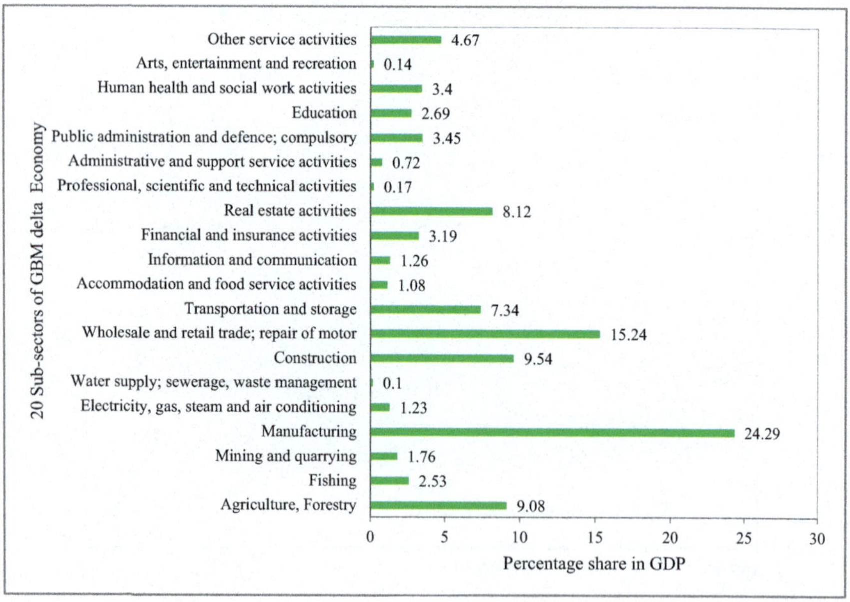

Fig. 4 Sub-sector wise contribution to Bangladesh's GDP in 2022. *Data source* BBS (2023)

et al. (2018), the textile industry of the nation contributes over $28 billion annually, or 82% of the entire export revenue of the nation. The major industries in Bangladesh include the textile industry (25%), the engineering industry (18%), the chemical industry (15%), the power industry (10%), and the food industry (9%) (Sarkar et al., 2017).

The service sector is the most significant contributor to the Bangladesh economy. The top service sub-sectors in the country include trade, transportation, public administration, real estate, and finance. In 2022, trade, transportation, public administration, real estate, and finance contributed 15.24%, 7.34%, 3.45%, 8.12%, and 3.19%, respectively, with growth rates of 8.46%, 5.75%, 4.91%, 3.70%, and 5.87% (Fig. 4). Health and education are two of the most basic human requirements, but their contributions in 2022 were 3.4% and 2.69%, respectively, with growth rates of 9.88% and 7.87% (Fig. 4). Because of the COVID-19 effect and its recovery investment, the health sub-sector had the highest growth rate, 9.88%, in 2022. Policymakers in the country should invest more in the health and education sub-sectors.

Table 1 shows Bangladesh's GDP at constant prices, as well as sectoral contributions to the national economy. Agriculture, industry, and services contributed 13.62%, 32.98%, and 53.4%, respectively, to the national GDP in 2017. The contributions of the agricultural, industrial, and service sectors turned to 11.61%, 36.92%, and 51.48%, respectively, after 6 years in 2022. Between 2017 and 2022, on average, the contributions of the sectors of agriculture, industry, and services were 12.59%, 34.95%, and 52.47%, respectively. The economy of Bangladesh is gradually expanding along with its 3 major economic sectors (i.e., agriculture, industry, and service).

Table 2 represents the year-wise GDP and sector-wise growth rate of the economy of Bangladesh. In 2017, the country's GDP growth rate was 6.59%, and in 2022, it increased to 7.1%. The COVID-19 shock caused the country's growth rate to drop to 3.45% in 2020. The country's economy has recovered successfully from the COVID shock. Another issue of concern for Bangladesh and the world economy is the Russia-Ukraine conflict. Since February 2022, Bangladesh has experienced a high dollar rate, high inflation rate, problems with supply chain management, and numerous other

Table 1 GDP of Bangladesh at constant prices with sector wise share

Year	GDP (billion USD)	Agriculture (%)	Industry (%)	Services (%)
2017	202.77	13.62	32.98	53.40
2018	229.68	13.14	33.85	53.01
2019	257.72	12.56	34.99	52.45
2020	279.00	12.52	34.94	52.54
2021	310.10	12.07	36.01	51.92
2022	348.62	11.61	36.92	51.48
Average	271.32	12.59	34.95	52.47

Data source BBS (2023)

Table 2 GDP and sector-based growth rate (%) of Bangladesh economy

Year	GDP	Agriculture	Industry	Services
2017	6.59	3.20	8.27	6.37
2018	7.32	3.54	10.20	6.55
2019	7.88	3.26	11.63	6.88
2020	3.45	3.42	3.61	3.93
2021	6.94	3.17	10.29	5.73
2022	7.10	3.05	9.86	6.26
Average	6.55	3.27	8.98	5.95

Data source BBS (2023)

economic issues as a result of this war. Therefore, the country economy may slow down in the coming years.

3 Current Issues and Challenges

Bangladesh is a country that faces various climatic hazards due to its geographical and geological features. It is located on the floodplains and deltaic land of three major rivers, which receive heavy monsoon rainfall and large amounts of sediments from the Himalaya (Ali et al., 1997; Kabir & Golder, 2017; Karim & Mimura, 2008). The near-flat topography, low river gradients, low elevation, and rising temperatures at land and sea increase the risk of floods, cyclones, storm surges, salinity intrusion, droughts, and sea level rise.

The climate in Bangladesh is changing and becoming more unpredictable in recent years. The lengths of the winter, summer, and rainy seasons have changed. Bangladesh ranks 6th among the countries with a population exposed to multi-hazards (cyclone, drought, and flood) and ranks second for flood hazards (Christenson et al., 2014). Bangladesh regularly faces different types of climatic hazards such as floods, extreme temperatures, droughts, storm surges, tornadoes, thunderstorms, hailstorms, shoreline recession, riverbank erosion, tidal surges, landslides, salinity intrusion, coastal erosion, and tropical cyclones (Alam & Dominey-Howes, 2015; Burkart & Kinney, 2017; Sarwar & Islam, 2013).

3.1 Coastal Erosion and Siltation

The Ganges-Brahmaputra-Meghna (GBM) Delta is one of the most dynamic and complex river systems in the world. It is constantly changing due to various natural and human-induced factors, such as erosion, deposition, channel shifting, tidal and wave actions, cyclones, storm surges, sea level rise, and monsoon flows. These factors affect the morphology, hydrology, and ecology of the delta, as well as the livelihoods of millions of people living in the region.

The GBM Delta has undergone significant geohistorical evolution and shifts since the 18th century, as shown in Fig. 5. The rivers bring in fresh sediments from the Himalayas and other sources, while the sea influences the shape and size of the coastline. The delta is also prone to frequent and severe natural disasters, such as floods, droughts, and cyclones, which redistribute sediments and alter the landforms.

One of the major challenges facing the GBM Delta is riverbank erosion, which is influenced by the slope and material composition of the riverbanks, as well as sea level rise and tidal pressure (Brammer, 2014; Emran et al., 2019). Riverbank erosion is more common and severe along the main rivers, such as the Brahmaputra/Jamuna, the Ganges/Padma, and the Meghna (Fig. 6). Riverbank erosion is often associated with monsoon flooding and the deltaic topography in Bangladesh (Nishat & Mukherjee, 2013). Riverbank erosion causes huge economic losses and social impacts, as it destroys agricultural land, infrastructure, and settlements, and displaces millions of people every year (Alam, 2017; Penning-Rowsell et al., 2013). Riverbank erosion affects approximately 20 of the country's 64 districts (Aktar, 2012).

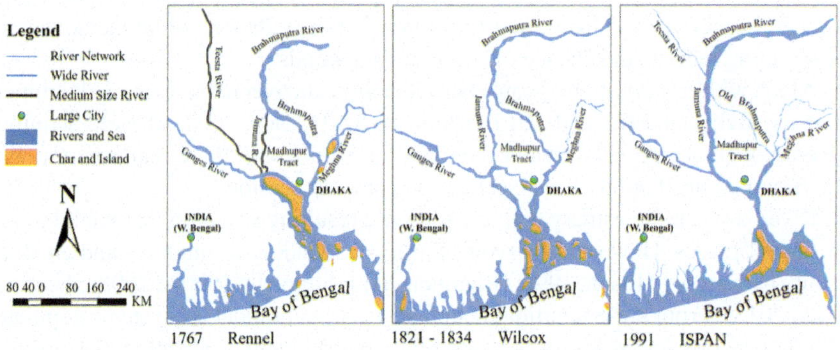

Fig. 5 Shift of major rivers channels in the GBM deltaic region (1767–1991)

Fig. 6 Riverbank erosion along the coast of lower Brahmaputra and lower Meghna River

Fig. 7 On June 21, 2022, people wade past stuck trucks on a flooded street in Bangladesh. *Source* Mamun Hossain/AFP/Getty Images

River bank erosion and land siltation are two dynamic processes that shape the GBM Delta, depending on the river flow and sediment transport. The Jamuna, Padma, and Lower Meghna rivers are eroding at rates of 1,770, 1,298, and 2,900 ha per year, respectively. The increased discharge in the rivers is a major cause of erosion. Flow statistics for the stations Bahadurabad (Brahmaputra/Jamuna), Bhairab Bazar (Meghna), and Hardinge Bridge (Ganges) reveal that peak discharge is changing in magnitude and timing over the years. Rising Ganges and Brahmaputra peak discharges increase the likelihood of future river degradation, while the advancing peak timing reduces the chance of simultaneous floods.

Along with erosion, there is some land accretion via river movements and sediment transfer. Between 1973 and 2015, a total of 52,313 acres of land were acquired. Overall, land accretion is greater than land erosion for all three major rivers, with the Brahmaputra/Jamuna River seeing the worst net erosion.

Riverbanks in the estuary actively migrate, resulting in significant bank erosion in several places. However, the river mouth is expanding as mudflats and channel-mouth bars grow (Allison, 1998; Brammer, 2014; Sokolewicz et al., 2008; Shearman et al., 2013). From these intertidal features emerge supratidal ecosystems populated by coastal grass and mangroves. Over time, active channels abandon these islands and erode them as the sediment supply decreases. This relative elevation decrease allows tides to re-inundate the land surface, creating a dense tidal channel network (Passalacqua et al., 2013; Perignon et al., 2020). This east–west trend in channel-network density shows that the GBM's main river mouths have migrated eastward since the Holocene (Allison et al., 2003; Goodbred et al., 2014). Land development around the GBM river mouth reflects the large riverine sediment supply that may balance the expected 21st-century sea-level rise (Brammer, 2014).

3.2 Flooding

Bangladesh is a flood-prone country where three major rivers, the Ganges, the Brahmaputra, and the Meghna, converge to form the world's largest delta. Most of the country consists of vast flood plains and deltas that are susceptible to heavy monsoon rains, especially in the highlands, and frequent tropical storms in the coastal region. These events trigger frequent floods in Bangladesh, affecting 20–25% of the country on average. The situation becomes extreme when the three major rivers reach their flood peak at the same time, inundating 55–60% of the country. Recent evidence indicates that the magnitude and frequency of mega floods are increasing as a result of climate change. Other human factors, such as dam construction in upper-riparian countries, unplanned urbanization in illegally encroached floodplains, lack of a combination of structural and non-structural measures, etc., are exacerbating the situation.

Floods in Bangladesh can be categorized into four types: riverine floods, coastal floods, flash floods, and urban floods. Riverine floods occur during the monsoon season (June to September) due to the overflow of the rivers. Coastal floods occur due to the coincidence of a high astronomical tide and a storm surge in low-lying coastal areas (Jisan et al., 2018). Flash floods occur in the hilly areas during the pre-monsoon season (March–May) due to heavy rainfall. Urban floods occur in cities due to inadequate drainage systems and water logging.

Floods have severe impacts on the socio-economic and environmental aspects of Bangladesh. Floods cause loss of life and damage to properties, crops, infrastructure, and the environment (Fig. 7). Floods also disrupt the normal functioning of society, affecting education, health, communication, and transportation. Floods also increase the risk of waterborne diseases, malnutrition, and food insecurity. According to the World Bank (2010), the total flooded area in Bangladesh may increase by 6%, 10%, and 20% for a sea level rise of 15 cm, 27 cm, and 62 cm, respectively. The flood risk area in Bangladesh would be 15.3% greater than the present risk area, and the depth of flooding would increase by as much as 22.7% within 20 km from the coastline if a storm surge happens with a 0.3 m sea-level rise (Karim & Mimura, 2008).

In 1998, two-thirds of Bangladesh's total land area was flooded (Table 3). The worst flood in recorded history—a combination of severe rains from within and outside the country, synchronization of major river peak flows, and a very strong back-water effect merged into a mix—lasted more than two months and caused significant damage.

3.3 Relative Sea Level Rise

The Bangladesh delta, located at a low elevation and vulnerable to flooding, faces the challenges of sea level rise and saline water intrusion. Climate change is likely to exacerbate these challenges, as sea levels in the Bay of Bengal are rising faster than the global average and are projected to increase by 0.2–1 m by 2100, depending on emission scenarios (IPCC, 2014). Several factors contribute to the relative sea level rise in the Bay of Bengal, such as the melting of Himalayan glaciers, land subsidence, and storm surges (Ali, 1999). Two main things cause land to sink in Bangladesh: active tectonic pressures and sediment loads from Himalayan upwelling, and the dryness and flattening of Proto-Bengal Fan shale and mud (Alam, 1996). The rate of

Table 3 Year wise flood affected area in Bangladesh since 1990–2020 (BBS, 2023)

	Flood affected area			Flood affected area	
Year	km^2	Percentage (%)	Year	km^2	Percentage (%)
1990	3,500	2.4	2006	16,175	11
1991	28,600	19	2007	62,300	42
1992	2,000	1.4	2008	33,655	23
1993	28,742	20	2009	28,593	19
1994	419	0.2	2010	26,530	18
1995	32,000	22	2011	29,800	20
1996	35,800	24	2012	17,700	12
1998	100,250	68	2013	15,650	10.6
1999	32,000	22	2014	36,895	25
2000	35,700	24	2015	47,200	32
2001	4,000	2.8	2016	48,675	33
2002	15,000	10	2017	61,979	42
2003	21,500	14	2018	33,941	23
2004	55,000	38	2019	45,747	31
2005	17,850	12	2020	59,028	40

relative sea level rise varies across Bangladesh's coastal zone, ranging from 2.8 to 21 mm/year (CEGIS and DOE, 2011; CCC, 2016; Ericson et al., 2006; Han et al., 2010; Pethick & Orford, 2013; Unnikrishnan & Shankar, 2007). The Chittagong coastal plain experiences the highest rate at 11–21 mm/year, followed by the Ganges tidal floodplain (7–8 mm/year) and the Meghna Estuarine flood plain (6–10 mm/year) (CCC, 2016). A sea level rise of 10 cm, 25 cm, and 1 m by 2020, 2050, and 2100 is expected to affect 2%, 4%, and 17.5% of the total land area of Bangladesh, respectively (World Bank, 2000). The land along the coastline is sinking at a rate of 4 mm/year (Sarwar & Islam, 2013), adding to the challenges posed by sea level rise.

These challenges pose serious threats to the coastal communities, ecosystems, and essential resources in the Bangladesh delta. Rising sea levels lead to more extensive and frequent coastal flooding, especially in the central and western parts of the coastal zone (Fig. 8). The flooding extent is expected to increase by up to 6% and 8% from the baseline (2005) by 2050 and 2100, respectively (CCC, 2016). Saltwater intrudes into surface water and soil as sea levels rise, reducing the availability of fresh water in coastal areas. The areas with 1 and 5 ppt salinity are projected to increase from 10 and 16% in the base year (2005) to 17.5% and 24% by 2050 in a particular scenario (CCC, 2016).

These impacts have negative consequences for agriculture and freshwater resources in the delta. Therefore, it is urgent to address climate change and implement adaptation measures in the Bangladesh delta. Adaptation strategies may include building resilient infrastructure, implementing coastal zone management plans, and developing strategies to mitigate the impacts of saltwater intrusion.

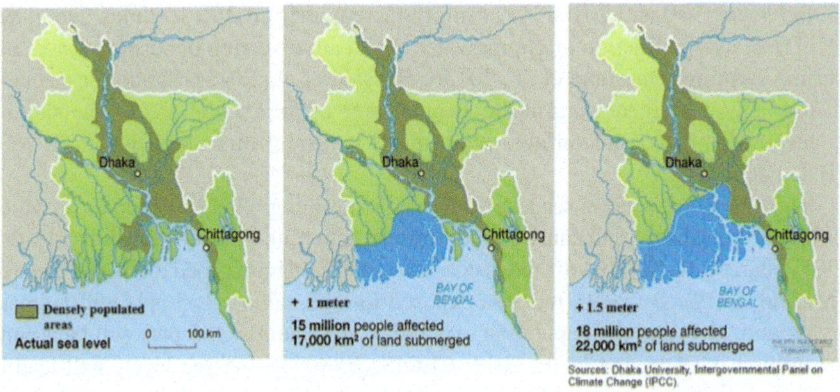

Fig. 8 Climate change may raise sea levels, causing widespread flooding. This 18-million-people region is prone to flooding due to its low elevation. *Source* Dhaka University, IPCC

3.4 Pollution

The rivers in the delta are heavily polluted, primarily due to industrial discharges and untreated sewage. This pollution is a significant threat to both aquatic ecosystems and human health. The country's surface water quality is expected to deteriorate further due to salinity intrusion and industrialization, which includes the mechanization of agriculture and urbanization.

Water pollution has three direct negative impacts:

1. It increases health problems among both rural and urban populations.
2. It leads to a loss of agricultural and industrial productivity due to pollution and saltwater intrusion.
3. It results in environmental degradation, which in turn degrades the environmental services provided.

Groundwater quality is also a concern, with key issues being the presence of arsenic and salinity. Industrial and domestic sources of pollution pose a clear risk to groundwater quality.

3.4.1 Saltwater Intrusion

From 2000 to 2009, saline water intrusion in coastal Bangladesh increased up to 15 km north of the coast and reached up to 160 km inland in the dry season due to lower flow from upstream rivers (SRDI, 2010). This was a result of climate change, which poses a major soil salinization risk for the region (Dasgupta et al., 2015a). Salt accumulation in the root zone of the soil affects plant growth and crop production. By 2013, around 30% of the coastal cultivable land was already affected by salinity (Minar et al., 2013).

Between 1970 and 2010, river salinity increased from 2 to 10 times due to sea level rise, over-extraction of groundwater, upstream diversion of surface water, and shrimp farming (Dasgupta et al., 2015b; Hossain et al., 2016). Salinity, especially in dry season, is increasing in the groundwater of the coastal areas. The salinity concentration of coastal surface water in 2015 was 1.6 times higher than in 2003 (Baten et al., 2015).

With the rise of present sea level, the river-sea water gradient will decrease, which will retard discharge and cause saline water to penetrate inward. This indicates that every year, nearly 4000 ha of land in coastal areas could be exposed to salinity hazards (Masum, 2019). By 2050, the area exposed to high salinity (>5 ppt) will increase by around 24%, and the population exposed to high salinity (>5 ppt) will be around 14 million (CEGIS and DOE, 2011; CEGIS, 2006). Rising sea levels will lead to increased salinity in both surface and ground water (Brammer, 2014).

3.5 Heavy Metal Pollution

In recent decades, heavy metals have been considered the main hazardous element and have become a major global concern due to their toxicity, persistency, and non-biodegradability (Vu et al., 2018). Heavy metals originate from both natural and anthropogenic sources but, anthropogenic activities in coastal areas have led to excessive discharges of heavy metals into coastal environments and caused adverse effects and risks to water, sediment, and biota. The GBM delta also faces heavy metal pollution due to various anthropogenic activities such as sewage discharge, untreated industrial wastewater emision, fertilizer leaching, urban construction, mining, damming, transportation, etc. The GBM delta is the most populated in the world with around 670 million people (Whitehead, 2018), and acts as a major source of water for domestic activities, agriculture, aquaculture, and industries. Bangladesh generates an average of about 22.53 million tons of municipal solid waste and 22×10^9 m^3 of industrial liquid waste per year, of which the collection coverage is 20% and the recycling rate is 0% (Waste Atlas, 2020). Consequently, large quantities of untreated industrial wastewater are discharged directly or indirectly into lowlands and water bodies in adjacent areas and ultimately discharged into the surrounding rivers of the GBM delta and transferred to the sediment by physical, chemical, and biological processes. The accumulated heavy metals in sediments, released into the aquatic system under favorable conditions (pH, Eh, salinity, etc.), enter the food web and pose potential risks to human health and ecosystem stability by bioaccumulation and biomagnification in the food chain. Toxic heavy metals (As, Cr, Co, Cu, Pb) concentrations are lower in modern sediment upstream than downstream of the delta due to the higher water flow and less industrial activities (Haque et al., 2023). Ahmed et al. (2021) stated that the heavy metal concentration in the waters, sediments, and fish species of the different rivers in the GBM delta vary with different anthropogenic activities as well as natural sources. The waters of Buriganga, Bangshi, Dhaleshawri, Halda, and Karnaphuli rivers are polluted with Cr, Cd, and Pb. On the other hand, concentrations of Zn, Cu, Ni, and Mn as well as Cr, Cd, and Pb are reported higher in the sediments of Buriganga, Bangshi, Dhaleshawri, Karnaphuli, Korotoa, Paira, Rusash, Shitalakha and Turag. The levels of Cr, Pb, and Ni surpass the maximum permissible levels (MPLs) in certain fish species (Gudusia chapra, Colisa fascitus, Channa punctatua) found in the Buriganga, Shitalakhya, Turag, and Karnaphuli rivers, as recommended by the Government of Bangladesh and various international organizations.

The arsenic problem in the GBM delta region of Bangladesh, first identified in 1993, is believed to stem from natural geological sources, specifically quaternary confined and semi-confined alluvial or deltaic aquifers. Numerous chemical and biological reactions actively contribute to the arsenic cycle in groundwater (Fig. 9). The issue arises from the younger sediments deposited by the GBM river system in the delta region, which contain minerals like biotite. Under groundwater's reducing conditions, these minerals undergo a slow "diagenetic" reaction when compacted, releasing toxic As^{3+}.

This process has led to significant arsenic contamination in Bangladesh's drinking water, affecting approximately 30 million people who consume water with arsenic levels exceeding the acceptable limit of 10 μg/L (Reddy et al., 2023). The country heavily relies on this contaminated groundwater for irrigating dry season rice (boro), which is rich in arsenic and contributes an estimated 1,360 tons of arsenic to the soil annually (Edmunds et al., 2015).

The main contamination process involves the oxidation of arsenic pyrites or ferrous hydroxides, which are usually stable under the water table. Anthropogenic activities like lowering the water table accelerate their oxidation, releasing arsenic. Some are absorbed into iron hydroxide but returns to the water table during recharge,

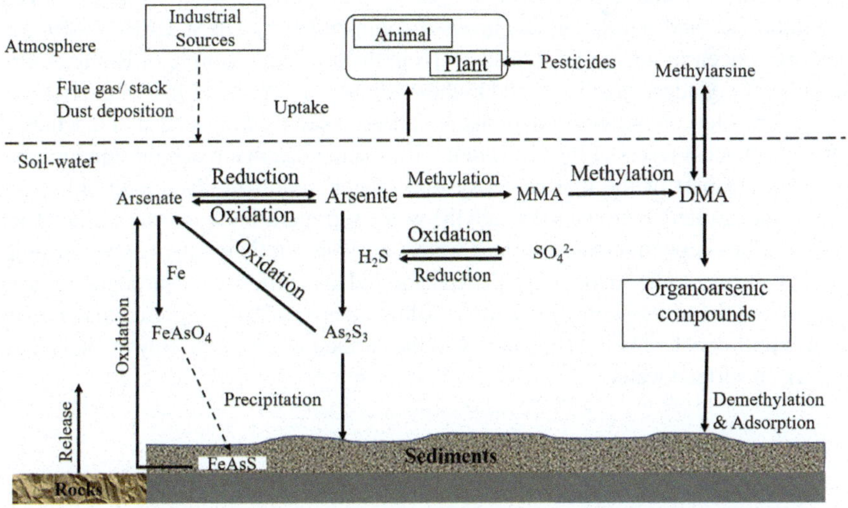

Fig. 9 Diverse chemical and biological reactions in the cycling of arsenic toxic element

causing water poisoning. This suggests that arsenic-rich groundwater is man-made and recent. Geological processes like weathering, erosion, sedimentation, and the use of irrigation and fertilizers also accelerate these processes.

3.6 *Radiological Impact*

The sources of natural radionuclides on the earth's surface are primitive sources and cosmic radiation. Primitive radionuclides (^{238}U, ^{232}Th, ^{40}K) have been present in the Earth's crust since its formation, and others (^{3}H, ^{7}Be, ^{14}C) formed by cosmic radiation. Natural radioactive elements are transferred and circulated in the environment by natural processes. However, Human activities such as agricultural inputs (for example, the use of fertilizers, fungicides, insecticides, and herbicides), burning fossil fuels, rapid urbanization, exploration of coal, petroleum, and natural gas removal of radioactive waste and industrial sewages, etc. can lead to redistribution of activity concentrations on the earth's crust. Natural radioactivity in the GBM delta of Bangladesh and the adjacent aquatic environment was increased through human activities mostly the use of phosphate fertilizer, coal mining, and building materials which generate and redistribute more and more radioactive materials, causing radioactive ecological pollution of these areas. Phosphate fertilizer plants cause environmental problems with emissions of process-specific chemicals into the air, releases into water, storage of solid waste, and released radioactivity can produce an unambiguous radioactive impact in the surrounding environment through radionuclides from phosphogypsum wastes (Samad et al., 2012). The soil around the Barpukuria coal mine area is heavily contaminated because radionuclides leaked during mining, surpassing the global average values for uranium (U), thorium (Th), and potassium (K) levels recommended by UNSCEAR (2000). The activity concentrations of ^{226}Ra and ^{232}Th in the sediment sample of Brahmaputra, Jamuna, Bangladesh are also found higher (Khalil et al., 2016). Most of the radioactivity deposited on surface sediments is carried away by rainwater and discharged into the adjacent rivers of the GBM delta. Radionuclides that reach the aquatic environment are part of the marine ecosystem (water, sediment, sand, and biota) and can be transmitted to humans through the water-sediment-biota interface. Most rural communities depend on these surface waters of rivers for their daily water consumption. Therefore, radionuclides are also transported to the food chain through irrigation water from the aforementioned sources. Bhuiyan et al. (2014) reported that activity concentrations of ^{238}U, ^{232}Th, and ^{40}K in the surface water of the Bay of Bengal are higher on surface water.

3.7 *Biodiversity Loss*

There used to be a great diversity of native fish species in the river, but sadly, many of them have died out. The Gangetic leaffish is one of many native fish species that have vanished from river systems. Once upon a time, Bangladesh was home to more than 300 species of native freshwater fish. However, due to factors such as habitat loss, overfishing, pollution, and climate change, many of these fish are now extinct or on the verge of extinction. Native fish species in open rivers and other bodies of water in Bangladesh are disappearing due to development interventions, unplanned urbanization, encroachment, and siltation. According to the International Union for the Conservation of Nature (IUCN), 30 freshwater fish species have gone extinct in the wild in the last 30–40 years, and the status of another 64 varies from vulnerable to critically endangered. The Bangladesh Fisheries Research Institute (BFRI) has revived 37 of 67 species that were once thought to be extinct in the wild through captive breeding programs and conservation efforts. Based on research conducted in 2018, it was determined that annually, the GBM river system moves between 1 billion and 2.4 billion tons of silt. During periods of low flow or a drop in depth below critical levels, rivers, floodplains, and beel bottoms that have accumulated silt are more likely to dry up. This creates an environment that is unhealthy for fish and makes it difficult for them to reproduce. The loss of fish diversity is largely attributable to encroachment, development interventions, and the diversion of wetlands and rivers into agricultural land. In addition, the construction of new roads across floodplain areas that provide inadequate passage for fish is severing wetlands, creating smaller and more isolated fish populations overall.

4 Management Measures

4.1 *Adaptation Strategies*

Bangladesh is categorized as one of the most vulnerable nations globally with regards to the impacts of climate change. Given the country's low-lying geographical features and the high concentration of inhabitants, it is especially prone to the adverse consequences of escalating sea levels, heightened occurrences of inundation, and intensified meteorological phenomena. In an effort to effectively respond to these repercussions, Bangladesh is actively participating in the formulation and execution of multiple climate-resilient frameworks and tactics. These initiatives encompass the following:

(i) Embankments: These structures consist of elevated earthen walls constructed along rivers and coastlines to safeguard against flooding. Embankments are a widely employed adaptation approach in Bangladesh and have proven successful in mitigating the impact of flooding in many regions.

(ii) Cyclone shelters: These robust buildings are specifically designed to serve as secure refuges during cyclones. Given the coastal nature of Bangladesh, cyclones pose a significant threat, making cyclone shelters of utmost importance.
(iii) Early warning systems: These networks consist of sensors and communication systems that enable the timely dissemination of warnings about impending hazards. Early warning systems play a vital role in enhancing preparedness and response to disasters such as floods and cyclones by providing individuals with advance notice.
(iv) Crop diversification: The technique of cultivating a wide variety of crops, known as crop diversification, has the potential to reduce the likelihood of crop failure caused by climate change. For example, if one crop is affected by drought, there is a possibility that other crops will be able to endure the unfavorable conditions.

Bangladesh is diligently taking measures to protect its people and economy from the detrimental impacts of climate change by making substantial investments in climate-resilient infrastructure and adopting strategic approaches.

4.2 Sustainable Agriculture

Bangladesh is a nation with a swiftly growing population and a limited amount of cultivable land. Therefore, it is crucial to embrace sustainable agricultural methods that can help meet the country's food needs while also protecting the environment. In Bangladesh, research on sustainable agriculture covers several important aspects, including organic farming, diversification of crops, integrated pest management, and climate-smart agriculture. The employment of organic farming techniques can augment soil health, reduce reliance on external inputs, and yield food of superior quality. The implementation of crop diversification and rotation holds the potential to enhance soil fertility, mitigate pests and diseases, and strengthen resilience to climate change. Techniques for integrated pest management can decrease dependence on pesticides and promote the utilization of natural predators for pests. The adoption of climate-smart agriculture practices can empower farmers to adapt to the repercussions of climate change.

A variety of sustainable agriculture practices are being implemented in Bangladesh. Agroforestry involves the integration of trees and shrubs into crop and livestock production systems. Agroforestry can enhance soil fertility, mitigate erosion, and provide shade for crops and livestock. Vermicomposting entails the utilization of earthworms to convert organic waste into a fertilizer rich in nutrients. Vermicomposting can help reduce waste and enhance soil fertility. Integrated rice-fish farming entails raising fish in rice paddies. Fish can aid in pest and disease control in rice crops, and their waste can supply nutrients to rice plants. Biogas production is the process of generating renewable energy from organic waste. Biogas can be utilized for cooking, lighting, and electricity generation.

Despite the progress that has been achieved, there are numerous challenges to sustainable agriculture in Bangladesh. Small landholdings make it challenging for farmers to embrace sustainable practices. Limited access to credit can impede farmers' ability to invest in sustainable practices. Insufficient awareness of the benefits of sustainable agriculture practices can also hinder their adoption. The government of Bangladesh is fostering sustainable agriculture through different initiatives, such as the National Agriculture Policy and the National Adaptation Plan for Climate Change. Non-governmental organizations (NGOs) are also actively working to promote sustainable agriculture in Bangladesh. International organizations are also contributing to support sustainable agriculture in Bangladesh.

4.3 Coastal Aquaculture

The future of aquaculture is looking bright, especially in regions that are largely protected from flooding due to embankments. Fish farming has become a year-round endeavor, with a large portion of embankment-enclosed ponds and tanks being used for this purpose. However, this method is not without its risks. High-intensity floods or breaches in the embankments can cause these ponds to overflow, releasing the fish into the open water. While the fish may survive, the financial impact on the farmers can be significant. Moreover, the threat of climate change, which could intensify monsoon floods, makes these fisheries increasingly vulnerable.

The strength of surges and tidal bores is increasing, raising the chances of coastal embankments being overtopped by saline water. Shrimp farms outside these embankments use small earthen enclosures, locally known as "ghers," for shrimp cultivation. This has become a major industry in coastal regions like Cox's Bazaar, Satkhira, Khulna, and Bagerhat. However, rising tides pose a significant risk to these ghers, both inside and outside the embankments (WB, 2000). The intrusion of saltwater into areas north of the current shrimp cultivation zones could potentially expand the shrimp industry (CEGIS, 2006).

In addition, an overall increase in surface water temperature could lead to heat stress in shrimp. Studies have shown that when water temperatures exceed a critical limit of 32 °C, young shrimp have a significantly higher mortality rate. This is particularly relevant considering that temperatures are already quite high in April. Concurrently, warmer water temperatures could lead to algal blooms that negatively affect shrimp growth. Consequently, climate change introduces an element of uncertainty to this thriving sector.

4.4 Mangrove Conservation

Bangladesh is the home of the largest mangrove forest in the world, known as the Sundarbans. The Sundarbans mangrove forest in the delta plays a crucial role in both biodiversity conservation and protection against storm surges. The enhancement of ecosystem resilience can be achieved through dedicated conservation efforts. Several research organizations are actively involved in the conservation of the Sundarbans. Some of the key areas of research include:

(i) Investigation of climate change impacts on mangroves: Researchers are conducting studies to understand the effects of climate change on mangroves, such as the rise in sea levels, alterations in salinity levels, and the occurrence of more extreme weather events.
(ii) Restoration of mangrove ecosystems: Researchers are currently involved in the endeavor to develop effective methods for the revitalization of deteriorated mangrove forests.
(iii) Evaluation of mangrove ecosystem services: Researchers are actively engaged in studying the various ecosystem services provided by mangroves, including carbon sequestration, coastal protection, and support for fisheries.
(iv) Development and assessment of mangrove management strategies: Researchers are committed to formulating and evaluating management strategies with the goal of conserving mangroves.

The government of Bangladesh has taken numerous measures to ensure the protection of the Sundarbans. These measures include:

(i) Establishment of protected areas: The government has established various protected areas within the Sundarbans, such as the Sundarbans Reserved Forest and the Sundarbans World Heritage Site.
(ii) Implementation of forest management plans: The government has put in place comprehensive forest management plans specifically designed to safeguard the Sundarbans from deforestation and other potential threats.
(iii) Raising awareness about the importance of mangroves: The government is actively engaged in increasing awareness about the importance of mangroves through educational initiatives and outreach programs.

4.5 Clean Energy

The utilization of renewable energy sources like solar and wind power has the ability to significantly reduce the region's dependency on fossil fuels and mitigate greenhouse gas emissions. Bangladesh has a significant potential for generating solar and wind power, with an estimated solar radiation potential of 5–6 $kWh/m^2/day$ and wind speeds that range between 3 and 6 m/s at a height of 50 m. The government of Bangladesh has implemented various measures to promote the adoption of renewable energy, including setting a target to generate 30% of the country's electricity from renewable sources by 2041 (BDP2100, 2015). Presently, there are numerous ongoing solar and wind power initiatives, and the nation is primed to witness substantial expansion in the generation of renewable energy in the forthcoming years.

The transition to renewable energy will yield several advantages for Bangladesh. It will effectively curtail the nation's reliance on imported fossil fuels, which can prove to be both costly and volatile. Additionally, it will serve to mitigate the prevalent issue of air pollution within the country. Furthermore, it will substantially contribute to the reduction of greenhouse gas emissions, thereby addressing the issue of climate change. Despite the challenges that may arise during the transition to a clean energy future, the undeniable advantages of renewable energy make it a worthwhile endeavor for Bangladesh. The country is well-equipped to navigate this transition and establish a clean energy economy.

4.6 Water Management

Efficient management of freshwater resources, such as the optimal utilization of rainwater collection and the implementation of highly effective irrigation methods, has the potential to effectively tackle the challenges posed by saltwater intrusion and water pollution. A comprehensive study carried out by the prestigious International Water Management Institute (IWMI) highlights the significant role that rainwater collection can play as a substantial water source for irrigation purposes in Bangladesh. The study reveals that rainwater collection has the capacity to provide ample water supply for the irrigation of up to 1.5 million hectares of land within the nation. Furthermore, a thorough investigation conducted by the renowned Bangladesh University of Engineering and Technology (BUET) reveals that the adoption of efficient irrigation practices can lead to a noteworthy 50% reduction in water consumption within the agricultural sector. This study underscores the benefits of utilizing drip irrigation and sprinkler irrigation methods instead of traditional flood irrigation techniques, as they enable farmers to achieve substantial water savings.

To enhance water management within the country, the Bangladesh government has introduced various programs, including but not limited to:

(i) The National Water Management Plan (NWMP), which strives to ensure that every citizen of Bangladesh has access to safe and affordable water.
(ii) The Water Resources Planning Organization (WARPO), which holds the responsibility of planning and managing water resources throughout the nation.
(iii) The Bangladesh Water Development Board (BWDB), which bears the responsibility of constructing and maintaining water infrastructure.

Additionally, the government has made significant financial commitments towards the development of rainwater collection initiatives, which have demonstrated their efficacy in supplying water for agricultural and other rural needs. Concurrently, measures are being taken to encourage the implementation of productive irrigation methods. To facilitate this objective, the government has provided subsidies to farmers to facilitate the procurement of water-efficient irrigation machinery.

4.7 International Cooperation

The GBM delta, comprising the Ganges-Brahmaputra-Meghna rivers, is an extensive and intricate system that spans across multiple nations. Consequently, the significance of international collaboration in safeguarding this delta cannot be undermined. A key area in which international cooperation is imperative lies in the management of transboundary water resources. The GBM delta is home to a multitude of rivers that traverse numerous countries, playing a pivotal role in irrigation, drinking water provisions, and hydropower generation. However, the management of these rivers can engender conflicts between nations. For instance, upstream countries may construct dams that diminish the downstream water flow or they may contaminate the water, thereby adversely impacting the downstream communities. International cooperation can effectively ensure the equitable and just management of transboundary water resources. This can be accomplished through the formulation of joint management agreements, the exchange of data and information, and the establishment of mechanisms for conflict resolution.

Another area necessitating international cooperation involves the development of climate-resilient infrastructure. The GBM delta is exceptionally susceptible to the ramifications of climate change, including rising sea levels, heightened occurrences of extreme weather events, and alterations in rainfall patterns. These impacts can result in flooding, erosion, and the intrusion of saltwater. International collaboration has the potential to assume a crucial position in funding the establishment of climate-resistant infrastructure, encompassing flood prevention embankments, early cautionary mechanisms, and crops capable of withstanding drought conditions. This can effectively shield the delta from the adverse effects of climate change and ensure its continued productivity and habitability.

Furthermore, international cooperation is indispensable in promoting sustainable development in the GBM delta. This encompasses investments in clean energy, the promotion of water conservation, and the protection of the environment. By collaborating, the nations in the region can guarantee that the GBM delta serves as a sustainable and prosperous home for millions of individuals in the coming generations. Some notable examples of international cooperation initiatives in the GBM delta include:

(i) The Bangladesh-India Joint Rivers Commission, which functions to oversee the management of shared water resources between the two nations.
(ii) The Global Environment Facility (GEF), which is actively supporting several projects aimed at fostering climate-resilient development in the delta.

(iii) The World Bank, in collaboration with the governments of Bangladesh, India, and Nepal, is engaged in formulating a regional plan for managing the delta's water resources.

In order to prioritize joint cooperation over individual development, it is crucial to establish closer collaboration with neighboring countries. The following strategies should be implemented (BDP2100, 2015):

(i) A comprehensive action plan should be developed, taking into account the water usage of upstream countries.
(ii) Prospective sites for constructing embankments should be selected, considering both the water flow from upstream and the understanding and cooperation of upstream countries.
(iii) Multi-track water diplomacy should be employed to prevent or peacefully resolve conflicts.
(iv) Ongoing efforts should be made to sign a Treaty that addresses the sharing of water for the Teesta and all other transboundary rivers.
(v) Initiatives for demand-based common river basin management schemes need to be initiated.
(vi) Third party involvement, whether it be a multilateral or bilateral development partner or country, should be sought to resolve transboundary water-related issues.
(vii) Enhanced basin-wide flood forecasting should be implemented.

These initiatives exemplify the myriad ways in which international cooperation is effectively safeguarding the GBM delta. By joining forces, the nations in the region can ensure the preservation of this vital ecosystem for future generations.

4.8 Education and Awareness

Education and awareness are crucial for community involvement and the implementation of sustainable practices in order to address ecological challenges and their consequences. In Bangladesh, numerous endeavors are currently in progress to enhance awareness about ecological challenges. These include:

(i) Inclusion of environmental education programs in schools and universities.
(ii) Execution of public awareness campaigns via media and community events.
(iii) Implementation of training programs for farmers and other rural communities.

These initiatives are instrumental in fostering a society that is more cognizant of environmental concerns. As individuals become increasingly conscious of the challenges confronting the environment, they are more inclined to endorse and partake in sustainable practices. For instance, farmers who possess knowledge of the hazards associated with pesticide usage are more inclined to embrace organic farming techniques. Similarly, communities that recognize the significance of water conservation are more inclined to embrace measures that curtail their water consumption.

5 Future Perspectives

In conclusion, the ecological crisis in Bangladesh's GBM River Delta is a complex issue that requires a multi-pronged approach. While the challenges are significant, there are opportunities for sustainable development and resilience building through innovative strategies, international cooperation, and community involvement:

1. Agricultural Potential: The delta's fertile soil is ideal for agriculture, and with proper management and investment, it can significantly contribute to food production in Bangladesh. Rice production has seen remarkable growth, from 12 million tonnes in 1973 to 36.3 million tonnes in 2018. There is an opportunity to modernize agricultural practices and increase yields.
2. Fisheries and Aquaculture: The delta's extensive network of rivers, lakes, wetlands, and estuaries provides abundant opportunities for fisheries and aquaculture. Marine fishing has gained prominence, and the agriculture landscape is gradually shifting from crops to fisheries. Sustainable management can enhance livelihoods and food security.
3. Renewable Energy: The delta's geographic features make it suitable for renewable energy sources like solar and wind power. Investing in these technologies can provide sustainable energy solutions.
4. Tourism and Biodiversity: The delta's unique ecosystems and natural beauty offer great potential for eco-tourism. Conservation efforts can protect biodiversity and generate income through tourism. Coastal tourism, exemplified by destinations like Kuakata in Patuakhali district, presents an alternative to traditional coastal resorts like Cox's Bazar.
5. Trade and Transportation: The delta's waterways serve as vital transportation routes. The extensive river network connecting various districts and growth centers, such as Dhaka, Chattogram, and Khulna, offers a sustainable and cost-effective means of transportation. Inland water transport not only benefits the country as a whole but also provides employment opportunities for rural communities. The open access to the sea is a significant advantage for Bangladesh. With proper investments, the country can become a regional hub for international sea transportation, following the examples of dynamic port cities like Mumbai, Shanghai, Singapore, and Hong Kong. This potential is further amplified by the growing energy demand in Bangladesh, necessitating port facilities for the handling of imported coal, LNG, and LPG.
6. Climate Resilience: Implementing climate-resilient infrastructure and practices is crucial for the delta's future. There are opportunities for innovative solutions to address climate change challenges.
7. Research and Education: The GBM Delta provides a valuable site for research in various fields, including environmental science, agriculture, and climate change. Investing in education and research facilities can yield significant benefits.
8. Community Empowerment: Engaging and empowering local communities in delta management and development projects can foster sustainable growth and reduce poverty.

9. International Collaboration: Collaborative efforts with neighboring countries, such as China, India, and Myanmar, can lead to joint projects that harness the delta's resources for mutual benefit.
10. Infrastructure Development: Investments in infrastructure, such as roads, bridges, and flood protection systems, are essential for the delta's development and resilience. Finally, Bangladesh's delta offers a diverse range of opportunities, from increasing agricultural productivity and developing fisheries to expanding inland water transport and capitalizing on the open access to the sea for trade, energy, and the blue economy. Careful planning and investment can transform these natural advantages into significant development gains for the country. Addressing these challenges is vital not only for the well-being of the delta's inhabitants but also for the global environment, given the delta's importance in climate regulation and biodiversity conservation.

References

Ahmed, M. K., Rani, S., Khan, M. I., Failler, P., Islam, M. M., & Watson, G. (2021). Trace metals in water, sediment and fishes of the Ganges-Bhahmaputra-Meghna Delta, A review (pp. 103–130). Bangladesh Maritime Journal Special issue.

Aktar, N. (2012). Prediction of river bank erosion along the Jamuna, the Ganges the Padma and the lower Meghna rivers in 2012. Center for Environmental and Geographic Information Services.

Alam, E., & Dominey-Howes, D. (2015). A new catalogue of tropical cyclones of the northern Bay of Bengal and the distribution and effects of selected landfalling events in Bangladesh. *International Journal of Climatology, 35*(6), 801–835.

Alam, G. M. (2017). Livelihood cycle and vulnerability of rural households to climate change and hazards in Bangladesh. *Environmental Anagement, 59*(5), 777–791.

Alam, M. (1996). Subsidence of the Ganges—Brahmaputra Delta of Bangladesh and associated drainage, sedimentation, and salinity problems. In J. D. Milliman & B. U. Haq (Eds.), *Sea-level rise and coastal subsidence: Causes, consequences, and strategies* (pp. 169–192). Springer.

Ali, A. (1999). Climate change impacts and adaptation assessment in Bangladesh. *Climate Research, 12*(2–3), 109–116.

Ali, M. M., Saheed, S. M., Kubota, D., Masunaga, T., & Wakatsuki, T. (1997). Soil degradation during the period 1967–1995 in Bangladesh: I. *Carbon and nitrogen. Soil Science and Plant Nutrition, 43*(4), 863–878.

Allison, M. A. (1998). Historical changes in the Ganges-Brahmaputra delta front. *Journal of Coastal Research, 14*(4), 1269–1275.

Allison, M. A., Khan, S. R., Goodbred, S. L., Jr., & Kuehl, S. A. (2003). Stratigraphic evolution of the late Holocene Ganges-Brahmaputra lower delta plain. *Sedimentary Geology, 155*(3–4), 317–342.

Baten, M. A., Seal, L., & Lisa, K. S. (2015). Salinity intrusion in interior coast of Bangladesh: Challenges to agriculture in south-central coastal zone. *American Journal of Climate Change, 4*(03), 248–262.

BBS. (2022). Bangladesh Bureau of Statistics. Bangladesh National Portal.

BBS. (2023). Bangladesh Bureau of Statistics. Bangladesh National Portal.

BDP2100. (2015). Bangladesh Delta Plan: Coast and Polder Issue. General Economic Division, Bangladesh Planning Commission.

Bhuiyan, M. K. A., Siddique, M. A. M., Zafar, M., & Kmal, A. H. M. (2014). Spatial distribution of radioisotope concentrations in the offshore water and sediment of Bay of Bengal (Indian Ocean), Bangladesh. *Isotopes in Environmental and Health Studies, 50*(1), 134–141.

Brammer, H. (2014). Bangladesh's dynamic coastal regions and sea-level rise. *Climate Risk Management, 1*, 51–62.

Burkart, K., & Kinney, P. L. (2017). What drives cold-related excess mortality in a south Asian tropical monsoon climate season versus temperatures and diurnal temperature changes. *International Journal of Biometeorology, 61*(6), 1073–1080.

Climate Change Cell (CCC). (2016). Assessment of Sea Level Rise on Bangladesh Coast through Trend Analysis. Department of Environment, Ministry of Environment and Forests.

Center for Environmental and Geographic Information Services (CEGIS) and Department of Environment (CEGIS and DoE). (2011). Final Report on Programmes Containing Measures to Facilitate Adaptation to Climate Change of the Second National Communication Project of Bangladesh.

Center for Environmental and Geographic Information Services (CEGIS). (2006). Impact of sea level rise on the land use suitability and adaptation options, component of SEMP.

Christenson, E., Elliott, M., Banerjee, O., Hamrick, L., & Bartram, J. (2014). Climate-related hazards: A method for global assessment of urban and rural population exposure to cyclones, droughts, and floods. *International Journal of Environmental Research and Public Health, 11*(2), 2169–2192.

Coleman, J. M. (1969). Brahmaputra River: Channel processes and sedimentation. *Sedimentary Geology, 3*(2–3), 129–239.

Dasgupta, S., Hossain, M. M., Huq, M., & Wheeler, D. (2015a). Climate change and soil salinity: The case of coastal Bangladesh. *Ambio, 44*(8), 815–826.

Dasgupta, S., Kamal, F. A., Khan, Z. H., Choudhury, S., & Nishat, A. (2015b). River salinity and climate change: Evidence from coastal Bangladesh. In M. Agarwal, J. Pan, & J. Whalley (Eds.), *World Scientific reference on Asia and the World Economy* (pp. 205–242). World Scientific.

Edmunds, W. M., Ahmed, K. M., & Whitehead, P. G. (2015). A review of Arsenic and its impacts in groundwater of the Ganges-Brahmaputra_Meghna delta. *Bangladesh. Environmental Science: Processes and Impacts, 17*(6), 1032–1046.

Emran, A., Rob, M. A., & Kabir, M. H. (2019). Coastline change and erosion-accretion evolution of the Sandwip Island, Bangladesh. In Information Resources Management Association (USA). (Ed.), *Environmental information systems: Concepts, methodologies, tools, and applications* (pp. 1497–1509). IGI Global.

Ericson, J. P., Vörösmarty, C. J., Dingman, S. L., Ward, L. G., & Meybeck, M. (2006). Effective sea-level rise and deltas: Causes of change and human dimension implications. *Global and Planetary Change, 50*(1–2), 63–82.

Goodbred, S. L., Jr., Paolo, P. M., Ullah, M. S., Pate, R. D., Khan, S. R., Kuehl, S. A., Singh, S. K., & Rahaman, W. (2014). Piecing together the Ganges-Brahmaputra-Meghna River delta: Use of sediment provenance to reconstruct the history and interaction of multiple fluvial systems during Holocene delta evolution. *Geological Society of America Bulletin, 126*(11–12), 1495–1510.

Han, W., Meehl, G. A., Rajagopalan, B., Fasullo, J. T., Hu, A., Lin, J., & Wallcraft, A. (2010). Patterns of Indian Ocean sea-level change in a warming climate. *Nature Geoscience, 3*(8), 546–550.

Haque, M. M., Selim Reza, A. H. M., & Hoyanagi, K. (2023). Anthropogenic and natural contribution of potentially toxic elements in southeastern Ganges-Brahmaputra-Meghna delta, Bangladesh. *Marine Pollution Bulletin, 192*, 115103.

Hossain, L., Sarker, S. K., & Khan, M. S. (2018). Evaluation of present and future wastewater impacts of textile dyeing industries in Bangladesh. *Environmental Development, 26*, 23–33.

Hossain, M. S., Dearing, J. A., Rahman, M. M., & Salehin, M. (2016). Recent changes in ecosystem services and human well-being in the Bangladesh coastal zone. *Regional Environmental Change, 16*(2), 429–443.

IPCC. (2014). Climate Change 2014: Impacts, adaptation and vulnerability. Part A: global and sectoral aspects. Contribution of working group II to the fifth assessment report of the intergovernmental panel on climate change. Cambridge University Press.

Jisan, M. A., Bao, S., & Pietrafesa, L. J. (2018). Ensemble projection of the sea level rise impact on storm surge and inundation at the coast of Bangladesh. *Natural Hazards and Earth System Sciences, 18*(1), 351–364.

Kabir, H., & Golder, J. (2017). Rainfall variability and its impact on crop agriculture in southwest region of Bangladesh. *Journal of Climatology and Weather Forecasting, 5*(1), 1–20.

Karim, M. F., & Mimura, N. (2008). Impacts of climate change and sea-level rise on cyclonic storm surge floods in Bangladesh. *Global Environmental Change: Human and Policy Dimensions, 18*(3), 490–500.

Khalil, M. I., Majumder, R. K., Kabir, M. Z., Deeba, F., Khan, M. N. I., Ali, M. I., Paul, D., Haydar, M. A., & Islam, S. M. A. (2016). Assessment of natural radioactivity levels and Identification of minerals in Brahmaputra (Jumuna) river sand and sediment, Bangladesh. *Radiation Protection and Environment, 39*(4), 204–211.

Kuehl, S. A., Hariu, T. M., & Moore, W. S. (1989). Shelf sedimentation off the Ganges-Brahmaputra river system: Evidence for sediment bypassing to the Bengal fan. *Geology, 17*(12), 1132–1135.

Manik, M. H. (2022). Current marine fisheries production status and opportunity in Bangladesh. *International Journal of Agriculture and Animal Production, 2*(03), 49–56.

Masum, J. H. (2019). Climatic hazards in Bangladesh: A literature review. Coastal Development Partnership (CDP), Bangladesh, Coastal Development Partnership (CDP).

Minar, M. H., Hossain, M. B., & Shamsuddin, M. D. (2013). Climate change and coastal zone of Bangladesh: Vulnerability, resilience and adaptability. *Middle-East Journal of Scientific Research, 13*(1), 114–120.

Nishat, A., & Mukherjee, N. (2013). Climate change impacts, scenario and vulnerability of Bangladesh. In *Climate change adaptation actions in Bangladesh* (pp. 15–41).

Passalacqua, P., Lanzoni, S., Paola, C., & Rinaldo, A. (2013). Geomorphic signatures of deltaic processes and vegetation: The Ganges-Brahmaputra-Jamuna case study. *Journal of Geophysical Research: Earth Surface, 118*(3), 1838–1849.

Penning-Rowsell, E. C., Sultana, P., & Thompson, P. M. (2013). The 'last resort'? Population movement in response to climate-related hazards in Bangladesh. *Environmental Science and Policy, 27*, S44–S59.

Perignon, M., Adams, J., Overeem, I., & Passalacqua, P. (2020). Dominant process zones in a mixed fluvial-tidal delta are morphologically distinct. *Earth Surface Dynamics, 8*(3), 809–824.

Pethick, J., & Orford, J. D. (2013). Rapid rise in effective sea-level in southwest Bangladesh: Its causes and contemporary rates. *Global and Planetary Change, 111*, 237–245.

Rahman, M. M., Ghosh, T., Salehin, M., Ghosh, A., Haque, A., Hossain, M. A., Das, S., Hazra, S., Islam, N., Sarker, M. H., & Nicholls, R. J. (2020). Ganges-Brahmaputra-Meghna delta, Bangladesh and India: A transnational mega-delta. In R. J., Nicholls, W. N. Adger, C. W. Hutton, & S. E. Hanson (Eds.), *Deltas in the anthropocene* (pp. 23–51). Springer.

Reddy, R. R., van Velden, G. A., Abedin, M. J., Karim, M. R., Hayes, K. F., Agrawal, A., & Raskin, L. (2023). Low-cost informational intervention reduced drinking water arsenic exposure in Bangladesh. *PNAS Nexus, 2*(3), 1–9.

Samad, M. A., Ali, M. I., Pau, D., & Islam, S. M. A. (2012). Assessment of radiological significance of the wastes generated from the tripal super phosphate (TSP) fertilizer factory, Chittagong, Bangladesh. *Journal of Bangladesh Academy of Sciences, 36*(1), 79–88.

Sarkar, P., Anjum, A., & Khan, E. A. (2017). Overview of major industries in Bangladesh. *Journal of Chemical Engineering, 30*(1), 51–58.

Sarwar, M. G. M., & Islam, A. (2013). Multi hazard vulnerabilities of the coastal land of Bangladesh. In R. Shaw, F. Mallick, & A. Islam (Eds.), *Climate change adaptation actions in Bangladesh* (pp. 121–141). Springer.

Schmuck, W. S. (2001). Facing the Jamuna River: Indigenous and engineering knowledge in Bangladesh. In *Bangladesh Resource Centre for Indigenous Knowledge (BARCIK)* (pp. 1–242). Bersha (Pvt) Ltd.,

Shearman, P., Bryan, J., & Walsh, J. P. (2013). Trends in deltaic change over three decades in the Asia-Pacific region. *Journal of Coastal Research, 29*(5), 1169–1183.

Sokolewicz, M., Zhang, G., & Louters, T. (2008). Planform development of the Meghna Estuary in Bangladesh. In C. M. Dohmen-Jenssen & S. J. M. H. Hulscher (Eds.), *River, Coastal and Estuarine Morphodynamics* (pp. 129–135). Taylor and Francis Group.

Soil Resources Development Institute (SRDI). (2010). Saline Soils of Bangladesh. Ministry of Agriculture.

Unnikrishnan, A. S., & Shankar, D. (2007). Are sea-level-rise trends along the coasts of the north Indian Ocean consistent with global estimates? *Global and Planetary Change, 57*(3–4), 301–307.

United Nations Scientific Committee on the Effects of Atomic Radiation (UNSCEAR). (2000). Report of UNSCEAR to the General Assembly (pp. 111–125).

Vu, C. T., Lin, C., Nguyen, K. A., Shern, C. C., & Kuo, Y. M. (2018). Ecological risk assessment of Heavy metals sampled in sediments and water of the Houjing River, Taiwan. *Environmental Earth Sciences, 77*(10).

Waste Atlas. (2020). Country data: Bangladesh.

Whitehead, P. G. (2018). Biophysical modelling of the Ganges, Brahmaputra and Meghna catchment. In R. J. Nicholls, C. W. Hutton, W. N. Adger, S. E. Hanson, M. M. Rahman, & M. Salehin (Eds.), *Ecosystem service for well-being in deltas, integrated assessment for policy analysis* (pp. 249–262). Palgrave Macmillan.

Wilson, C. A., & Goodbred, S. L., Jr. (2015). Construction and maintenance of the Ganges-Brahmaputra-Meghna delta: Linking process, morphology, and stratigraphy. *Annual Review of Marine Science, 7*, 67–88.

Woodroffe, C. D., Nicholls, R. J., Saito, Y., Chen, Z., & Goodbred, S. L. (2006). Landscape variability and the response of Asian megadeltas to environmental change. In N. Harvey (Ed.), *Global change and integrated coastal management: The Asia-Pacific region* (pp. 277–314). Springer.

World Bank. (2000). Bangladesh: Climate change and sustainable development. Report No. 21104 BD.

World Bank. (2010). Economics of adaptation to climate change: Bangladesh.

Indus River Delta

Waqar Ahmed and Samina Kidwai

1 Physical Setting

1.1 The Indus River

The Indus River basin spans around 2,900 km over four nations (Pakistan, Afghanistan, India, and China) in a region that is 30% barren (Wong et al., 2007). The Indus River begins at Lake Manasarovar, north of the Himalayas, on China's Kailash Parbat Mountain, at an altitude of 5500 m. It then travels through India to Pakistan, where it flows the entire length of the country (Fig. 1). The Indus River forms the Indus Basin, which covers 65% of Pakistan's land area. It is one of the oldest rivers known to man. The Indus River was created when the Indian and Eurasian plates collided more than 45 million years ago (Clift et al., 2001). Pakistan comprises ~56% of the total catchment area of the Indus River, which is 970,469 km^2 (Kidwai et al., 2019). The catchment area of the Indus River includes Pakistan and neighboring nations, as well as its tributaries (Table1). The two major catchment zones of Pakistan are the Indus basin and the dry region of Balochistan, with the basin covering 55,3413 km^2 (NDMA-UNDP, 2010).

The milder winters, reduced water from the glaciers over years, and low precipitation during the common dry spells influences the fresh water and sediment flows. Constructions of agriculture canals and drains diverts the water and water intensive crops and industry around the catchment and coast contributes to the country's agroeconomy but has serious implications to the delta and its adjacent coastal zones in terms of stability of the coastline or integrity of the ecosystems.

W. Ahmed (✉) · S. Kidwai
National Institute of Oceanography, Karachi 75600, Pakistan
e-mail: waqar25@gmail.com

W. Zhang and H. de Vriend (eds.), *Delta Sustainability*,
https://doi.org/10.1007/978-981-97-7259-9_7

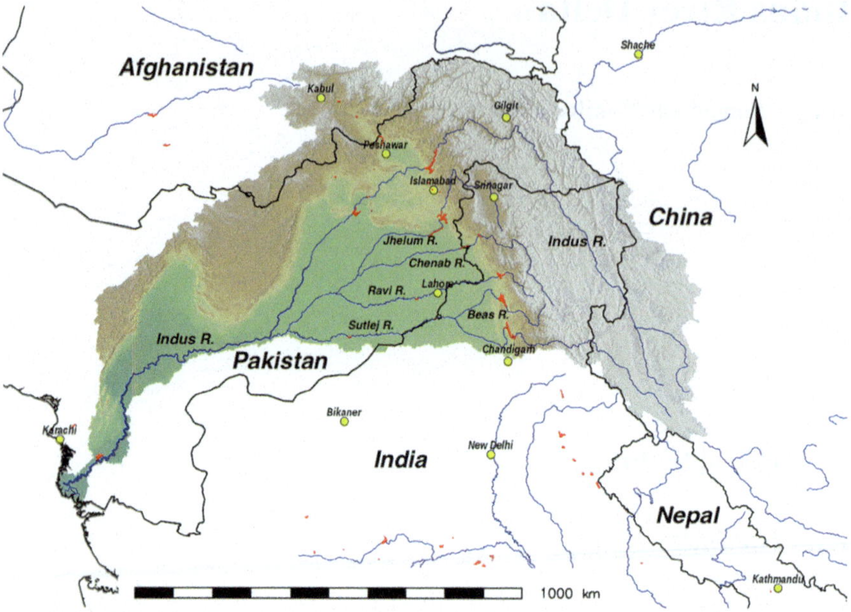

Fig. 1 Map of the Indus River Basin with major tributaries. Red prominent areas are representing reservoirs. Modified from Adams (2019)

Table 1 Details of the catchment area of the Indus River within Pakistan

Rivers	Catchment area (km^2)	Main tributaries in Pakistan
Sutlej	75,369	(8 tributaries, with only 1 in Pakistan) Rohi Nullah, River Beas (Largest river)
Ravi	24,960	Deg Nulah
Chenab	41,760	(12 tributaries) Palku Nullah is the largest
Jhelum	39,200	10 tributaries

Modified from Kidwai et al. (2019)

1.2 The Indus Delta

The main Indus River flows through the entire country and into the Arabian Sea, forming the Indus Delta (20° E and 24° N). The delta (Fig. 2) is located in a dry subtropical zone with little precipitation, with an average annual rainfall of 160 mm, the Indus River, through glacier melt and excessive rainfall, is the primary supplier of freshwater to the delta (Inam et al., 2022; Kidwai et al., 2019). The annual sediment flux in 2016 was 400 million tons, while the average discharge of water downstream of Kotri Barrage was 416 $m^3\ s^{-1}$ (Fig. 3; Sindh Irrigation Department, personal communication). In 2017, the largest water discharge from the Indus River to offshore (2960 $m^3 s^{-1}$) was reported in August and the SW monsoon period, when precipitation (30–85 mm/month) was at its maximum during the sample period (Fig. 4; Pakistan Meteorological Department).

The Indus River's discharge has fallen significantly (Fig. 3) over the past few years due to the construction of dams and channels, which has resulted in a decrease in sediments reaching the delta (Fig. 3). The Indus Delta which was originally nourished by 18 creeks (Syvitski et al., 2014), currently has one functional freshwater channel (the Khobar Creek). In addition, some portions of the Indus Delta have been eroded (Kidwai et al., 2019) and mangrove forests have been lost as a result of this (Saifullah, 2017). However, in some months, the delta and the downstream region receive little to no water.

The drastic decrease in freshwater flow has decreased the deltaic region, and nutrients carried by the river can no longer reach the North Arabian Sea (Ahmed et al., 2021; Inam et al., 2007). There are distinct and unpredictable wet and dry seasons (Fig. 4). The latest flood year was in 2021, and there were distinct patterns of normal and extreme flood years. Generally, river flow is high during the summer

Fig. 2 The Indus Delta and the major creeks along the coast. *Source* National Institute of Oceanography, Pakistan

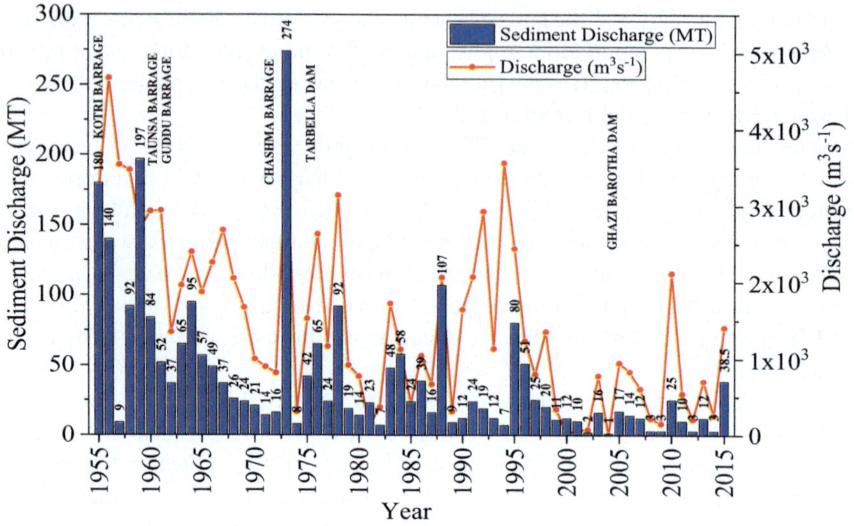

Fig. 3 Sediment discharge (blue bars) and river discharge (red dots and line) from downstream Kotri Barrage from 1955 to 2015. Modified from Kidwai et al. (2019)

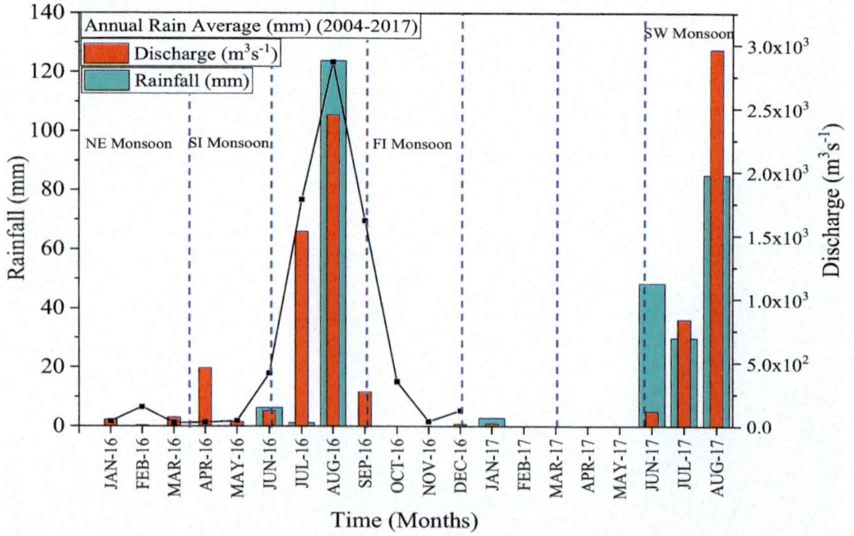

Fig. 4 Rainfall in the sampling period; rain data was collected from the Pakistan Metrological Department, and water discharge data was collected from the Sindh Irrigation Department Pakistan. Modified from Ahmed et al. (2021)

monsoon and low during the winter monsoon and other seasons (Ahmed et al., 2021). The delta is characterized by strong seasonal variability in salinity, 38–40 ppt in the dry season and 32–35 ppt in the rainy season (July–August).

There can be up to a 30 km intrusion of seawater into the estuary during low flow periods (Kidwai et al., 2016), but freshwater is still well mixed, and no stratification has been seen until it enters the estuary (Ahmed et al., 2021; Harrison et al., 1994).

2 Socio-economic Characteristics

2.1 Urbanization and Infrastructures at the Delta

Urbanization and infrastructures in the Indus Delta region of Pakistan have witnessed significant changes and challenges. The areas around the Indus Delta have experienced rapid urban growth and population expansion in recent decades. As urbanization progresses, there is an increasing demand for adequate infrastructures to support the growing population. One of the key infrastructural challenges in the Indus Delta is the provision of reliable and sustainable water supply and sanitation systems. The region heavily relies on the Indus River for its water needs, but the availability and quality of water is adversely affected by factors such as upstream water diversions, pollution, and climate change. Consequently, ensuring clean drinking water and efficient sanitation services for the urban areas of the delta remains a pressing concern.

Transportation infrastructure is another crucial aspect affected by urbanization in the Indus Delta. The expansion of urban centers has resulted in increased traffic congestion, especially in major cities like Karachi and Hyderabad. To address this issue, the government has been investing in the development of road networks, flyovers, and public transportation systems. These efforts aim to improve connectivity, reduce travel times, and enhance the overall mobility within and around the delta region.

Furthermore, urbanization has led to the growth of residential and commercial areas in the Indus Delta. As more people migrate to urban centers in search of better opportunities, there is a need for the construction of housing units, commercial complexes, and other urban facilities. Balancing urban development with environmental preservation is crucial to protect the delicate ecosystem of the delta, which serves as a habitat for various flora and fauna.

In conclusion, urbanization in the Indus Delta region of Pakistan has brought about both opportunities and challenges. Adequate infrastructures, including water supply, sanitation systems, transportation networks, and sustainable urban development, are essential to support the growing population and ensure a high quality of life. It is crucial for the government and relevant stakeholders to prioritize sustainable planning and management practices to preserve the unique natural environment of the delta while meeting the needs of the urban population.

2.2 *Major Industries*

The Indus Delta region in Pakistan has been home to various industries, mainly centered around the coastal areas and urban centers. These industries have developed to take advantage of the delta's natural resources and strategic location. The delta's coastal areas support a significant fishing industry. Fishing communities rely on the delta's waters for various marine species. The seafood processing industry involves activities such as fish cleaning, processing, and packaging for domestic consumption and export.

The fertile soils of the deltaic region have traditionally supported agricultural activities, including the cultivation of rice, cotton, sugarcane, and other crops. However, water availability and quality can be challenges due to factors like water salinity and irrigation practices. Urban centers in the Indus Delta, such as Karachi, have a substantial textiles and garments industry. This includes manufacturing clothing, textiles, and related products for both domestic consumption and export. The Karachi Port and Port Qasim, both located in the deltaic region, are among the largest and busiest ports in Pakistan. These ports play a crucial role in facilitating trade and cargo movement. The delta's coastal areas and cultural sites offer potential for tourism and hospitality industries to thrive.

It's important to note that while these industries offer economic opportunities, they also pose challenges to the environment and communities in the region. The growth of industries can lead to environmental degradation, pollution, habitat loss, and conflicts over resource use. Balancing economic development with environmental conservation and sustainable practices is a key consideration for the industries in the Indus Delta.

3 Current Issues

3.1 *Sea Level Rise and Coastal Erosion*

Global sea levels have been rising steadily for over a century, and the rate of sea level is continuously rising. This is due to a complex interplay of factors, including the thermal expansion of the warming oceans, the melting of glaciers and ice sheets, dynamic geophysical processes, and changes in the Earth's gravitational field. The impacts of rising sea levels are felt around the world, but perhaps nowhere more acutely than in coasts. Erosion, flooding, and inundation are all common in the coastal zones, and their impacts on human populations and ecosystems become more severe. Sea level rise and coastal erosion have far-reaching implications universally, but more so for poorer countries. Climate change in its various forms will exacerbate these challenges in the coming years. With over 1,000 km of coastline along the Arabian Sea and a rapidly growing population concentrated in urban coastal areas, Pakistan is a high-risk country, with intense consequences expected due to sea level rise and coastal erosion (Kanwal et al., 2020).

Land subsidence, as a result of the sinking of land due to geological processes, can also contribute to relative sea level rise (Mentaschi et al., 2018). Natural drivers, such as plate tectonics and changes in ocean currents also affect sea levels. Nevertheless, these factors are not contributing significantly to the current rate of sea level rise (Cazenave & Cozannet, 2014). According to the RCP 8.5, the Global Mean Sea Level (GMSL) increased by 1.5 mm yr^{-1} between 1901 and 1990 rising to 3.6 mm yr^{-1} between 2005 and 2015. If the global Green House Gases (GHG) emissions do not reduce, this is predicted that at this rate the GHG will increase to 0.61–1.10 m by the year 2100 thereby, increasing the rise in the seas by 2 m. Depending on the quantity of GHG emissions and the response of the Antarctic Ice Sheet (AIS), both of which are unpredictable, the sea levels may rise to more than 3 m by the year 2300 (Vitousek et al., 2017). Considering that the mitigation measures will be successful, Extreme Sea Level (ESL) events, which have become uncommon over the past century, will become more frequent before 2100 and as soon as 2050 in many places. Radical adaptation is critical, the combined effects of risks like coastal storms and extremely high tides would significantly increase the frequency and intensity of inundation in low-lying areas (Oppenheimer et al., 2019).

The effects of sea level rise are far-reaching and varied. Its impacts on coastal communities are devastating. As sea levels rise, coastal areas are subjected to more frequent and severe flooding, erosion, and saltwater intrusion. Coastal erosion is yet another by-product of sea level rise, as higher water levels cause waves and tides to erode shorelines and cliffs. This could damage infrastructure, homes, and businesses, as well as displace populations. Sea level rise can also lead to seawater intrusion into freshwater resources, which can have a significant impact on agriculture and drinking water supplies. The effects of sea level rise are felt most acutely in low-lying areas, such as small island nations and coastal cities.

Rising sea levels and increased erosion are already beginning to threaten the country's coastline. Infrastructure, including roads, bridges, and buildings near the coast are the first line to experience the grunt (Zia et al., 2017). Coastal spatial planning and proactive efforts to address the challenges faced due to sea level rise (SLR), coastal erosion (CE), and sea water intrusion (SWI) are critical to sustaining coastal development. If not paid attention, the situation is most likely to worsen and have significant economic, social, and environmental impacts to coastal infrastructure, populations and livelihoods of the average Pakistanis. Millions of people living in the low-lying areas close to the sea are at risk of displacement due to potential flooding and storm surges. A study conducted by World Bank (Christophe et al., 2022) estimates that more than 10% of Pakistan's population lives in coastal areas that are likely to be affected acutely by climate change. Single one factor contributing to this problem is the ongoing trend of global warming caused by human activity. According to recent studies, carbon emissions in the atmosphere have increased global temperatures and have subsequently led to a rise in sea levels due to glaciers melting (Arnell & Freeman, 2022). With approximately one-third of all large cities located near coastlines worldwide, rising sea levels pose significant threats globally—not just for Pakistan alone. The situation becomes even more intense when the impact is directly on the trade route that is the lifeline to a country's economy. Karachi, the major port city of Pakistan and business hub handles over two-thirds of Pakistan's total shipping traffic; coastal erosion or loss to the port infrastructure and road connections to the rest of the country consequently disrupts supply chains feeding into Pakistan's economy. Every year, natural factors such as the changes in the weather patterns in the form of unpredictable monsoon rains disrupts the business activities and the supply chains up country. Human communities, catchment areas, coastal ecology bear serious significant consequences the important coastal wetlands and mangrove forests provide critical habitat for a range of biodiversity be them, migratory birds and marine turtles. The rising seas are causing rapid erosion of these ecosystems, and threaten the existence and long-term viability of these important habitats.

3.2 Sea Water Intrusion Impacts in Pakistan

Sea water intrusion in Pakistan has dire consequences for the country's coastal regions. As saline water infiltrates freshwater aquifers, it renders groundwater unsuitable for drinking and irrigation, leading to crop damage, drinking water shortages, and increased costs for both farmers and local communities. This phenomenon also poses environmental threats, harming coastal ecosystems like mangrove forests and disrupting biodiversity. Moreover, it can force displacement and migration of communities in search of suitable water sources, exacerbating urbanization challenges. During the Southwest monsoons, the sea water circulates towards the Indus River Estuary effecting the natural habitat of the delta, thereby affecting the brackish and freshwater lakes around the delta making them hypersaline. Sediment deposition behind the barrages and dams on Indus River affecting the natural flow of water and sediment, use of groudwater for irrigation purposes around the delta, natural land subsidence due to tectonic activity, as well as the natural flatness of delta have resulted the sinking and compaction of Indus Delta (Kidwai et al., 2016). As result of seawater intrusion, land subsistence and less sediment flowing into the delta than one century ago, the Indus delta has lost one fifth of its size since 1932 (Giosan et al., 2014).

The vulnerability of the delta to these coastal hazards has intensified significantly in recent years. While there isn't a single cause for this heightened vulnerability, seawater intrusion is a prominent factor. The low water tables in the area create an opening for seawater intrusion, making the intrusion of salinity even more problematic. Overall, the Indus Delta is grappling with multiple challenges, and addressing seawater intrusion (Fig. 5) is crucial for the long-term health and sustainability of the region.

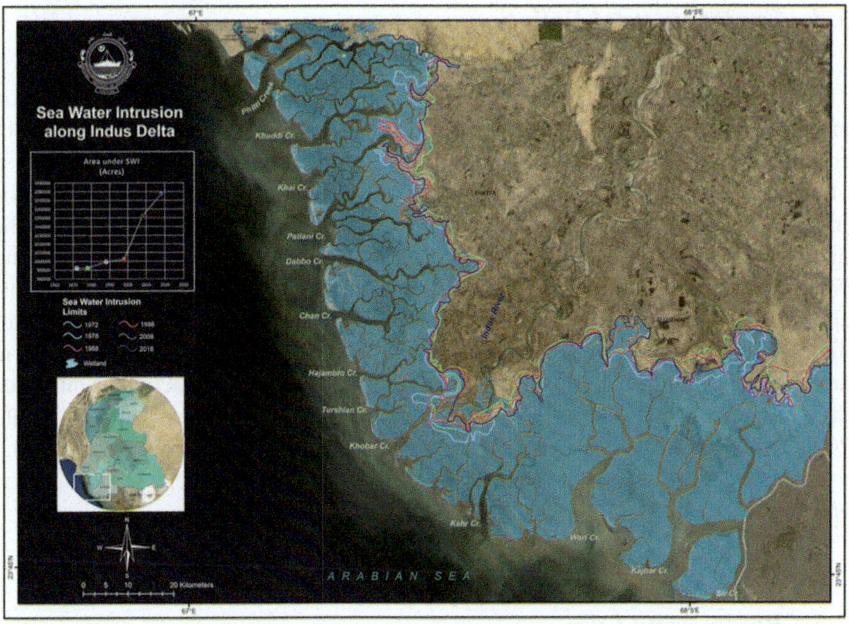

Fig. 5 Sea water intrusion along Indus Delta (From National Institute of Oceanography)

3.3 Extreme Events—Storm Surges and Flooding

The Indus Delta region in Pakistan is highly vulnerable to extreme events such as storm surges and flooding. Storm surges, caused by tropical cyclones and intense storms in the Arabian Sea, pose a significant threat to the low-lying coastal areas of the delta. These surges, characterized by a rapid rise in sea level, can result in the inundation of coastal communities, erosion of shorelines, and damage to infrastructure. In addition to storm surges, the Indus Delta experiences frequent flooding, primarily driven by heavy monsoon rains and the release of water from upstream reservoirs. The vast network of rivers and tributaries that converge in the delta make it susceptible to overflow and the subsequent displacement of communities residing along the riverbanks. The flooding not only leads to the loss of lives and destruction of homes but also impacts agriculture, which is a vital livelihood for many people in the region (Zia et al., 2023).

These extreme events have been further exacerbated by factors such as climate change, deforestation, and land-use changes. Rising sea levels, changing rainfall patterns, and increased intensity of cyclones contribute to the heightened risks faced by the Indus Delta. Efforts are being made to improve early warning systems, enhance coastal defenses, and implement flood management strategies to mitigate the impacts of storm surges and flooding. However, the challenges are complex, requiring long-term planning, coordination, and investment in resilient infrastructure and adaptive measures to protect the vulnerable communities and ecosystems of the Indus Delta.

3.4 Environmental Pollution

The surplus concentration of nutrients causes eutrophication, hypoxia, and anoxia in the coastal waters, which results in an adverse impact on the biogeochemistry of coastal waters and on the ecosystems (Herbeck et al., 2013). Hypoxia and the presence of harmful algal blooms (HAB) due to eutrophication have been recorded from the Indus Delta shelf area (Chaghtai & Saifullah, 2006), and are often associated with the mass mortality of fish. Sediment balance, vegetation change, fertilizer use, dam and barrage construction, industrialization, and land utilization for human settlements all have an impact on the quality and quantity of water in the Indus. Loading of dissolved nutrients in rivers beyond primeval concentrations is linked to changes in anthropogenic emissions and land-use (Dan et al., 2009). Human activities influence the nitrogen and phosphorus concentrations from land to oceans (Howarth et al., 2002). Nitrogen enters the ocean via atmospheric deposition, waste discharge, ground water, rivers, nitrogen fixation, and sedimentation, and nitrogen exits via anammox and denitrification.

3.5 Biodiversity Loss

Fisheries output in the Indus Delta and its nearby coastal waters is approximately 4000 metric tons per year; 2000 metric tons per year for shrimp; and 2500 metric tons per year for crabs (Hoekstra et al., 1997). With the current level of exploitation from decreasing area, this leads to a signficantly increasing fishing pressure. The vast majority of fishing for small fish, shrimp, and crabs occur within the mangrove habitat and in the open sea immediately adjacent to it. The Fishery Resource Appraisal Program's (2013–14) findings from the deltaic creeks and (2009–15) findings from the shelf areas indicate a drop in the fish catch per effort, the size of the fish and shrimp taken, and the number of commercial fish and shellfish species (Fanning et al., 2009, 2015; Kidwai et al., 2016).

The Indus Delta is characterized by tidal creeks and small over-bank plains that cover an area of approximately 2500 km^2. The lower deltaic plain is lined with stunted mangroves on a sand/silt substrate (Kristensen et al., 1992). *Avicennia marina*, *Ceriops tagal*, and *Aegiceras corniculata* are the three species that are now naturally present in the Indus Delta's four open entrances to the sea streams (Hajamro, Khobar, Dabbo, and Waddi Khuddi). *Avicennia* marina is the dominant species in the Indus Delta and on a portion of Shah Bundar and Keti Bundar, and *Rhizophora macronata* is transplanted.

The mangrove forests in the Indus Delta once covered the entire delta region, and it was the fifth-largest mangrove stand in a semiarid location in the world. It was recently ranked as the fifteenth mangrove stand in the world in terms of size, composed of fewer species (Saifullah, 2017). The rate of degradation of mangrove forests in the delta was estimated to be 6% per year between 1980 and 1995, and only a small percentage is now considered to be healthy. The mangrove forests serve as major breeding grounds for shrimp, prawns, crabs, and other economically important species of seafood (Bouillon et al., 2008). A healthy mangrove forest helps to keep this complex ecosystem in ecological balance and can provide a sustainable supply of products and services to meet the growing demands of the coastal communities and the adjacent urban settlements. The health of the ecosystem shows how people are doing in coastal communities. 70% of the shrimp fishery is dependent on the mangroves of the Indus Delta, and they nurture 23 species of shrimps and about 155 species of fish.

Waddi Khuddi Creek (WKC; total length (TL) 32 km), Dabbo Creek (DC; TL 38 km), Hajamro Creek (HC; TL 15 km), and Khobar Creek (KC; TL 40 km) are among the major creeks of the Indus Delta (Fig. 2). WKC is a virgin tidal stream with healthy natural mangrove coverage that is tidally dominated, with no freshwater intake from other sources other than the sea and other smaller connecting creeks or possibly seepage from groundwater during floods. DC is also a tidal creek. The end of the creek connects to a small creek called the Ambro, which gets runoff from farms and plantations of bananas, betel leaves, and vegetables upstream. DC also features a strong mangrove strand interspersed with mixed vegetation (*Temerax* and *Salvadora* spp.). HC is a fishing trawler navigation channel. There is a sizable human community in the vicinity of Keti Bundar (population of 58,832, according to the 2017 Census). This area has the most anthropogenic activity, with minimal mangrove cover and mangrove replanting programmes (Kidwai et al., 2019). KC is the principal river channel, which is estuarine, impacted by tides, and has a 36 km^2 basin area. It is characterized as a somewhat mixed estuary and a coastal plain estuary (Inam et al., 2007). In earlier environmental studies (Kalhoro et al., 2015), HC, Isaro, Gizri, and Gharo are mentioned as lesser tributary streams west of the delta; they are tiny creeks that empty into larger main creeks, with the exception of HC. The river or the sea has no direct influence on these creeks.

4 Measures to Solve the Problems

Saline intrusion and water scarcity in riverine and deltaic systems are caused by river water diversion and groundwater extraction for agricultural activities and arid conditions. These issues are due to poor management because of the difficulties in determining a management plan for the catchment. In these regions, not only are future environmental changes influenced by global warming and infrastructure (such as dams) necessitated by a growing population, but there is also a high potential for water-related conflicts (Kidwai et al., 2019).

4.1 Mitigation Strategies

The global response to sea level rise has been mixed. While some countries have taken significant strides in mitigating the impacts of sea level rise, others have been slow to act. The United Nations Framework Convention on Climate Change (UNFCCC) has been instrumental in bringing together nations to work towards limiting global warming and mitigating the effects of sea level rise. Additionally, the Intergovernmental Panel on Climate Change (IPCC) has published several reports outlining the impacts of sea level rise and strategies to mitigate its effects. The Paris Agreement, signed in 2015, aims to limit global warming to well below 2 °C above pre-industrial levels and to pursue efforts to limit the temperature increase to 1.5 °C. The agreement also calls for countries to regularly report on their greenhouse gas emissions and their efforts to reduce them. There are a number of strategies that can be employed to mitigate the effects of sea level rise. These include measures to reduce greenhouse gas emissions, such as transitioning to renewable energy sources and improving energy efficiency. In addition, coastal communities can implement measures such as mangrove reforestation, sustainable land-use practices, beach nourishment, construction of sea walls, and relocation of infrastructure and populations from high-risk areas (Oppenheimer et al., 2019).

4.2 The Role of Government and Policy Makers in Addressing the Issue

Governments and policy makers have a critical role to play in addressing the issue of sea level rise. This includes developing and implementing policies to reduce greenhouse gas emissions, as well as providing funding and resources for mitigation and adaptation measures. In addition, governments can work with coastal communities to develop comprehensive adaptation plans that consider the specific needs and vulnerabilities of each community. Research and technology development are critical for addressing the issue of sea level rise. This includes improving our understanding of the causes and effects of sea level rise, developing new mitigation and adaptation technologies, and improving our ability to model and predict sea level rise. Governments, universities, and private sector organizations can all play a role in supporting research and technology development in this area. There is no single solution to these issues, and it will require a collaborative effort between governments, businesses, communities, and individuals worldwide to mitigate their effects on coastal areas. The anthropogenic impact of upstream water and sediment blockage has resulted in the shrinkage of active delta and stunted growth of mangrove forest. The Indus delta has been forced to face severe problem of coastal erosion due to unplanned coastal development in the area. The wellbeing of Indus Delta demands a realistic assessment of the minimum quantity of fresh water and sediments required to prevent total disappearance of the delta. There is need for a certain amount of water and sediment to be discharged in to the delta on year-round basis. It is also important that the management of the delta should become a part of an integrated coastal zone management practice.

4.3 *Efforts to Restore, Sustain, and Revive the Indus Delta*

In the context of national efforts there are various solutions to restore and revive the Indus Delta. Among these, two are important to be addressed for the Indus Delta to sustain as part of Pakistan's heritage, wealth and healthy ecosystem. The first is the role of scientific consideration, community-based management and local action to reduce the impact of direct exploitation of the delta ecosystem within the creeks by external users and delta communities. Day to day implementation of effective management and commitment based on sound scientific understanding is vital for the long-term success. The second solution is the inflow of sufficient freshwater every year—not just in flood years—to nourish the delta. This will support mangrove and fish production. Downstream areas depend on a balanced upstream river system, for irrigation, freshwater supply, sediment management, as well as a balanced estuarine system. These existential challenges can only be addressed at the highest political wills and levels. For that there is need of a national policy to ensure the basic water needs of the delta ecosystem in consultation with all stakeholders; regulators, communities, resource users and environmental groups.

5 Conflicting Perspectives

Conflicting perspectives regarding the Indus Delta often arise from various stakeholders, including environmentalists, policymakers, local communities, and industrial interests. Here are some common conflicting viewpoints related to the Indus Delta:

Environmental Conservation versus Economic Development

Conservation Perspective: Environmentalists and conservationists emphasize the importance of protecting the delta's fragile ecosystem, including its mangrove forests, biodiversity, and wetlands. They argue that unsustainable development, such as urbanization, industrialization, and agricultural expansion, could lead to irreversible damage to the ecosystem and its services.

Development Perspective: Economic development proponents often prioritize infrastructure projects, agriculture, and industrial growth for the region. They might argue that development activities can lead to job creation, economic growth, and improved living standards, outweighing potential environmental concerns.

Local Livelihoods versus Industrial Growth

Livelihood Perspective: Local communities, including fisherfolk and those dependent on delta resources, often prioritize maintaining traditional livelihoods and sustaining their way of life. They may oppose industrial projects that could pollute water bodies, disrupt fishing activities, or displace communities.

Industrial and Economic Perspective: Industries and businesses seeking to establish or expand operations in the region may focus on economic growth and employment opportunities. They might argue that industrial development can contribute to the overall prosperity of the region.

Navigating these conflicting perspectives requires a balanced approach that considers environmental, social, and economic factors, as well as the long-term sustainability of the Indus Delta ecosystem and the well-being of the people who rely on it.

References

Adams III, T. E. (2019). Water resources forecasting within the Indus River Basin: A call for comprehensive modeling. In S. I. Khan, & T. E. Adams III (Eds.), *Indus River Basin* (pp. 267–308). Elsevier.

Ahmed, W., Wu, Y., Kidwai, S., Li, X., Zhang, G., & Zhang, J. (2021). Spatial and temporal variations of nutrients and chlorophyll a in the Indus River and its deltaic creeks and coastal waters (Northwest Indian Ocean, Pakistan). *Journal of Marine Systems, 218*, 103525.

Arnell, N. W., & Freeman, A. (2022). The impact of climate change on policy-relevant indicators of temperature extremes in the United Kingdom. *Climate Resilience and Sustainability, 1*, e12.

Bouillon, S., Connolly, R. M., & Lee, S. Y. (2008). Organic matter exchange and cycling in mangrove ecosystems: Recent insights from stable isotope studies. *Journal of Sea Research, 59*, 44–58.

Cazenave, A., & Cozannet, G. L. (2014). Sea level rise and its coastal impacts. *Earth's Future, 2*(2), 15–34.

Chaghtai, F., & Saifullah, S. M. (2006). On the occurrence of green Noctiluca scintillans blooms in coastal waters of Pakistan, North Arabian Sea. *Pakistan Journal of Botany, 38*(3), 893–898.

Christophe, C., Jabeen, R., & Kondo, S. (2022). Pakistan's coastal ecosystem and opportunities to tackle climate change. World Bank Blogs.

Clift, P. D., Shimizu, N., Layne, G. D., Blusztajn, J. S., Gaedicke, C., Schlüter, H.-U., Clark, M. K., & Amjad, S. (2001). Development of the Indus fan and its significance for the erosional history of the Western Himalaya and Karakoram. *Geological Society of America Bulletin, 113*(8), 1039–1051.

Dan, S. F., Liu, S. M., Yang, B., Udoh, E. C., Umoh, U., & Ewa-Oboho, I. (2009). Geochemical discrimination of bulk organic matter in surface sediments of the Cross River estuary system and adjacent shelf, South East Nigeria (West Africa). *Science of the Total Environment, 678*, 351–368.

Fanning, P. L., Kalhoro, M., Hanif, T., Khan, M. W., & Ansari, M. A. (2009). Cruise Report—Pakistan Demersal Survey 2009001 29 October–6 November, 2009. Report Prepared for the Fisheries Resources Appraisal in Pakistan Project UTF/PAK/108/PAK (p. 34).

Harrison, P. J., Snedaker, S. C., Ahmed, S. I., & Azam, F. (1994). Primary producers of the arid climate mangrove ecosystem of the Indus River Delta, Pakistan: An overview. *Journal of Tropical Ecology, 35*(2), 155–184.

Giosan, L., Syvitski, J., Constantinescu, S., & Day, J. (2014). Climate change: Protect the world's deltas. *Nature, 516*(7529), 31–33.

Herbeck, L. S., Unger, D., & Wu, Y. (2013). Effluent, nutrient and organic matter export from shrimp and fish ponds causing eutrophication in coastal and back-reef waters of NE Hainan, tropical China. *Continental Shelf Research, 57*, 92–104.

Hoekstra, D. A., Mahmood, N., Shah, G. R., Domki, M. A., & Ali, Q. M. (1997). Diagnostic study—Indus Delta mangrove ecosystem. Main subsystem characteristics, problems, potentials, proposed interventions and pilot sites. Sub-project. RRIDM (World Bank/GoS funds).

Howarth, R. W., Sharpley, A., & Walker, D. (2002). Sources of nutrient pollution to coastal waters in the United States: Implications for achieving coastal water quality goals. *Estuaries, 25*, 656–676.

Inam, A., Clift, P. D., Giosan, L., Tabrez, A. R., Tahir, M., Rabbani, M. M., & Danish, M. (2007). The geographic, geological and oceanographic setting of the Indus River. In A. Gupta (Ed.), *Large Rivers: Geomorphology and management* (pp. 333–345). Wiley.

Inam. A., Clift. P. D., Giosan. L., Zai. A. A., Kidwai. S., Shahzad. M. I., Zia. I., Nazeer. M., Khan. M. J., Ali. S. S., Shaheen. A., Qari R., & Kehkashan. S. (2022). The geographic, geological and oceanographic setting of the Indus River-an update. In A. Gupta (Ed.), *Large Rivers: Geomorphology and management* (2rd ed., pp. 488–520). Wiley.

Kanwal, S., Ding, X., Sajjad, M., & Abbas, S. (2020). Three decades of coastal changes in Sindh, Pakistan (1989–2018): A geospatial assessment. *Remote Sensing, 12*(1), 8.

Kalhoro, N. A., Saleem, M., Muhammad, F., & Bhutto, A. H. (2015). Dynamics of a polluted tidal estuary, Gizri Creek during south west monsoon. *Scholars Academic Journal of Biosciences, 3*(1A), 46–51.

Kidwai, S., Fanning, P. L., Ahmed, W., Tabrez, S. M., Zhang, J., & Khan, M. W. Practicality of marine protected areas—Can there be solutions for the River Indus delta. *Estuaries, Coastal and Shelf Science, 183*(Part B), 349–359.

Kidwai, S., Ahmed, W., Tabrez, S. M., Zhang, J., Giosan, L., Clift, P., & Inam, A. (2019). The Indus Delta—Catchment, river, coast, and people. In E. Wolanski, J. W. Day, M. Elliott, & R. Ramachandran (Eds.), *Coast and estuaries: The future* (pp. 213–232). Elsevier.

Kristensen, E., Devol, A. H., Ahmed, S. I., & Saleem, M. (1992). Preliminary study of benthic metabolism and sulfate reduction in a mangrove swamp of the Indus Delta, Pakistan. *Marine Ecology Progress Series, 90*, 287–297.

Mentaschi, L., Vousdoukas, M. I., Pekel, J.-F., Voukouvalas, E., & Feyen, L. (2018). Global long-term observations of coastal erosion and accretion. *Scientific Reports, 8*(1), 12876.

NDMA-UNDP. (2010). Pakistan Indus River system. 5-Day training course on "Flood Mitigation".

Oppenheimer, M., Glavovic, B. C., Hinkel, J., van de Wal, R., Magnan, A. K., Abd-Elgawad, A., Cai, R., Cifuentes-Jara, M., DeConto, R. M., Ghosh, T., Hay, J., Isla, F., Marzeion, B., Meyssignac, B., & Sebesvari, Z. (2019). Sea level rise and implications for low-lying islands, coasts and communities. In H.-O. Pörtner, D. C. Roberts, V. Masson-Delmotte, P. Zhai, M. Tignor, E. Poloczanska, K. Mintenbeck, A. Alegría, M. Nicolai, A. Okem, J. Petzold, B. Rama, & N. M. Weyer (Eds.), *IPCC special report on the ocean and cryosphere in a changing climate* (pp. 321–445). Cambridge University Press.

Saifullah, S. M. (2017). The effect of global warming (climate change) on mangroves of Indus Delta with relevance to other prevailing anthropogenic stresses—A critical review. *European Academic Research, V, 4*, 2110–2138.

Syvitski, J. P. M., Kettner, A. J., Overeem, I., Giosan, L., Brakenridge, M., & Billam Roger, M. H. (2014). Anthropocene metamorphosis of the Indus Delta and lower floodplain. *Anthropocene, 3*, 24–35.

Vitousek, S., Barnard, P. L., Fletcher, C. H., Frazer, N., Erikson, L., & Storlazzi, C. D. (2017). Doubling of coastal flooding frequency within decades due to sea-level rise. *Scientific Reports, 7*(1), 1399.

Wong, C. M., Williams, C. E., Pittock, J., Collier, U., & Schelle, P. (2007). World's top 10 rivers at risk. WWF International.

Zia, I., Zafar, H., Shahzad, M. I., Meraj, M., & Kazmi, J. H. (2017). Assessment of sea water inundation along Daboo creek area in Indus Delta Region, Pakistan. *Journal of Ocean University of China, 16*, 1055–1060.

Zia, I., Zafar, T., & Khan, M. (2023). *Analyzing seawater inundation extent during the recent extremely severe cyclone Biparjoy throughout scenario modeling approach, Pakistan.* Unpublished data.

Irrawaddy River Delta

Qinzi Cheng and Xing Li

1 Physical Setting

1.1 The Irrawaddy River

The Irrawaddy (Ayeyarwady) River originates from the confluence of the N'mai and Mali Rivers in the glaciers of the southeastern Himalayas (Fig. 1). It then flows to the southwest and combines with the Chindwin River in the middle of the Irrawaddy Basin. The river continues to meander through the Central Dry Zone, and then flows through the container valley between Minbu and Pyay to enter the Irrawaddy Delta (Fig. 1). The river has a length of 2010 km (Vogel et al., 2022).

The Irrawaddy Basin has an area of 415,700 km^2, covering 60% of Myanmar's territory. It has a primarily tropical monsoon climate with distinct dry and rainy seasons. In the dry season, the northeast (NE) monsoon is active between November and February, while in the rainy season, the southwest (SW) monsoon is active between May and October. Although its mean annual precipitation varies widely from 500 to 4000 mm, the heaviest rains occur during the rainy season, accounting for 90% of the annual precipitation in the area (Furuichi et al., 2009). The annual average water discharge was estimated to be 422 km^3/yr (Robinson et al., 2007). The heavy seasonal rain-induced floods carry large amounts of sediment into the coastal sea. The annual Irrawaddy suspended sediment load has been reported around 261–364 Mt (Gordon, 1879; Milliman & Meade, 1983; Robinson et al., 2007).

Q. Cheng (✉)
State Key Laboratory of Estuarine and Coastal Research, East China Normal University, Shanghai 200241, China
e-mail: qz_cheng@foxmail.com

X. Li
School of Geography, Geomatics and Planning, Jiangsu Normal University, Xuzhou 221116, China

W. Zhang and H. de Vriend (eds.), *Delta Sustainability*,
https://doi.org/10.1007/978-981-97-7259-9_8

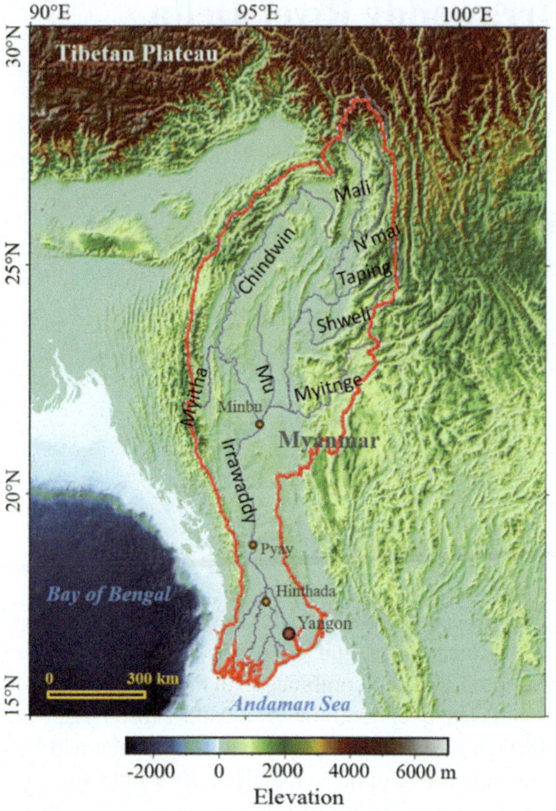

Fig. 1 Location of the Irrawaddy River basin

1.2 The Irrawaddy Delta

The Irrawaddy River branches into two main distributaries at the apex of the Irrawaddy Delta, about 90 km north of Hinthada: the Pathein River in the west and the Yangon River in the east. In this section, there is a conspicuous meander belt with alluvial ridges along active and abandoned channels, especially along the former Daga River channel (Giosan et al., 2018) (Fig. 2). In the Lower Irrawaddy, many tertiary branches form a complex network of channels that terminate in the shallow Andaman Sea, creating a lobate lower delta with eleven estuaries. Seven of these estuaries are located on the western delta coast (Chen et al., 2020).

The Irrawaddy Delta is around 40,000 km^2. It is a mud-silt system dominated by semi-diurnal tides with a mean tidal range of 3–6 m (Brakenridge et al., 2017). The tidal range increases from west to east along the delta coast. According to Kravtsova et al. (2009), tidal ranges during the spring and neap tides are 2.2 m and 1.8 m in the west of the delta coast, whereas they are 5.7 m and 4.0 m in the east of the delta coast, respectively. The spring tide influences extend nearly 300 km inland to the apex of the delta area (Hedley et al., 2010). In the northern Andaman Sea, during the spring

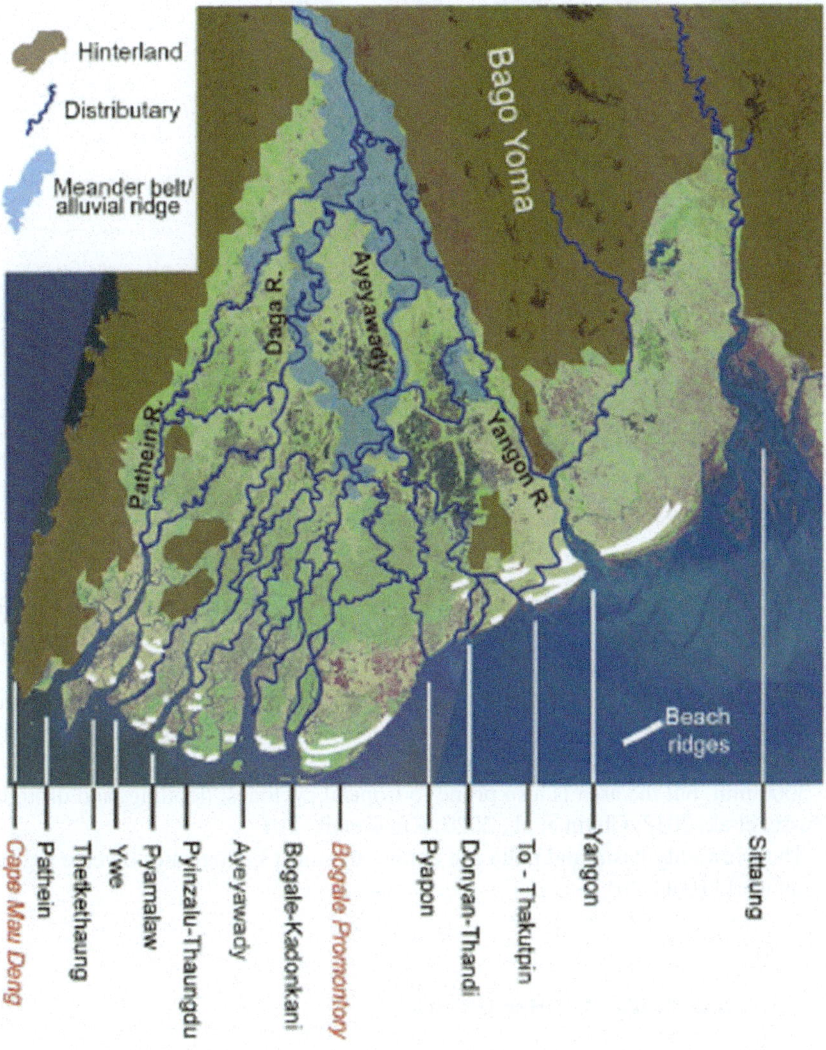

Fig. 2 Distributaries of the Irrawaddy River Delta (Giosan et al., 2018)

tide, the tidal current velocity reaches up to 3 m/s, which causes significant sediment suspension and re-suspension, resulting in a persistent high-turbidity zone in the gulf (Ramaswamy et al., 2004). Wave heights range from <1 m in winter to 1–2 m in summer (Kravtsova et al., 2009). The shore-parallel beach ridges near the coastline reflect periods of strong wave activity in the late Holocene (Giosan et al., 2018). About 90% of the sediment is transported eastward into the Gulf of Martaban by tidal and prevailing SW monsoon currents. A recent geophysical survey indicated that little modern sediment is accumulating on the shelf immediately off the Irrawaddy River

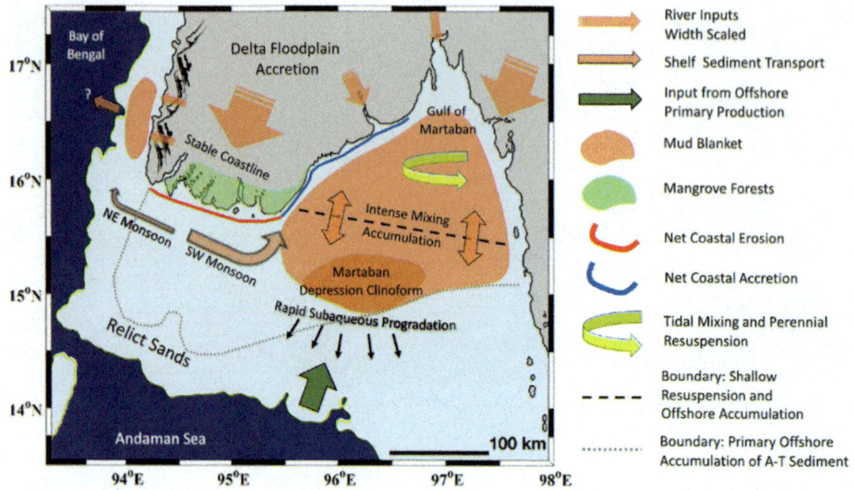

Fig. 3 Sediment transport and deposition in the Irrawaddy River Delta (Flynn et al., 2022)

mouths. In contrast, the modern Irrawaddy River derived sediment has accumulated in the Gulf of Martaban at a rate of 215 Mt/yr and formed a 60 m-thick mud depocenter on the shelf (Fig. 3) (Flynn et al., 2022; Liu et al., 2020).

The climate of the delta is determined by the Indian Monsoon with hot and humid months in the rainy season from May to October and relatively cool and dry months from December to March. The mean annual rainfall in the delta ranges from 2000 to 3000 mm, but the area is also prone to tropical cyclones, flooding, and droughts (Besset et al., 2017; Chhin et al., 2020; Kraas et al., 2017).

The Irrawaddy basin and delta are among the most biologically diverse regions in the world (HIC, 2017).

2 Socioeconomic Characteristics

2.1 Population and Economy

Myanmar is one of the poorest nations in Southeast Asia, with a total GDP of 59.36 billion current US $ and a GDP per capita of 1095.7 current US $ in 2022 (World Bank, 2023). It has experienced a great recession since 2020. The country also suffers from inadequate infrastructure. Most goods are transported across the Thai border and along the Irrawaddy River. Energy shortages are common throughout the country, including in Yangon. Only 25% of the country's population has access to electricity. Agriculture is a major sector of the Myanmar economy, accounting for 41% of the GDP and 68.9% of the labour force (Schmitt-Degenhardt, 2013).

The Irrawaddy River catchment hosts over 90% of the population of Myanmar and connects leading cities and economic centres (Furuichi et al., 2009; Kraas et al., 2017). The Irrawaddy Delta is one of the most populous areas in Myanmar. About 15 million people lives on the delta, about 26% of the population of Myanmar (Vogel et al., 2022). The delta includes the megacity Yangon and numerous regional and smaller cities and towns within the urbanized delta triangle of Yangon, Pathein, and Pyay (Kraas et al., 2010, 2017).

The Irrawaddy River and its delta have remained largely inaccessible for many years (Webb et al., 2014). Under British colonial rule, the delta became the world's largest exporter of rice (Adas, 2011). The delta is still among the important rice-producing areas of Southeast Asia and constitutes the most important agricultural region in Myanmar, with significant proportions of aquaculture (Kraas et al., 2017; Soe, 2020; Vogel et al., 2022).

2.2 *Land Use and Land Cover*

Agricultural areas are the dominant land cover in the Irrawaddy Delta (Fig. 4). Significant mangrove area has been converted to agricultural land and aquaculture areas have expanded rapidly (Fig. 4). Until 2001, the proportion of urban and built-up areas in the entire study area was below 2%. By 2005, there was a significant increase of urban areas. This expansion is concentrated mainly in the southeast of the delta (around Yangon city), and along the river courses and in agriculturally dominated rural areas, such as Myaungmya in the southeast of Pathein or Kyonpyaw in the upper delta region (Fig. 4; Vogel et al., 2022).

3 Current Issues

Recent decades have witnessed considerable human-induced disturbances in the Irrawaddy Basin, such as dam construction, agricultural development, deforestation, and terrestrial mining, which have altered the pristine state of the Irrawaddy River (Hennig, 2016). These activities have resulted in noticeable changes in the hydrological regime of the basin (HIC, 2017). The Irrawaddy Delta, which is characterized by its low-lying and flat topography, is highly vulnerable to flooding from tropical cyclones and storms. Syvitski et al. (2009) classified the Irrawaddy Delta as "in peril" based on its exposure to multiple threats. The delta is expected to become more susceptible to extreme events and coastal erosion under the combined effects of climate change and human activities.

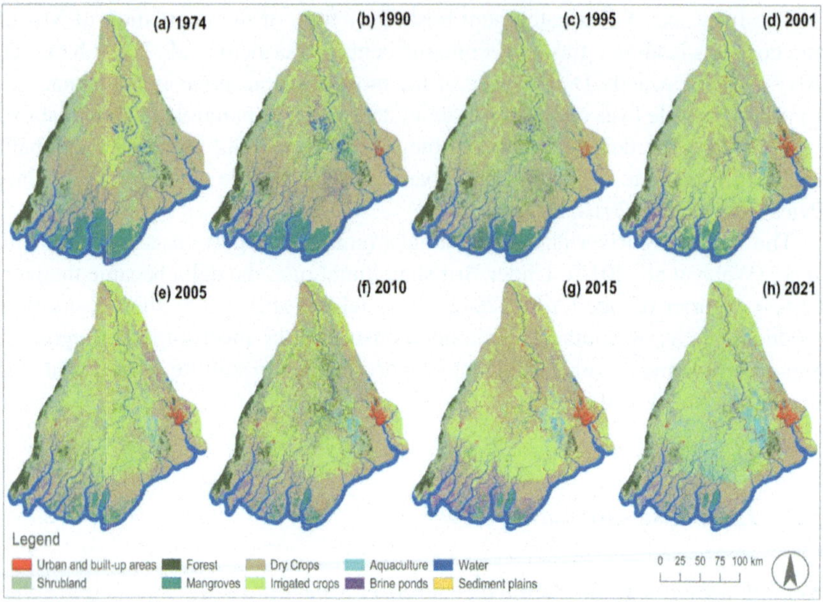

Fig. 4 Land use changes for the Irrawaddy Delta from 1974 to 2021 (Vogel et al., 2022)

3.1 Deforestation

The mangrove cover in the Irrawaddy Delta declined drastically from 8% (3000 km^2) of the total delta area in 1974 to 2% (800 km^2) in 2021, representing a loss of 2200 km^2 (Fig. 4; Vogel et al., 2022). The mangrove distribution also became more fragmented, with the maximum patch size decreasing from 606 km^2 in 1990 to 103 km^2 in 2021 (Vogel et al., 2022). The mangrove forests are projected to disappear completely within the next decade (Webb et al., 2014). The main drivers of this deforestation are consumption of wood for energy and the expansion of aquaculture, brine ponds, and agricultural areas (Frenken, 2012; Webb et al., 2014; Yang et al., 2019).

3.2 Coastal Erosion

The delta system has been relatively stable over the past ~150 years, with a seaward migration of 3.4 m/yr since 1925 (Hedley et al., 2010). Since 1974, however, the delta coast is undergoing significant changes, with an average accretion rate of 10.4 m/year (Chen et al., 2020). The coast is eroding (~4.7 m/yr on average) in the west and is accreting (~21.4 m/yr on average) in the east. Between 1974 and 2018, 58% of the total shoreline length accreted and 42% eroded (Fig. 5a). The evolution of the delta channels is illustrated in Fig. 5b. Between 1974 and 2018, the net land gain in the study area was 194.0 km^2, among which the channel bar, channel bank, mouth bar, and delta front changes contributed 11%, 5%, 8% and 76%, respectively (Chen et al., 2020).

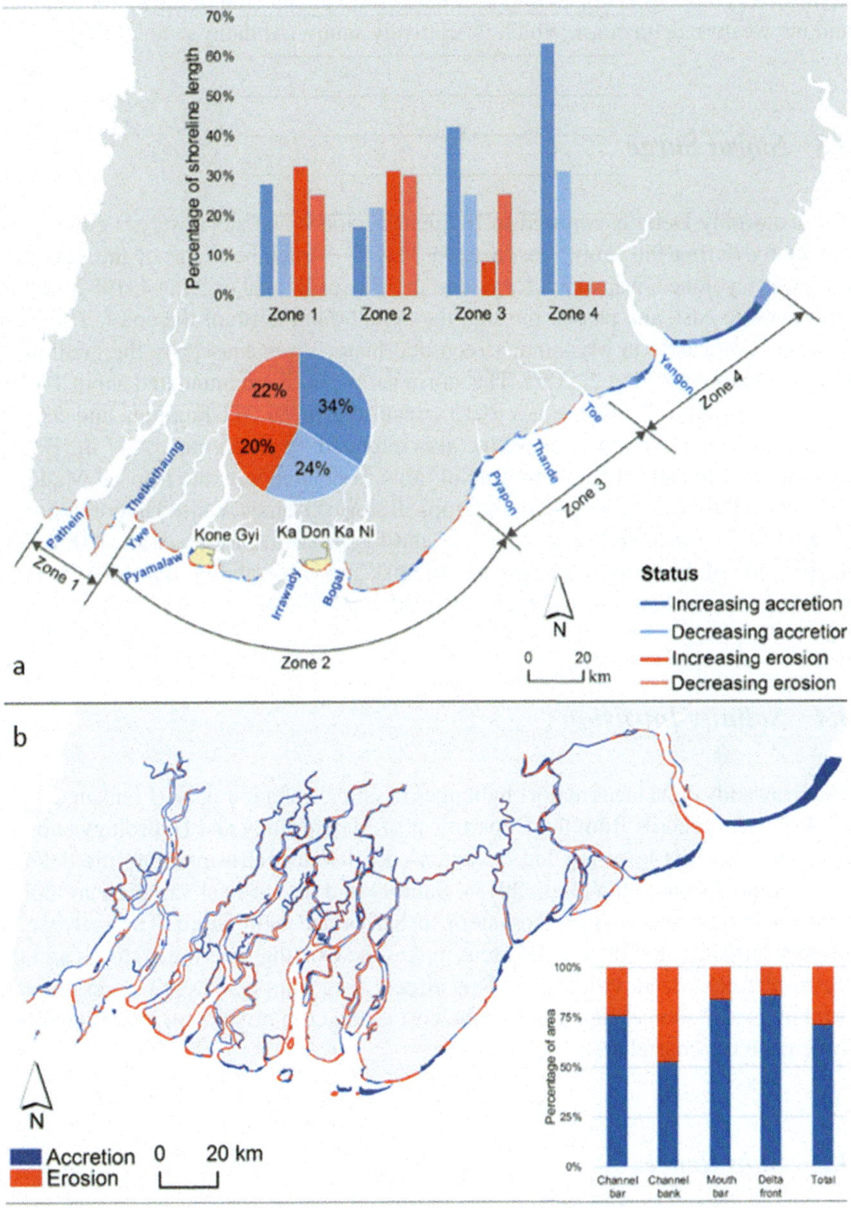

Fig. 5 The Irrawaddy River Delta shoreline changes between 1974 and 2018 (Chen et al., 2020)

Various human activities have affected the sediment supply to the delta, such as deforestation, mining, dam construction, and sediment mining. These activities have increased the fine sediment load (Anthony et al., 2019; Giosan et al., 2018; Hedley et al., 2010; Rao et al., 2005), but reduced the coarse sediment transport and delivery to the delta (Lazarus et al., 2018). This has resulted in erosion of the major channels and the western delta coast, which is relatively sandy (Anthony et al., 2019).

3.3 Storm Surge

The Irrawaddy Delta is exposed to frequent cyclones and storm surges every year, especially during the rainy season from May to September. One of the strongest ones was Cyclone Gwa, which formed in the Bay of Bengal on May 4, 1982, moved from west to east, and passed through the delta 60 km north of the coast. The most devastating disaster in Myanmar's recorded history was caused by the landfall of Cyclone Nargis on May 2, 2008. The storm surge and flood inundated about 14,400 km^2 in the delta (Tasnim et al., 2015), resulting in 84,537 fatalities and 53,836 missing people. The delta's shoreline also retreated by an average of 47 m (Besset et al., 2017). In 2015, heavy monsoonal rains associated with the tropical Madden-Julian Oscillation and a slow-moving tropical storm (Komen) caused major flooding, about 130 fatalities, and severe damage and losses (Brakenridge et al., 2017). The degradation of the mangroves increases the risk and vulnerability of the delta to cope with large tropical storms and cyclones (Osti et al., 2009).

3.4 Salinity Intrusion

The Irrawaddy delta faces major challenges in saline water intrusion (Frenken, 2012). Hydrological records from the Department of Meteorology and Hydrology showed that the rising sea level has led to increased saltwater intrusions into the delta in recent years (SeinnSeinn et al., 2015). Salinity and its seasonal variation are dominant factors for the coastal ecosystem, fisheries and agriculture. The increase of salinity intrusion into the coastal areas makes the existing freshwater stocks and the freshwater ecosystem vulnerable. This affects the local population's access to safe drinking water sources and increase the cost of the crop production and aquaculture systems in the coastal area.

3.5 Subsidence

The delta may be undergoing land subsidence at unknown rates (Higgins, 2016). A recent study revealed that the subsidence rate exceeded 10 mm/year in half of

Yangon City from 2014 to 2017, and reached up to 110 mm/year in some areas (van der Horst et al., 2018). Groundwater extraction, which amounts to 12% of the current recharge, was the main cause of the subsidence in Yangon. The increasing water demand is likely to further enhance groundwater extraction, which could worsen the land subsidence (HIC, 2017). Hydrocarbon extraction was another factor that contributed to the delta's subsidence (Ericson et al., 2006).

3.6 Climate Change

Sea level rise and climate change also potentially pose threats to the sustainability of the Irrawaddy Delta. The projected sea level rise will reach 14–56 cm in the 2050s, which would have devastating effects on the delta shoreline (Horton et al., 2017). In the period 1988–2000, the average monsoon duration in northern Irrawaddy, one of the most influential climate factors, shortened by about three weeks (late onset and early withdrawal), compared with the 1951–2000 average (McKinley et al., 2015). Other effects include rising temperatures, fewer rainy days per year, and a higher frequency of intense rainfall events (McKinley et al., 2015; SeinnSeinn et al., 2015). These climate change-related events will certainly impact the hydrological regime and sediment discharge of the Irrawaddy River, and ultimately affect the stability of the delta.

4 Future Perspectives

As one of the most underdeveloped river deltas, the Irrawaddy River Delta is facing the challenges of rapidly increasing population, intensification of agriculture, but lacking adequate infrastructures to protect the people from natural hazards including tropical cyclones and floods. To gain a more comprehensive understanding and better management of the delta, knowledge of evolution of human-natural coupled deltaic system is badly needed. In spite of recent studies on geomorphological change and land use change, the Irrawaddy Delta is one of the least studied mega-deltas in the world. More efforts should be invested to acquire accurate data with high spatiotemporal resolution. International cooperation and financial assistance should be sustained and strengthened to effectively cope with the challenges of climate change and economic development.

References

Adas, M. (2011). The Burma Delta: Economic development and social change on an Asian Rice frontier, 1852–1941, new perspectives in Southeast Asian studies. University of Wisconsin Press.

Anthony, E. J., Besset, M., Dussouillez, P., Goichot, M., & Loisel, H. (2019). Overview of the monsoon-influenced Ayeyarwady River Delta, and delta shoreline mobility in response to changing fluvial sediment supply. *Marine Geology, 417*, 106038.

Besset, M., Anthony, E. J., Dussouillez, P., & Goichot, M. (2017). The impact of Cyclone Nargis on the Ayeyarwady (Irrawaddy) River delta shoreline and nearshore zone (Myanmar): Towards degraded delta resilience? *Comptes Rendus Geoscience, 349*(6–7), 238–247.

Brakenridge, G. R., Syvitski, J. P. M., Niebuhr, E., Overeem, I., Higgins, S. A., Kettner, A. J., & Prades, L. (2017). Design with nature: Causation and avoidance of catastrophic flooding, Myanmar. *Earth-Science Reviews, 165*, 81–109.

Chen, D., Li, X., Saito, Y., Liu, J. P., Duan, Y., Liu, S., & Zhang, L. (2020). Recent evolution of the Irrawaddy (Ayeyarwady) Delta and the impacts of anthropogenic activities: A review and remote sensing survey. *Geomorphology, 365*, 107231.

Chhin, R., Shwe, M. M., & Yoden, S. (2020). Time-lagged correlations associated with Interannual variations of pre-monsoon and Postmonsoon Precipitation in Myanmar and the Indochina Peninsula. *International Journal of Climatology, 40*, 3792–3812.

Ericson, J. P., Vörösmarty, C. J., Dingman, S. L., Ward, L. G., & Meybeck, M. (2006). Effective sea-level rise and deltas: Causes of change and human dimension implications. *Global and Planetary Change, 50*(1–2), 63–82.

Flynn, E. R., Kuehl, S. A., Harris, C. K., & Fair, M. J. (2022). Sediment and terrestrial organic carbon budgets for the offshore Ayeyarwady Delta, Myanmar: Establishing a baseline for future change. *Marine Geology, 447*, 106782.

Frenken, K. (2012). Irrigation in Southern and Eastern Asia in figures: AQUASTAT Survey—2011. *FAO Water Reports 37* (p. 349). Food and Agriculture Organization of the UN.

Furuichi, T., Win, Z., & Wasson, R. J. (2009). Discharge and suspended sediment transport in the Ayeyarwady River, Myanmar: Centennial and decadal changes. *Hydrology Process, 23*(11), 1631–1641.

Giosan, L., Naing, T., Tun, M. M., Clift, P. D., Filip, F., Constantinescu, S., Khonde, N., Blusztajn, J. S., Buylaert, J.-P., Stevens, T., & Thwin, S. (2018). On the Holocene evolution of the Ayeyawady megadelta. *Earth Surface Dynamics, 6*, 451–466.

Gordon, R. (1879). Report on the Irawadi River (p. 550). PW Secretar Press.

Hedley, P. J., Bird, M. I., & Robinson, R. A. J. (2010). Evolution of the Irrawaddy Delta Region since 1850: Evolution of the Irrawaddy Delta Region since 1850. *Geographical Journal, 176*, 138–149.

Hennig, T. (2016). Damming the transnational Ayeyarwady basin. Hydropower and the water-energy nexus. *Renewable and Sustainable Energy Reviews, 65*, 1232–1246.

HIC. (2017). Ayeyarwady State of the Basin Assessment (SOBA) 2017: Synthesis Report. Hydro-Informatics Centre, Yangon.

Higgins, S. A. (2016). Advances in delta-subsidence research using satellite methods. *Hydrogeology Journal, 24*(3), 587–600.

Horton, R., De Mel, M., Peters, D., Lesk, C., Bartlett, R., Helsingen, H., Bader, D., Capizzi, P., Martin, S., & Rosenzweig, C. (2017). Assessing climate risk in Myanmar: Technical Report. Center for Climate Systems Research at Columbia University, WWF-US and WWF-Myanmar.

Kraas, F., May, Y., & Myint, Z. N. (2010). Yangon/Myanmar: Transformation processes and mega-urban developments. *Geographische Rundschau International, 6*(2), 26–37.

Kraas, F., Aung, K., & Oo, N. W. (2017). Agricultural regions. In F. Kraas, R. Spohner, & A. A. Myint (Eds.), *Socio-economic atlas of Myanmar* (pp. 112–115). Franz Steiner.

Kravtsova, V. I., Mikhailov, V. N., & Kidyaeva, V. M. (2009). Hydrological regime, morphological features and natural territorial complexes of the Irrawaddy River Delta (Myanmar). *Water Resources, 36*(3), 243–260.

Lazarus, K. M., Corbett, M., Cardinale, P., Lin, N. S., & Noeske, T. K. H. (2018). Strategic environmental assessment of the Myanmar hydropower sector: Final Report. World Bank Group.

McKinley, J., Adaro, C., Pede, V. O., Setiyono, T., Aung, N. M., Hom, N. H., Htwe, N. M., Hein, Y., Than, S. M., Swe, K. L., Quicho, E., Sheinkman, M., & Wassmann, R. (2015). The current state of climate change perceptions and policies in Myanmar: 2014 report. CGIAR Research Program on Climate Change. Agriculture and Food Security.

Milliman, J. D., & Meade, R. H. (1983). World-wide delivery of river sediment to the oceans. *The Journal of Geology, 91*(1), 1–21.

Osti, R., Tanaka, S., & Tokioka, T. (2009). The importance of mangrove forest in tsunami disaster mitigation. *Disasters, 33*(2), 203–213.

Ramaswamy, V., Rao, P. S., Rao, K. H., Thwin, S., Rao, N. S., & Raiker, V. (2004). Tidal influence on suspended sediment distribution and dispersal in the northern Andaman Sea and Gulf of Martaban. *Marine Geology, 208*(1), 33–42.

Rao, P. S., Ramaswamy, V., & Thwin, S. (2005). Sediment texture, distribution and transport on the Ayeyarwady continental shelf, Andaman Sea. *Marine Geology, 216*(4), 239–247.

Robinson, R. A. J., Bird, M. I., Oo, N. W., Hoey, T. B., Aye, M. M., Higgitt, D. L., Lu, X. X., Swe, A., Tun, T., & Win, S. L. (2007). The Irrawaddy River sediment flux to the Indian Ocean: The original nineteenth-century data revisited. *Journal of Geology, 115*(6), 629–640.

Schmitt-Degenhardt, S. (2013). A regional perspective on poverty in Myanmar. UNDP.

SeinnSeinn, M., Ahmad, M., Thapa, G., & Shrestha, R. (2015). Farmers' adaptation to rainfall variability and salinity through agronomic practices in lower Ayeyarwady Delta, Myanmar. *Journal of Earth Science and Climatic Change, 6*(2), 1000258.

Syvitski, J., Kettner, A., Overeem, I., Hutton, E. W., Hannon, M. T., Brakenridge, G. R., Day, J., Vörösmarty, C., Saito, Y., Giosan, L., & Nicholls, R. J. (2009). Sinking deltas due to human activities. *Nature Geoscience, 2*(10), 681–686.

Tasnim, K. M., Shibayama, T., Esteban, M., Takagi, H., Ohira, K., & Nakamura, R. (2015). Field observation and numerical simulation of past and future storm surges in the Bay of Bengal: Case study of cyclone Nargis. *Natural Hazards, 75*, 1619–1647.

van der Horst, T., Rutten, M. M., van de Giesen, N. C., & Hanssen, R. F. (2018). Monitoring land subsidence in Yangon, Myanmar using Sentinel-1 persistent scatterer interferometry and assessment of driving mechanisms. *Remote Sensing of Environment, 217*, 101–110.

Webb, E. L., Jachowski, N. R. A., Phelps, J., Friess, D. A., Than, M. M., & Ziegler, A. D. (2014). Deforestation in the Ayeyarwady Delta and the conservation implications of an internationally-engaged Myanmar. *Global Environmental Change, 24*, 321–333.

World Bank. (2023). National GDP per capita (current US $) in 2020, derived from World Bank DataBank on June 1, 2023.

Yang, R., Luo, Y., Yang, K., Hong, L., & Zhou, X. (2019). Analysis of forest deforestation and its driving factors in Myanmar from 1988 to 2017. *Sustainability, 11*(11), 3047.

Lena River Delta

Sergey Chalov, Kristina Prokopeva, and Dmitry Magritsky

1 Physical Setting

1.1 The Lena River Basin

The Lena River is one of the world's largest rivers, the second after the Yenisei River by runoff, and third by its watershed area in Russia. The Lena originates in the mountains surrounding Lake Baikal, about 10 km from the western shore of the lake and at an altitude of 1,470 m. The river flows into the Laptev Sea with a total length of approximately 4,400 km. The watershed area of the Lena is 2.49 million km^2; and its shape is fairly compact, with the length of its basin from south to north being approximately 2,400 km and from west to east nearly 2,000 km (Fig. 1).

The Lena River basin has a complex terrain, as it located within the Siberian Platform, which is characterized by plateaus, plains, and mountains, including the Baikal and Verkhoyano-Kolyma Mountain regions. More than two-thirds of the basin's land area is covered by continuous permafrost, which has the world's thickest permafrost layer at 1,500 m (Brown et al., 2002). The rest of the area, with the exception of the southwest corner, is covered by intermittent or isolated permafrost.

The basin is located in the temperate and subarctic climate zones. The climate is highly continental with cold, snowless winters and hot summers. In the winter, a region of high pressure (the Siberian anticyclone), with dry and frosty conditions, dominates the entire area. Air temperatures can reach −60 °C. In the summer, the average air temperature varies from +8 to +35 °C in July (the warmest month), with the maximum air temperature being even higher. The average annual precipitation is around 200–400 mm per year, with most of it occurring in the mountains in the south where the monsoons penetrate. The maximum precipitation occurs during the

S. Chalov (✉) · K. Prokopeva · D. Magritsky
Faculty of Geography, Lomonosov Moscow State University, Moscow 129626, Russia
e-mail: srchalov@geogr.msu.ru

W. Zhang and H. de Vriend (eds.), *Delta Sustainability*,
https://doi.org/10.1007/978-981-97-7259-9_9

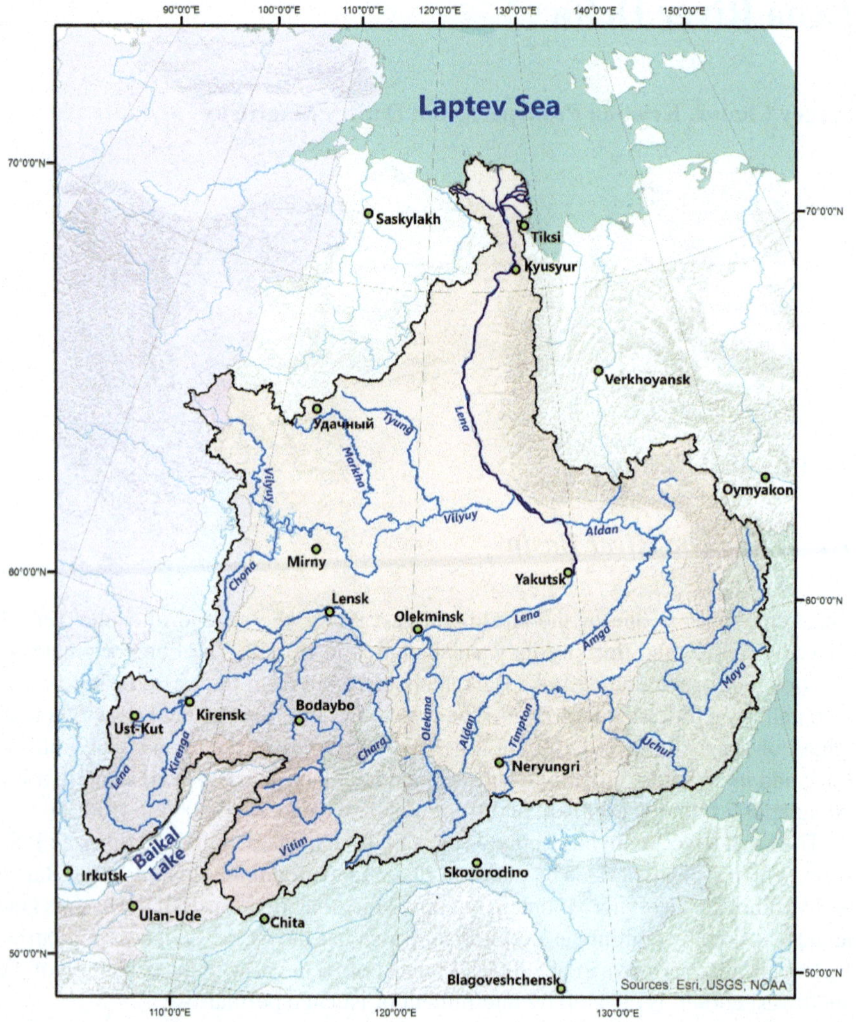

Fig. 1 Lena River Basin map

warmer season. The basin is located in the Taiga zone, with the exception of a small area to the north of 71°N, which is located in the Tundra and Forest-Tundra zones.

There are 242,500 rivers in the basin (Domanitsky et al., 1971). The main tributaries are Kirenga, Vitim, Olekma, and Aldan (the largest tributary) rivers, as well as the Vilyuy River, which is regulated by a series of reservoirs. The Lena River in upstream (up to the confluence with the Vitim), is a typical mountainous river. Below the mouths of the Vitim and Olekma rivers, the Lena becomes a full-flowing river, flowing through a well-defined valley with an anabranching patterns in semi-mountain conditions. The width of the channel with islands reaches 10 km and the width of the river valley is up to 25 km within the central Yakutsk lowlands. Then, the river valley gradually narrows, and in a section that is 150 km long the valley width decreases to 2–2.5 km prior to the delta (Alexeevsky, 2007).

The rivers of the majority of the Lena River basin have an almost exclusively snow-fed (>80% of annual runoff) or predominantly snow-fed (>50%) hydrology. The rainfall plays a less significant role, mainly in the southern part of the basin. Groundwater contributions in permafrost regions are very low. Flooding events caused by melted snow water are observed in the latter part of spring and summer.

The territory of the catchment is poorly populated and developed, practically not urbanized. Large-scale mining (oil, gas, black and brown coal, gold and silver, diamonds and semi-precious stones, ferrous metals, materials for the chemical industry and for building), as well as energy, logging, shipping and pipeline transport, northern livestock and crafts of the peoples of the North are the main developments in the area.

1.2 The Lena River Delta

Lena River Delta is the biggest delta in the Arctic region. Lena Delta extends to 28,000 (Alexeevsky, 2007), 29,630 km^2 (Bolshiyanov et al., 2013) or 32,000 km^2 (Antonow, 1967; Mikhailov, 1997). The different estimates are due to the various understanding of the delta upper border which is shown at the Tit-Ary Island or at the Stolb Island (Fig. 2).

The Lena Delta has complicated drainage network with more than 6000 branches with a total length of 14,600 km (Ivanov et al., 1983) and approximately 29,500 lakes (Lopatin & Fedorov, 1947). The number of lakes, oxbow lakes, and other genesis lakes as estimated by Ivanov et al. (1983) is 58,728. However, according to (Bolshiyanov et al., 2013) data, there are approximately 29,483 lakes in total. The area of these lakes ranges from 3,196 to 6,175 km^2 (Ivanov et al., 1983; Schneider et al., 2009). Lakes connected with river channels (oxbows and residuals) prevail in the eastern part of the delta. On the other hand, thermokarst lakes are found in the western, and particularly northwestern part (on relics of old plains and sea terraces). There are 1,614 lakes in the Lena Delta, which are composed of various islands of different origins, morphologies, and elevations (Fig. 3).

The four main distributaries represent the main sectors of the channel network: the largest is the Trofimovskaya branch (total length is 142 km); the second-largest branch by water discharge is the Bykovskaya branch (105 km) that turns sharply to the east after Sardakh Island and flows into the Buor Khaya Gulf. The third and fourth-largest distributaries are the Olenekskaya branch (210 km) which flows west into the Kuba Gulf and the system of Tumatskaya channels (155 km) which flow north into the Laptev Sea. The shoreline length from Olenekskaya branch mouth to Bykovskaya branch mouth is approximately 1,930 km (Lopatin & Fedorov, 1947).

Lena River Delta has been formed as river sediments slowly filled in small bays that were created by Laptev sea-level fluctuations over a long period of time over

Fig. 2 The Lena River Delta map with the positions of main gauge stations of discharges (black circles) and water levels (white circles) measuring and permanent gauging cross-sections presented by: 1: "4.7 km upper Stolb Island" (the end of Main channel); 2: "Olenekskiy" (Olenekskaya branch); 3: "Tumatskiy" (Tumatskaya branch); 4: "Trofimovskiy" (Trofimovskaya branch), in the top of branches

the past 7,000 years (Fig. 4). Different parts of the Lena River Delta were formed at different times and its individual parts have different origins (Bolshiyanov et al., 2013; Korotaev et al., 2007). The amplitude of wind surges can reaches 2.5 m. The maximum rise of sea level by tides not exceeds 0.5 m. Storm surges or tides never reach the top of the delta.

The entire delta is located within the Arctic continental climate zone. The average annual temperature, as measured by the Tiksi meteorological station, from 1930 to 2019 was −12.8 °C, with an average winter temperature of −30.9 °C in January and an average summer temperature of 7.7 °C in July. The temperature range in some years can be as high as 100 °C. The average annual precipitation varies between

Fig. 3 The surface of the Lena Delta Island. Photo by authors

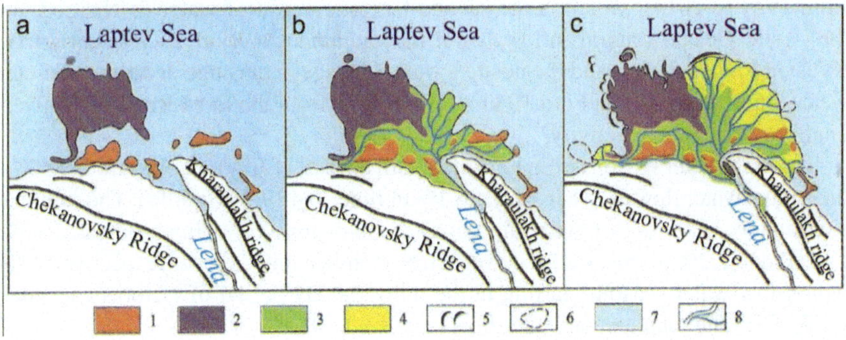

Fig. 4 Stages of formation of the Lena River Delta (Korotaev et al., 2007): (**a**) The maximum phase of postglacial transgression, (**b**) 5–4 thousand years ago, (**c**) less than 1000 years ago. Legend: 1—Ice complex of the lake-alluvial plain (Yedoma), 2—Karginskaya abrasive marine terrace (Arga-Muora-Sise Island), 3—Old delta floodplain, 4—Young delta floodplain, 5—Coastal marine bars, 6—Estuarine bars, 7—Marine area, 8—Hydrographic network

200 and 400 mm, peaking in summer season and falling off in the October-May. The number of days with snow cover averages 250 per year. The duration of the frost-free season is no longer than 50 days on average. North and northeast winds are prevalent during the summer season, while southerly winds are more common during the winter months.

In the condition of cold Arctic climate, the continuous permafrost has a thickness of 300–650 m, and the active (annually thawed) layer during summer has a depth of 0.3–0.5 m. The ice content in rocks can reach 20–40% or even more (up to 60–75%).

Repeated-wedge ice is also widely distributed (Boike et al., 2013; Kotlyakov et al., 1997; Kotlyakov, 2007; Sakha Republic, 2000). Under the main channel of the Lena River (the section of the Lena River between Tit-Ary Island and Stolb Island), as well as in large delta branches and lakes, various-sized taliks (unfrozen ground layers) are often formed.

The entire delta located in the tundra zone except for the southern part. The predominant landscape of the delta is a wet tundra, dominated by sedges and mosses (28.5% of the total area). These cover a large portion of the high floodplain, with wet areas accounting for 7.5%. The area occupied by shrubby tundra and other subspecies of tundra is approximately 16%, mainly located on Agra Island. Low and medium floodplain areas that are not occupied by vegetation account for approximately 10%, both with sandy banks and in-channel bar and along coasts. In 2006, based on the analysis of satellite images from the Landsat-7 ETM+, larch islets were identified in the southern part of the Delta (Tit-Ary Island).The images were taken on July 27th, 2000 and July 26th, 2001, which fell at the height of the growing season (Fig. 5).

The average annual water discharge of the Lena River at Kyusyur (nearest downstream gauging stations ~145 km upstream of the delta apex and 315 km from the Laptev Sea via the Bykovskaya branch) is 17,200 $m^3\ s^{-1}$ or 543 km^3 (for the period from 1927 to 2014), which is the second-largest amount of water discharged into the Arctic Ocean. Lateral inflow (from the catchment area of 27,900 km^2) from Kyusyur toward the sea and the positive water balance in the huge delta area increase water runoff approximately to 10 km^3. Water flow from the Lena River is relatively unaffected by human activity.

The high-water season (from mid-May until the end of July) accounts for 59.5% of the annual water flow (4.1% in May, 36.1% in June, and 19.3% in July). The summer-autumn low-water season with rain floods accounts for 32.7%. In particular, 13.3%, 12.2%, and 7.2% of this occurs in August, September, and October respectively. The winter low-water season (starting in the third ten-day period of October) accounts for 7.8% of the total annual water flow.

Over the last 20 years, flow partitioning between main distributaries has remained mostly unchanged: of the Lena's flow input to the delta, approximately 24.9–25.0% is directed to the Bykovskaya branch, 58–59.2% to Trofimovskaya, 6.0% to Tumatskaya, and 6.6% to Olenekskaya (Table 1). The difference ratio between various discharges less than 0.3%.

The remote location of the large Lena Delta explains data scarcity regarding sediment transport. The mean annual suspended sediment concentration (SSC) at the Kyusyur gauge station is approximately 40 g m^{-3}. SSC varies from 35 to 90 g m^{-3} during the high-flood water period, the average values during the winter low-water period is 1.5–4.5 g m^{-3}. The sediment load estimated by different authors ranges from 11.8 to 22.5 million tons per year (Alabyan et al., 1995; Doronina, 1962; Gordeev, 2006; Alexeevsky, 2007; Ivanov & Piskun, 1999; Magritsky et al., 2018; Milliman & Meade, 1983). Main data regarding sediment was based on the single observations and grab measurement and provided contradictory information. According to (Korotaev et al., 1990) only 10–30% of the sediments entering the Lena Delta

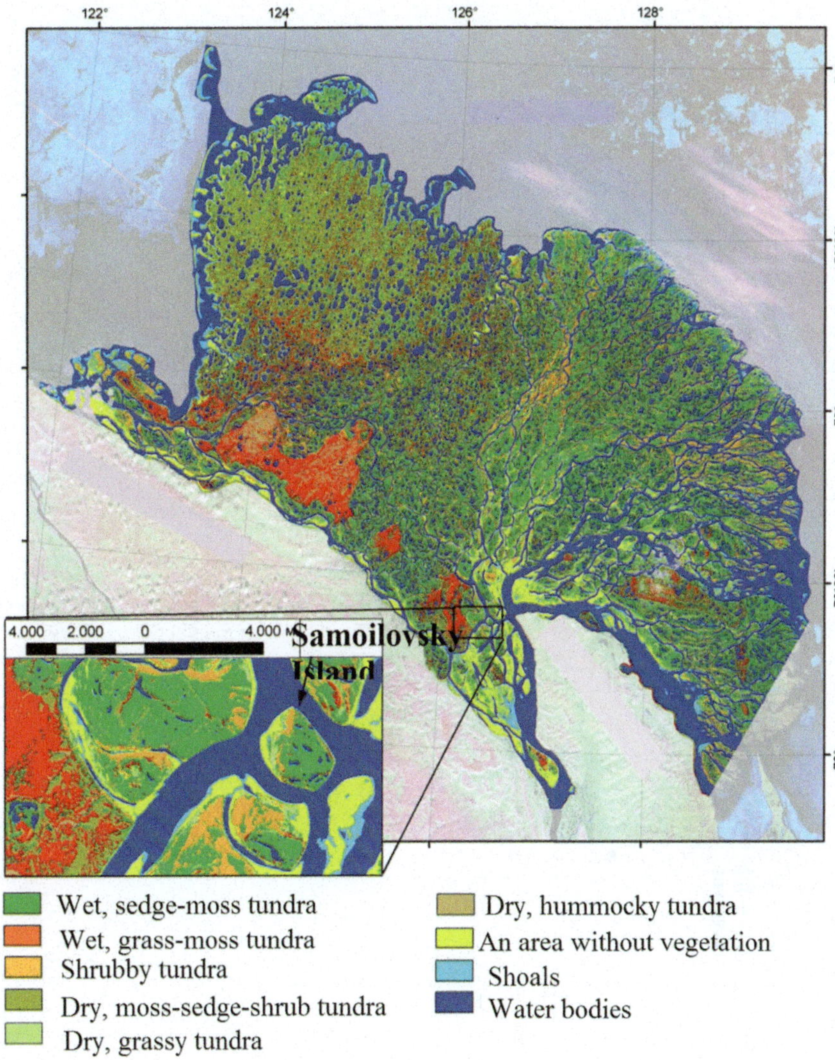

Fig. 5 Landscapes of the Lena River Delta (Schneider et al., 2009)

reach the Laptev Sea. Similary, Charkin et al. (2011) and Fedorova and Makushin (2016) argued that most of the sediment are stored within the delta area. However, according to (Rachold et al., 1996), significant parts of the sediments are transported directly to the sea. Additionally, some studies (Grigoryev, 1993) point at a longitudinal increase of the sediment concentrations along the main distributaries. The pioneer systematic assessment was done based on Landsat images collection since 1999–2021, which revealed contrasting patterns in the left and right sectors of the

Table 1 Distribution of flow discharges in the central part of Lena River Delta based on the monitoring data and field measurements (Magritsky et al., 2018)

Lena River		Top of the delta			Main delta branch point						
Kyusyur gauge		Main channel, 4.7 km upstream Stolb Island**	Bulkurskaya branch mouth	Discrepancy $\sum Q_i$ (towards Q at Kyusyur gauge)	Bykovskaya branch source, polar station Yu.A. Khabarova (Stolb)		Trofimovskaya branch source	Tumatskaya branch source	Olenekskaya branch downstream the confluence with Bulkurskaya branch	Discrepancy $\sum Q_i$ (toward Q at gauge rounded off)	
										Kyusyur	4.7 km upstream Stolb Island
Q, m^3/s	H, cm	Q_i, m^3/s	Q_i, m^3/s	%	Q_i, m^3/s	H, cm	Q_i, m^3/s	Q_i, m^3/s	Q_i, m^3/s	%	%
Open channel period											
10000	360	900	*−0*	−10	200	120	6800	200	400	−6	4.5
20000	620	19000	*−0*	−5,0	4800	267	13000	825	1040	−1.5	3.5
30000	828	28800	*100*	−3,5	7500	370	18900	1670	1800	−0.5	3.5
40000	1025	38600	*320*	−2,5	10,200	445	24200	2640	2680	−0.5	3.0
50000	1200	48500	*950*	−1,0	13,500	518	30000	3800	3750	2.0	5.5
60000	1360	57500	*1900*	−1,0	16,800	578	35300	4890	4880	3.0	7.5
80000	1640	76000	*5000*	1,0	23,500	685	46500	7700	7800	7.0	12.5
100000	*1930*	*92500*	–	–	*29500*	*805*	*55500*	*10700*	*10800*	*6.5*	*15.0*
120000	*2240*	*108000*	–	–	*34300*	*900*	*63000*	*13300*	*13900*	*4.0*	*15.5*
140000	*2560*	*122500*	–	–	*39000*	*990*	*70000*	*15000*	*15000*	*0*	*14.5*
160000	*2840*	*136000*	–	–	*45000*	*1050*	*77000*	–	–	–	–
Freezing-over period											
1000	*170*	900	–	−10	220	*55*	850	*0*	*0*	*7.0*	*19.0*
2000	*250*	1800	–	−10	310	*70*	1500	*0*	*0*	*−9.5*	*0.5*

(continued)

Table 1 (continued)

Lena River		Top of the delta			Main delta branch point						
Kyusyur gauge		Main channel, 4.7 km upstream Stolb Island**	Bulkurskaya branch mouth	Discrepancy $\sum Q_i$ (towards Q at Kyusyur gauge)	Bykovskaya branch source, polar station Yu.A. Khabarova (Stolb)		Trofimovskaya branch source	Tumatskaya branch source	Olenekskaya branch downstream the confluence with Bulkurskaya branch	Discrepancy $\sum Q_i$ (toward Q at gauge rounded off)	
										Kyusyur	4.7 km upstream Stolb Island
Q, m^3/s	H, cm	Q_i, m^3/s	Q_i, m^3/s	%	Q_i, m^3/s	H, cm	Q_i, m^3/s	Q_i, m^3/s	Q_i, m^3/s	%	%
3000	*340*	2700	–	−10	510	*100*	2150	*−0*	*−0*	*−11.5*	*−1.5*
4000	*415*	3750	–	−6	700	*140*	2850	*20*	*60*	*−9.0*	*−3.0*
5000	*505*	4400	–	−12	900	*170*	3200	*60*	*110*	*−14.5*	*−3.0*
7000	*655*	6600	–	−6	1500	*280*	4600	*190*	*270*	*−6.5*	*−0.5*
10000	*750*	10000	–	0	2450	*350*	6850	*400*	*480*	*2.0*	*2.0*
15000	***	15500	–	3	3750	–	*10400*	*730*	*840*	*5.0*	*1.5*
20000	–	22000	–	10	5400	–	*14400*	*1100*	*1250*	*11.0*	*0.5*

* Data does not include period of high-water wave rise; ** For the relation to gauge Kyusyur and taking into account a time lag of 1 to 3 days; *** Dashed figures mean lack of data or their unreliability. Very approximate data, especially for the freezing period, in view of the ambiguity of relation, are put in italics

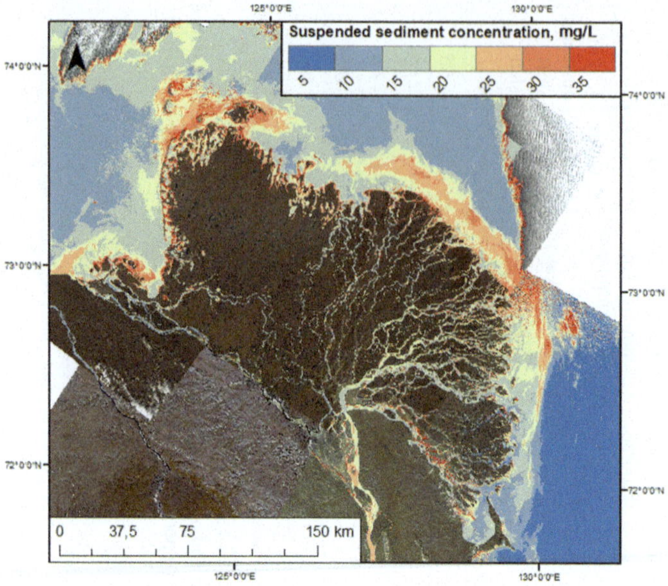

Fig. 6 Sediment concentration distribution over the Lena River Delta

delta (Chalov & Prokopeva, 2021, 2022), explained by the difference in distributary hydromorphology. A seaward increase of the sediment concentration increase is typical for the largest Lena Delta branches which distributes over 86% of the Lena total discharge. This is also evidenced by gravity cores from the Laptev Sea shelf, which show that Lena River-derived sediment bedload can be found up to 34.5–47.5 km from the delta edge of the Bykoskaya branch. This shows that sediment from the Lena River flows directly into the Laptev Sea (Rivera et al., 2006).

A significant drop in sediment concentrations is observed in the smaller distributaries, mostly Tumatskaya, which distributes less than 10% of the total Lena discharge and is characterized by the largest channel length and lowest channel slopes. Increase of sediment concentrations along the left banks of the large Eastern distributary (right down corner on the map) and nearshore increase of sediment concentrations of the Lena Delta are possibly due to upwelling and erosion of frozen coastal banks (based on Landsat 8 composite images from 29 August 2020 to 8 September 2020). Figure 6 illustrates the sediment distribution along delta.

2 Socioeconomic Characteristics

The vast territory of the Lena Delta remains largely unpopulated. The principal indigenous population of the area are the Evenk (formerly called Tungus) and—to a lesser extent—Sakha. The nearest settlement is Tiksi located about 115 km to the Southeast, and home to approximately 4,000 inhabitants. The single settlements along the banks of the delta were related to few meteorological and hydrological gauging stations: Tit-Ary from 1948 to 1990 (delta apex at the Tit-Ary Island), Malyshev Island (1953–1989), Sagyalah-Ary (1962–1991) and polar stations Yu.A. Khabarova (previously called Stolb) at the Bykovskaya branch. The latter have provided important data for the expeditions and the research community as a whole.

At the Samoylov Island in the upper part of the delta, there is a Russian-German research station since 2006, which is followed by a new large modern Russian station owned and run by the Trofimuk Institute of Petroleum Geology and Geophysics Siberian Branch of the Russian Academy of Sciences. The station can accommodate up to 25 people. Another year-round research facilities is related to the International Biological Station Lena-Nordenskjold and was opened by Prince Philip in 1995. The station is located on the right bank of Bykovskaya branch. The station's mission is to undertake complex research of Arctic ecosystems, observe their biodiversity and conduct ecological monitoring.

The intensive development of the North had an adverse impact on the delta environment at the begging of 20th century. The establishment of a fish processing plants network in the Lena Delta and its lower stretches in the period of the Great Patriotic War (World War II) in 1942–1943 and unlimited fishing caused a disastrous decline of Coregonus fish stocks by 1950. Between 1963 and 1983, the populations of Bewick's (tundra) swan and geese populations have decreased by a factor 7 and 3.5, respectively. There has been a decline in the number of reindeers, and some walrus breeding grounds have been lost. Considering these adverse changes in the environment and in the interests of its preservation and restoration, the Lena Delta Nature Reserve was established on the Northwest and Southeast parts of the delta in 1985 (Fig. 7). The majority of the delta lies within the Lena Delta Reserve and is expansion in 1995 made it the largest protected area in Russia. It is one of richest areas in the Arctic north of 71°N for both species diversity and breeding densities. The reserve is divided into two sub areas, and has a total land area of 14,330 km^2. The main aim of the Lena Delta Nature Reserve is protect of main nesting habitat for migratory birds and local mammals. It is also an important fish spawning site.

3 Current Issues

3.1 *Climate Change Influence*

Present climate changes influence on water regime and sediment transport and water and sediment quality in the delta branches. ERA5-Land reanalysis data along with the perennial monitoring data at Tiksi and Kyusyur weather stations show a significant change in air temperature and solar radiation which have occurred at the beginning of the 21 century.

Fig. 7 Lena Delta Nature Reserve map and its inhabitants. Red line is boundaries of the reserve; yellow line is boundaries of the protected area of the reserve. Photos by authors and Nature Reserve staff

According to (Chalov et al., 2023) air temperatures in the Lena Delta increased significantly (p-value < 0.05) (Fig. 8a). An increase in warming rate was observed between the periods 1950–1999 and 2000–2021, i.e., the most significant increase in air temperature occurred in the last 20-year period. Average air temperatures were 4.1 °C for the period 1950–1999, whereas for the last 20-year period (2000–2021) it reached 6.0 °C. The rate of air temperature rise was 0.86 °C per decade from 1979 to 2021. The subsequent period (2000–2021) warmed at 1.61 °C per decade. These trends are significantly higher compared to the Lena River basin, and to other Arctic regions.

Surface net solar radiation significantly increased from 110 W m^{-2} (from 1950 to 1992) to 118 W m^{-2} (from 1993 to 2021). Rising surface solar net radiation was 2.64% per decade from 1992 to 2021 and 6.99% per decade from 2004 to 2021

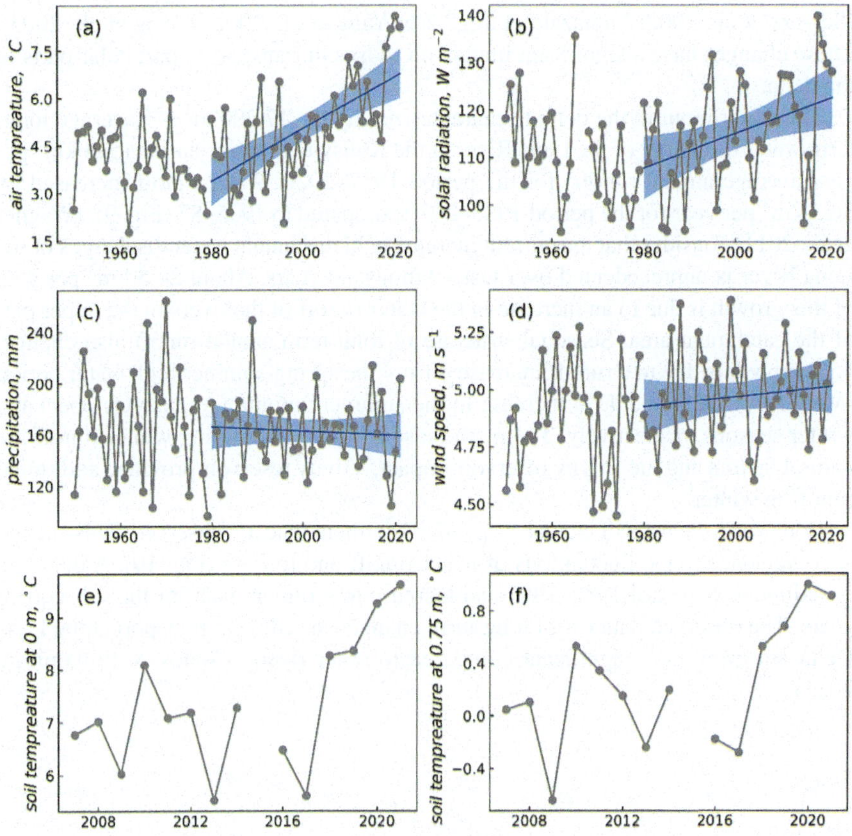

Fig. 8 Long-term changes in mean air temperature (**a**) and surface net solar radiation (**b**) in the Lena Delta. The overall linear trend for 1979–2020 is shown (blue line), with its 95% confidence interval (Chalov et al., 2023)

(Fig. 8b). There were no significant trends in rainfall and wind speed values (Fig. 8c, d). However, non-significant trends were recorded: one upward at the end of the 1980s and one downward from the end of the 2000s. Also, significant statistical changes have not been established for precipitation series at Tiksi and Kyusyur weather stations. Soil temperature from the observations on the Samoylov Island (Boike et al., 2019) displayed a general increasing tendency, with the higher values after 2017. Compared to temperature at 0 m, soil temperature at 0.75 m exhibited more stable values over time (Fig. 8e, f).

The water flow regime and its values in the lower reaches of the Lena River have undergone significant changes in the last 30–40 years, responding to climatic factors. Warming rates of annual temperature range from 0.25 to 0.36 °C per decade for cold-season (November–April) during 1936–2019 (Chalov et al., 2023) in the Lena River basin. Permafrost thawing and melting of ground ices due to increase of the air temperatures can also contribute in the river runoff in the permafrost zone

(Serreze et al., 2002; Yamazaki et al., 2006; Yang et al., 2002; Zhang et al., 2003). These changes have a significant impact on sediment transport in particular parts of the delta.

The water input to the delta is characterized by the 17,200 $m^3 \cdot s^{-1}$ average long-term water discharge of the Lena River at the Kyusyur gauging station (543 km^3 per year average annual runoff) for the period 1927–2020. Water runoff increased by 56.3 km^3 per year for the period 1988–2020 compared to 1935–87 (Fig. 9). (Wagner et al., 2011) consider that the current increasing in the annual water discharges of the Lena River is unprecedented over last ~9 thousand years. About 24.5 km^3 per year of this growth is due to an increase in the water runoff of the rivers in the upper part of the catchment area. Seasonal structure of long-term annual runoff also change: 45% growth is due to a runoff increase during the spring-summer high water period (May–July), 12% and 43% increase to the summer-autumn (August–October) and winter seasons, respectively. The main reasons for the increasing winter runoff are natural factors and the Vilyuy reservoir impact. Vilyuy reservoir provided additional runoff in winter.

Long-term changes of annual suspended sediment load are observed. This change is coincidence to the fluctuations of water runoff and follows (Fig. 10). A decrease in sediment load until 1986–1987 and increase in sediment load for the subsequent years. The observed patterns of long-term changes in sediment transport at the Lena Delta are even more significant compared to other rivers (Chalov & Prokopieva, 2021).

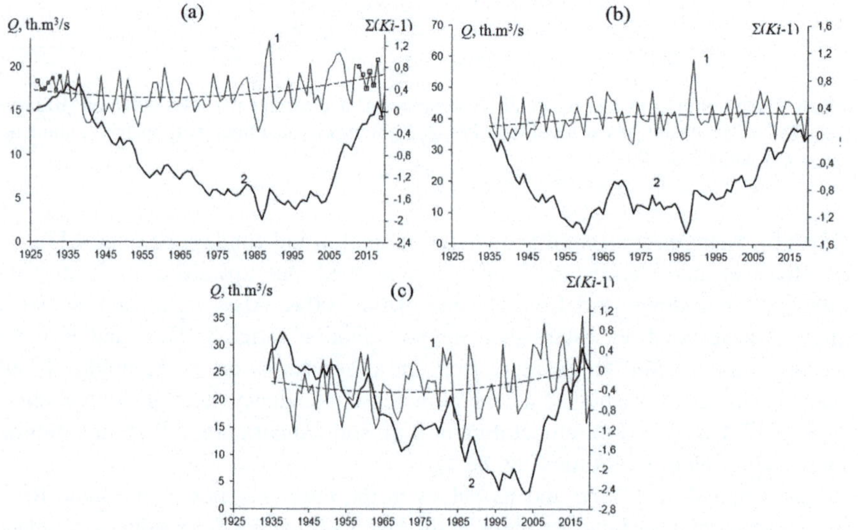

Fig. 9 Long-term fluctuations in average annual water discharges of the Lena River (**a**), high water (**b**, V–VII) and summer-autumn inter-flood season (**c**, VIII–X) discharges at the Kyusyur gauge (1—water discharge curve, 2—difference integral curve)

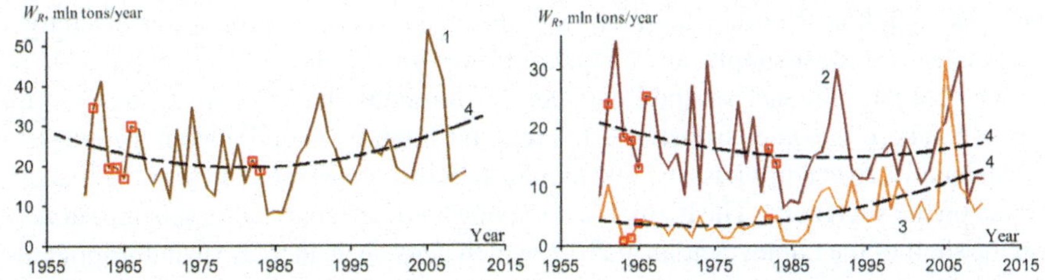

Fig. 10 Long-term changes of suspended sediment yield in the lower reaches of the Lena River (gauge Kyusyur). 1: annual sediment yield; 2: sediment yield averaged over snow-melt flood season; 3: sediment yield averaged over summer-autumn season, 4: the squares show the years, for which the data have been restored

A statistically insignificant increase in suspended sediment load since the end of 1988 (by 6.1 million tons per year) coincides with the increase of Lena River flow. Annual suspended sediment yield increased mainly due to the summer-autumn season. This increasing is caused by suspended sediment processes sensitivity to air and water temperature increasing which leads to the permafrost thawing (Chalov & Prokopeva, 2022; Costard et al., 2003; Chalov et al., 2021). The high values of SSC and its increase along the Lower Lena are observed downstream the confluence of the Aldan River and especially the Vilyuy River according to the field measurement data (Charkin et al., 2011; Magritsky, 2015; Rachold et al., 1996).

Degradation of permafrost have been reported here and led many hydrological alterations including old organic carbon release (Shakhova et al., 2007; Wild et al., 2019). The recent "permafrost erosion" due to heat transfer by lateral fluvial denudation and sunlight exposure as well as "thermokarst"-features within the Lena Delta areas transport large sediment masses into the river system and then in Artic Ocean. The Lena River still exhibits unchanged river patterns and a natural distributary deltaic system. Hydrological evolution over the past 100 years has been driven mainly by natural factors at the river scale. The impact of dam construction and reservoir impoundment on the coastal area is negligible. This means that the influence of other controls resulting in channel planform changes will be observed in near future (Alexeevsky et al., 2013).

3.2 *Pollution*

Given the absence of any human activities on delta and the establishment of the Natural Reserve, the issues of environmental pollution are likely related to sedimentation processes. Grab measurement indicate possible changes of metals concentrations and organic carbon due to thermal erosion processes along delta. Semiletov et al. (2011) indicated increase of methane along delta due to degradation of permafrost in the Lena, especially in the fault zones, and drainage of thaw lakes enriched in dissolved methane. The degradation of the ice complex, erosion of river terraces, and river bank erosion are the main drivers of the increasing of water discharge, suspended sediment supply, and chemical flow within delta.

The human influence on the Lena Delta is minimal, due to its status as a nature reserve. There is some subsistence fishing and hunting by indigenous people and occasionally, river ships pass by. The human activities in the areas are mostly concentrated on the shelf zone. The Rosneft's company has been conducting seismic surveys on the shelf of the Laptev Sea since 2017, which leads to an increase in anthropogenic load.

The Lena River Delta is largely navigable through the Bykovskaya branch, allowing for the transport of cargo, including excavated minerals, fur, food, and industrial products, from continental part of Yakutia and to Tiksi and the rest of the world by way of the Arctic Ocean routes. Threats from oil spills, however, do occur, as a large number of vessels carrying cargo ply on this river regularly. In fact, 25,000 tons of oil from the Lena River pollutes the Arctic Ocean each year.

The possible environmental risks in the Lena Delta are also related to observed climatic increase in air temperatures. Since the region is located in a permafrost zone, the climate changes induce a high ecological risk due to permafrost thawing. In the lower reaches of the Lena River so-called marginal zones can form, large areas where fresh and saline waters mix, like in the lower reaches of other large Arctic rivers. In the area impacted by seawaters intrusion, organic pollution due to a sharp increase of the number of diatoms has been reported (Gabyshev et al., 2022).

3.3 Hydrological Hazards

Some particular impacts on the environment and life of people have been observed in the delta. The water level of the river during ice jams can rise 5 to 10 m above winter level. However, in the gently sloping delta area floodwaters will spread out, the floodwaters spread out over a wide floodplain. The largest floodwater rise in recent decades occurred in 2008 and partially destroyed the Lena-Nordenskjold station (Fig. 11). As a result of an ice jam formed at the mouth of the Bulkurskaya branch, most of the flood flow of the Lena River flow into the Trofimovskaya and Bykovskaya branches.

The delta coast is significantly impacted by thermal erosional processes (Fig. 12). The role of coastal erosion is clearly indicated by sediment plumes (Shakhova et al., 2007) and increase of sediment concentrations in the most downstream parts of the Lena distributaries. According to (Semiletov et al., 2011) the Lena's particulate organic export is two orders of magnitude less than the annual input of eroded terrestrial carbon onto the shelf of the Laptev and East Siberia. Additional impacts on the sediment budget in the Arctic deltas is related to coastal upwelling events which regularly increased sea surface turbidity in certain areas adjacent to the Lena Delta in the Laptev Sea and the Kolyma deltas in the East-Siberian Sea (Osadchiev et al., 2020). These events are formed under strong easterly and southeasterly wind forcing and are estimated to occur during up to 10%–30% of the ice-free periods.

Fig. 11 Lena-Nordenskjold station after flood water level rise in 2008. Photo from Bolshiyanov et al (2013)

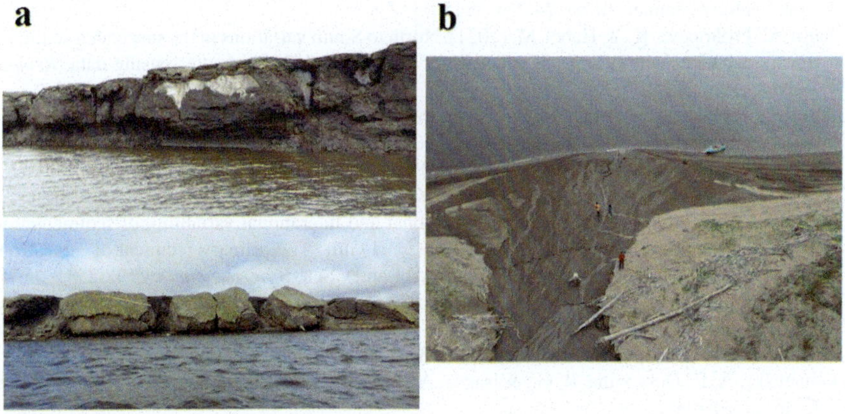

Fig. 12 Eroded river banks in the Lena Delta: (**a**) Exposed ice wedge and slump blocks associated bank erosion (Tumatskaya branch); (**b**) alluvial fan from thawing permafrost gully from the bank of Kurungnakh Island (Olenekskaya branch). Photos by authors. August 2022

References

Alabyan, A. M., Chalov, R. S., Korotaev, V. N., Sidorchuk, A. Y., & Zaitsev, A. A. (1995). Natural and technogenic water and sediment supply to the Laptev sea. *Berichte Zur Polar- und Meeresforschung, 182*, 264–271.

Alexeevsky, N. I. (Ed.). (2007). Geoecological state of Russian Arctic coast and their safety of nature management. GEOS.

Alexeevsky, N. I., Chalov, R. S., Berkovich, K. M., & Chalov, S. R. (2013). Channel changes in largest Russian rivers: Natural and anthropogenic effects. *International Journal of River Basin Management, 11*(2), 175–191.

Antonow, V. S. (1967). The Mouth Area of the Lena River. Gidrometeoizdat.

Boike, J., Kattenstroth, B., Abramova, K., Bornemann, N., Chetverova, A., Fedorova, I., Fröb, K., Grigoriev, M., Grüber, M., Kutzbach, L., Langer, M., Minke, M., Muster, S., Piel, K., Pfeiffer, E.-M., Stoof, G., Westermann, S., Wischnewski, K., Wille, C., & Hubberten, H.-W. (2013). Baseline characteristics of climate, permafrost and land cover from a new permafrost observatory in the Lena River Delta, Siberia (1998–2011). *Biogeosciences, 10*, 2105–2128.

Boike, J., Nitzbon, J., Anders, K., Grigoriev, M., Bolshiyanov D., Langer M., Lange S., Bornemann N., Morgenstern A., Schreiber P., Wille C., Chadburn S., Gouttevin I., Burke E., & Kutzbach L. (2019). A 16-year record (2002–2017) of permafrost, active-layer, and meteorological conditions at the Samoylov Island Arctic permafrost research site, Lena River delta, northern Siberia: An opportunity to validate remote-sensing data and land surface, snow, and permafrost models. *Earth System Science Data, 11*, 261–299.

Bolshiyanov, D. Y., Makarov, A. S., Schneider, V., & Stoof, G. (2013). Origin and Development of the Lena River Delta, AARI.

Brown, J., Ferrians, O., Heginbottom, J. A., & Melnikov, E. (2002). Circum-Arctic map of permafrost and ground-ice conditions, Version 2 [Электронный ресурс].

Chalov, S. R., & Prokopeva, K. N. (2021). Assessment of suspended sediment budget of the Lena River Delta based on the remote sensing dataset. *Izvestiya, Atmospheric and Oceanic Physics, 57*(9), 1051–1060.

Chalov, S. R., & Prokopeva, K. N. (2022). Sedimentation and erosion patterns of the Lena River anabranching channel. *Water, 14*(23), 3845.

Chalov, S., Prokopeva, K., & Habel, M. (2021). North to South variations in the suspended sediment transport budget within large Siberian River deltas revealed by remote sensing data. *Remote Sensing, 13*, 4549.

Chalov, S., Prokopeva, K., Magritsky, D., Grigoriev, V., Fingert, E., Habel, M., Juhls, B., Morgenstern, A., Overduin, P. P., & Kasimov, N. (2023). Climate change impacts on streamflow, sediment load and carbon fluxes in the Lena River delta. *Ecological Indicators, 157*, 111252.

Charkin, A. N., Dudarev, O. V., Semiletov, I. P., Kruhmalev, A. V., Vonk, J. E., Sánchez-García, L., Karlsson, E., & Gustafsson, Ö. (2011). Seasonal and interannual variability of sedimentation and organic matter distribution in the Buor-Khaya Gulf: The primary recipient of input from Lena River and coastal erosion in the southeast Laptev Sea. *Biogeosciences, 8*(9), 2581–2594.

Costard, F., Dupeyrat, L., Gautier, E., & Carey-Gailhardis, E. (2003). Fluvial thermal erosion investigations along a rapidly eroding river bank: Application to the Lena River (Central Siberia). *Earth Surface Process and Landforms, 28*, 1349–1359.

Domanitsky, A. P., Dubrovina, R. G., & Isaeva, A. I. (1971). Rivers and Lakes of the Soviet Union. Gidrometeoizdat.

Doronina, N. A. (1962). The rivers. Severnaya Yakutia.Tr. AANI (Proceedings AARI).

Fedorova, I., & Makushin, I. (2016). Hydrological part of expedition report "Lena-2016". Saint-Petersburg.

Gabyshev, V., Barinova, S., Ivanova, A., Gabysheva, O., & Curtean-Bănăduc, A. (2022). The indicator role of algae in assessing the organic pollution in the Lena River Delta, the Russian Arctic. *Frontiers in Environmental Science, 10*, 921819.

Gordeev, V. V. (2006). Fluvial sediment flux to the Arctic Ocean. *Geomorphology, 80*(1–2), 94–104.

Grigoryev, M. N. (1993). Cryomorphogenesis in the Lena Delta. Permafrost Institute Press.

Ivanov, V. V., Piskun, A. A., & Korabel, R. A. (1983). Distribution of runoff through the main channels of the Lena River Delta. *Tr. AANI (proceedings AARI), 378*, 59–71. (in Russian).

Ivanov, V. V., & Piskun, A. A. (1999). Distribution of river water and suspended sediment loads in the deltas of rivers in the basins of the Laptev and East Siberian seas. In H. Kassens, H. A. Bauch, I. Dmitrenko, H. Eicken, H.-W. Hubberten, M. Melles, J. Thiede, & L. Timokhov (Eds.), *Land-ocean systems in the Siberian Arctic: Dynamics and history* (pp. 287–299). Springer.

Korotaev, V. N., Mikhailov, V. N., Babich, D. B., Zayats, G. M., & Bogomolov, A. L. (1990). Hydrological and morphological processes, dynamics of the hydrographic network and channel deformations in the Lena River delta. In R. Charov (Ed.), *Land and water resources: erosion protection and regulation of riverbeds*. Lomonosov Moscow State University.

Korotaev, V. N., Mikhailov, V. N., Babich, D. B., Li, C. X., & Liu, S. G. (2007). Estuarine-deltaic systems of Russia and China: The hydrological-morphological processes, Geomorphology and Evolution Forecast (p. 445). Lomonosov Moscow State University.

Kotlyakov, V. M. (2007). The National Atlas of Russia, Vol. 2, Nature, Ecology. Roscartography.

Kotlyakov, V. M., Kravtsova, V. I., & Dreyer, N. (1997). Atlas of snow and ice resources of the world. Russian Academy of Sciences.

Lopatin, G. V., & Fedorov, M. K. (1947). Physical and geographical description of the Lena River Delta. AARI.

Magritsky, D. V. (2015). Factors and patterns of long-term changes in water flow, suspended sediments and heat of the Lower Lena and Vilyui. *Vestn. Moscow. un-ta. Ser.5. Geogr.*, 85–9.

Magritsky, D. V., Alexeevsky, N. I., Aybulatov, D. N., Fofonova, V. V., & Gorelkin, A. (2018). Features and evaluations of spatial and temporal changes of water runoff, sediment yield and heat flux in the Lena River delta. *Polarforschung, 87*(2), 89–110.

Mikhailov, V. N. (1997). The river mouths of Russia and contiguous countries: Past, present, and future. GEOS.

Milliman, J. D., & Meade, R. H. (1983). World-wide delivery of river sediment to the oceans. *The Journal of Geology, 91*(1), 1–21.

Osadchiev, A., Silvestrova, K., & Myslenkov, S. (2020). Wind-drivencoastal upwelling near large river deltas in the Laptev and East-Siberian Seas. *Remote Sensing, 12*(5), 844.

Rachold, V., Alabyan, A. M., Hubberten, H. W., Korotaev, V. N., & Zaitsev, A. A. (1996). Sediment transport to the Laptev Sea—Hydrology and geochemistry of the Lena River. *Polar Research, 15*(2), 183–196.

Rivera, J., Karabanov, E. B., Williams, D. F., Buchinskyi, V., & Kuzmin, M. (2006). Lena River discharge events in sediments of Laptev Sea, Russian Arctic. *Estuarine, Coastal and Shelf Science, 66*(1–2), 185–196.

Sakha Republic. (2000). Geographic Atlas. Roskartografiya.

Schneider, J., Grosse, G., & Wagner, D. (2009). Land cover classification of tundra environments in the Arctic Lena Delta based on Landsat 7 ETM+ data and its application for upscaling of methane emissions. *Remote Sensing of Environment,* 113(2): 380-391.

Semiletov, I. P., Pipko, I. I., Shakhova, N. E., Dudarev, O. V., Pugach, S. P., Charkin, A. N., Mcroy, C. P., Kosmach, D., & Gustafsson, Ö. (2011). Carbon transport by the Lena River from its headwaters to the Arctic Ocean, with emphasis on fluvial input of terrestrial particulate organic carbon versus carbon transport by coastal erosion. *Biogeosciences, 8*(9), 2407–2426.

Serreze, M. C., Bromwich, D. H., Clark, M. P., Etringer, A. J., Zhang, T., & Lammers, R. (2002). Large-scale hydro-climatology of the terrestrial Arctic drainage system. *Journal of Geophysical Research: Atmospheres, 107*(D2), ALT 1-1–ALT 1-28.

Shakhova, N., Semiletov, I., & Bel'cheva, N. (2007). The great Siberian rivers as a source of methane on the Russian Arctic shelf. *Doklady Earth Sciences, 415*(5), 734–736.

Wagner, A., Lohmann, G., & Prange, M. (2011). Arctic river discharge trends since 7 ka BP. *Global and Planetary Change, 79*(1–2), 48–60.

Wild, B., Andersson, A., Bröder, L., Vonk, J., Hugelius, G., McClelland, J. W., Song, W., Raymond, P. A., & Gustafsson, Ö. (2019). Rivers across the Siberian Arctic unearth the patterns of carbon release from thawing permafrost. *Proceedings of the National Academy of Sciences, 116*(21), 10280–10285.

Yamazaki, Y., Kubota, J., Ohata, T., Vuglinsky, V., & Mizuyama, T. (2006). Seasonal changes in runoff characteristics on a permafrost watershed in the southern mountainous region of eastern Siberia. *Hydrological Processes, 20*(3), 453–467.

Yang, D., Kane, D. L., Hinzman, L. D., Zhang, X., Zhang, T., & Ye, H. (2002). Siberian Lena River hydrologic regime and recent change. *Journal of Geophysical Research: Atmospheres, 107(D23)*, 4694.

Zhang, Y., Chen, W., Cihlar, J. (2003). A process-based model for quantifying the impact of climate change on permafrost thermal regimes. *Journal of Geophysical Research: Atmospheres, 108*(D22), 2002JD003354.

Mekong River Delta

Cong Mai Van, Xuefei Mei, and Tri Cao Mai

1 Physical Setting

1.1 The Mekong River

Mekong River (MR) is the 12th longest river in the world with a total length of 4800 km (Li et al., 2017). From its source in the Tibetan Plateau, MR flows across six different countries, namely China, Myanmar, Laos People's Democratic Republic (Laos PDR), Thailand, Cambodia and finally Vietnam before draining into the South China Sea (Fig. 1). The river has a drainage area of 795000 km^2 (Hoang et al., 2016).

In its natural and undisturbed state, MR annually transports approximately ~470 billion m^3 of water and 160 million tons of sediment to the delta (Li et al., 2017; Wild & Loucks, 2014). Following various human interferences, the delta presently receives 400 billion m^3 of water and only 37 million tons of sediment (Thi Ha et al., 2018). MR basin is dominated by Asian monsoon climate, which generates distinct wet (June–November) and dry (December–May) seasons. Tropical typhoons in the Pacific Ocean also significantly contribute to the rainfall during the later parts of the wet season (August to early October).

C. M. Van
Faculty of Civil Engineering, Thuyloi University, Hanoi, Vietnam

X. Mei (✉)
State Key Laboratory of Estuarine and Coastal Research, East China Normal University, Shanghai 200241, China
e-mail: xfmei@geo.ecnu.edu.cn

C. M. Van
Faculty of Civil Engineering and Geosciences, Delft University of Technology, Delft 2628 CN, The Netherlands

T. C. Mai
Faculty of Coastal and Offshore Engineering, Hanoi University of Civil Engineering, Hanoi, Vietnam

W. Zhang and H. de Vriend (eds.), *Delta Sustainability*,
https://doi.org/10.1007/978-981-97-7259-9_10

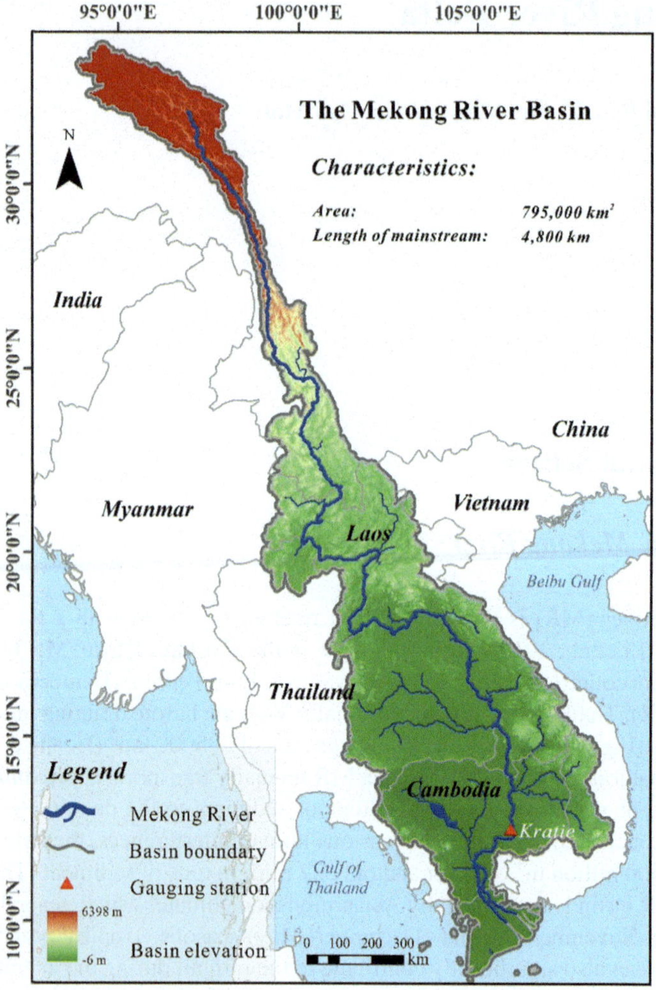

Fig. 1 Map of Mekong River basin

1.2 The Mekong Delta

The Vietnamese part of the Mekong Delta (MD, also known as the *Cuu Long* or "Nine Dragons") is the region in the far south of Vietnam (Fig. 2). The delta roughly forms a triangle of over 39,000 km^2 stretching from Mỹ Tho in the east to Châu Đốc and Hà Tiên in the northwest, down to Cà Mau at the southernmost tip of Vietnam (12.7% of the total land area of Vietnam) (Hoa, 2008; Ta et al., 2002).

The MD is a wave-influenced tide-dominated delta. The mean tidal range is 2.5 m and the maximum tidal range is 3.2–3.8 m (Nguyen et al., 2000). The mean wave height is 0.9 m (Ta et al., 2002). Southwestward coastal and longshore currents

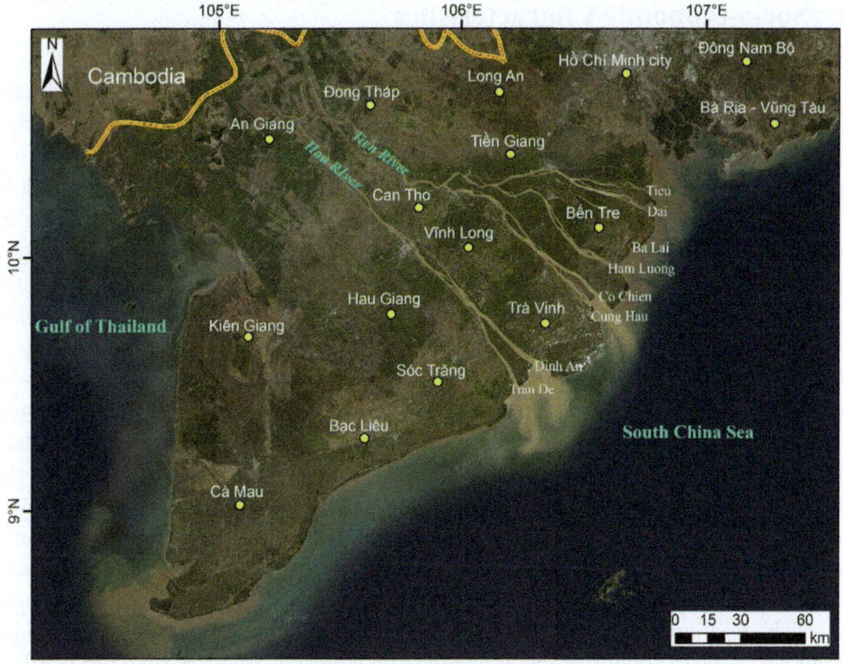

Fig. 2 Schematised map of the Mekong Delta

generated by winter monsoons are dominant and they are the cause of the formation of a beach ridge and spit system.

The subaerial delta plain of the MD is composed of two parts: an upper (inner) delta plain dominated by fluvial processes, and a lower (outer) delta plain characterized by a well-developed beach-ridge system and mainly influenced by marine processes (Nguyen et al., 2000; Szczuciński et al., 2013). In addition, the MR-system has a relatively young subaqueous delta ($\leq$1000 yr), a shallow rollover at 4–6 m water depth, gentle foreset gradients (0.03–0.57°), and a short cross-shelf dimension of 15–20 km within 20 m water depth (Liu et al., 2017a; Tamura et al., 2012).

The MD comprises highly fertile soils. The coastal wetlands are featured by mangrove forest and peatland Melaleuca forest wetlands. These nature wetlands, along with manmade ones, such as rice fields and aquaculture ponds, provide vital ecosystem services to the region (Dang et al., 2021; Leinenkugel et al., 2011).

2 Socio-economic Characteristics

Home to nearly 18 million people (approximately 20% of the Vietnamese population), the Vietnamese MD covers 13 provinces and plays a key role in the demographic fabric of the Vietnam region (Schmitt et al., 2019). With its catchment, the delta has a comparatively high population density of 435 people/km^2, of which about 20% is urbanised, leaving an 80% rural population (GSO, 2020; Vo, 2012). Despite its economic significance, the MD belongs to the regions with the lowest population growth in Vietnam, due to a high out-migration (Berchoux et al., 2023). Since the 1990s, population growth has been a steady 1.8–2% per year. Approximately 85% of the population lives from agricultural activities (GSO, 2011).

The economies and societies along the MD are strongly linked to good water access and highly fertile land. As the "rice bowl" of Asia, the MD is a major rice-growing region, accounting for not less than 54% of Vietnam's paddy production in 2018 (23.6 million tons) and over 95% of Vietnam's rice exports over the last decade. It also produces around 70% of Vietnam's fruit, including oranges, tangerines, bananas and mangoes. The Mekong Delta is also a major contributor to Vietnam's fisheries and aquaculture industries (56.0% and 69.9%, respectively in 2018). It accounts for some 60% of national fish exports, with shrimp, pangasius fish and tilapia as key aquaculture products, making Vietnam the world's third largest exporter of aquatic products and the fifth largest producer of rice (FAO, 2016). The fishing and aquaculture industries in the region alone employ over 2.8 million people—10% of Vietnam's entire labour force.

The Mekong Delta houses 11.1% of Vietnam's industry and construction (I&C) sector. It is the third largest industrial region in Vietnam after the metropolitan areas of Ho Chi Minh City (HCMC) and Hanoi. Activities in the region's services sector account for 14.8% of the national services output. Annual tourism-related revenues alone were estimated at around $400 million in 2019. The region is a net supplier of labour to the rest of the country. The net out-migration from the Mekong Delta has averaged 0.53% per annum in the years 2005–2018 (GSO, 2020, with over 90% of these migrants reportedly relocating to the neighbouring South East region of the country (particularly to HCMC). Overall, the Delta has close economic ties with the Southern Economic region (which includes HCMC) and the dependence on national road connections means that almost all Mekong Delta's output presently passes through HCMC before being traded within Vietnam or abroad. As economic growth centred on HCMC has expanded in recent years, there has been an expansion of activities (particularly light industrial) out of the formal industrial areas of HCMC and towards the south—effectively coalescing the two economic regions, such that provinces such as Long An may be de facto in both regions. This physical encroachment is in part facilitated by the over-development of industrial parks by local authorities and contributes to the urban fragmentation and low economic density which is described later.

The 2018 composition of the region's GDP by sector in comparison with Vietnam as a whole is shown in Fig. 3. Due to the rural nature of the Mekong Delta's economy,

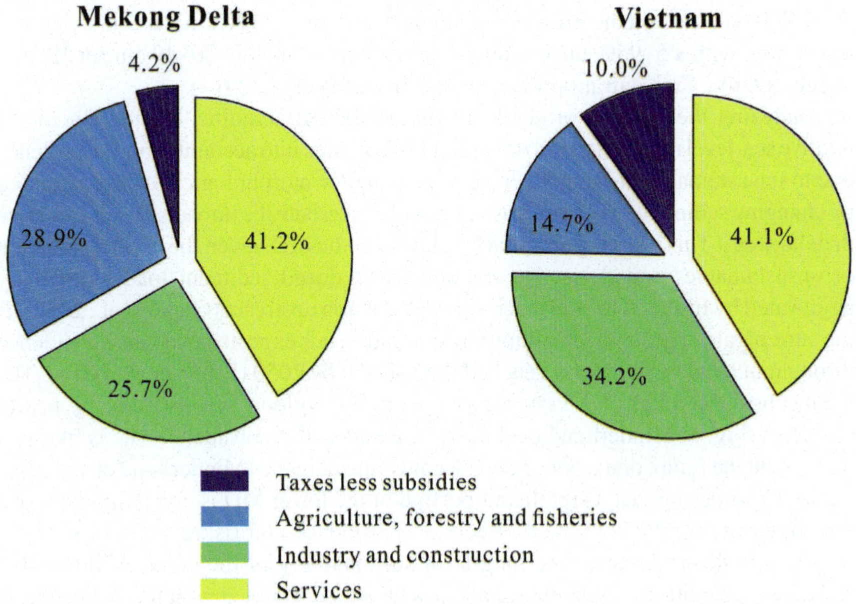

Fig. 3 GDP composition of Mekong Delta and Vietnam. *Data source* GSO (2020)

the share of agriculture, fisheries and forestry in the regional GDP is very high (at 28.9%) compared to the national average of 14.7% of GDP, while I&C's share is low (at 25.7% regionally vs. 34.2% nationally). However, with the industry and construction and services sectors now growing much more quickly than agriculture, fisheries and forestry (i.e., at 7.2% per annum for industry and construction and 8.5% per annum for services but only 2.4% per annum for agriculture, fisheries and forestry during the 5 years 2013–2018), the region is now undergoing a process of industrialisation and structural transformation, although later and at a slower pace than the rest of the country.

3 Current Issues

3.1 Relative Sea Level Rise and Land Subsidence

Continuing sea level rise in the coming centuries potentially has large impacts on the MD, in particular by increasing salinization, inundation and flood risks. The rate of sea-level rise, however, is still uncertain depending atmospheric temperatures rises and the rate of glacier and snow melt particularly in Greenland. According to the IPCC Fourth Assessment Report, a prediction has been made for the coast of southern Vietnam that the sea level will experience a rise of 30 cm by the year 2050 (Wassmann et al., 2004). Based on the Fifth Assessment Report by the IPCC have confirmed this prediction, with a 5–95% uncertainty range of approximately 20–40 cm for IPCC's Representative Concentration Pathway 8.5 (Katzfey et al., 2014).

Analyzing the impacts of sea-level rise in the MD requires calculation of the relative sea-level rise, which also takes land subsidence into account. Land subsidence due to unsustainable ground water extraction and/or morphological changes caused by changing sediment load from upstream will exacerbate the impacts of sea-level rise in delta areas. This issue is particularly relevant in the MD, given the current situation of unsustainable ground water extraction and reduced sediment load, potentially aggravated by further dam and reservoir construction upstream (Gupta et al., 2012). If the current rate of groundwater pumping continues, it is expected that land subsidence throughout the delta will have reached 0.35–1.4 m by 2050 (Erban et al., 2014). The local subsidence rates of 2.5 cm per year exceed the global sea level rise by nearly ten times (Fig. 4; Minderhoud et al., 2017). Besides, this subsidence rate is likely to increase in the future due to increased groundwater demand (Minderhoud et al., 2017, 2020). Considering that a significant portion of the lower MD is only slightly above 2 m in elevation, the combined effects of subsidence and rising sea levels would pose a significant threat to the long-term sustainability of the delta. Additionally, with the acceleration of subsidence and sea level rise, coupled with the construction of planned dams and ongoing sediment mining, the MD will face the risk of almost complete disappearance by the end of this century (Schmitt et al., 2017).

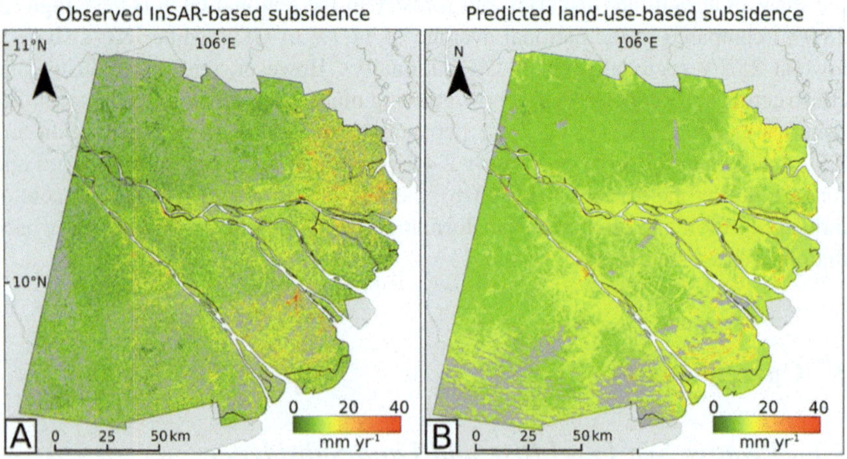

Fig. 4 **a** Observed InSAR-based subsidence rate. **b** Predicted subsidence rates for the period 2006–2010 (from Minderhoud et al., 2017)

3.2 Floods and Storm Surges

The MD experiences annual widespread flooding, which constitutes the region's primary natural hazard (Heiko et al., 2012). The delta has been significantly impacted by a series of major floods this century, notably those in 2000, 2001, and 2002 (Pengel et al., 2007). These floods have led to loss of lives (nearly 500 casualties), livestock and property, inflicted damage on agriculture and rural infrastructure, and disrupted socio-economic activities. The total economic cost of the flood damage was estimated at $400 million (Weichselgartner, 2005). The flood regime in the MD is shaped by a complex interplay of four influencing factors. Type 1 floods are primarily triggered by the MD's flood pulse and overland flow. Type 2 floods are artificial floodwaters distributed via a dense network of canals and controlled by dykes and sluice gates. Type 3 floods are short-term floods resulting from extreme local precipitation events. Type 4 floods are primarily associated with high tides during storms. Typically, no single flood occurrence in the MD can be exclusively attributed to one of these four factors. Instead, flood occurrences are invariably triggered by a combination of several of these sources (Hoang et al., 2016).

Storm surge, associated with tropical storms and typhoons, is a recurrent issue in the MD (MRC, 2015). Extreme wind speeds, along with the associated storm surges, cause most of the damage in coastal regions because of its low topography, high population density and inadequate protection (Tung et al., 2020). Typhoon Linda in 1997, considered to be the worst storm in recent times, claimed more than 3000 lives and caused severe damage in the delta and the remote islands (Anh et al., 1997, 2022). An area of 9800 km^2 of the MD is estimated to be susceptible to storm surges with potentially serious damage on agriculture, transportation, and threats to lives of the local inhabitants (Syvitski et al., 2009). Furthermore, the situation can be exacerbated during floods due to sea-level rise (SLR) induced by climate change (Gupta et al., 2003).

3.3 Sediment Starvation, Sand Mining and Coastal Erosion

The MR is one of the world's most crucial sediment-carrying rivers, continuously transporting sediment to the MD. For the past 3000 years, the average sediment discharge of the MR was nearly 150.000 Mt/yr, almost equivalent to the present value. Consequently, the delta's leading edge continuously advanced towards the sea, and its steepness increased during this period (Ta et al., 2002). However, in recent decades, due to human activities and climate change, the sediment load of the MR is rapidly decreasing, leading to intensified erosion in the MD (Kondolf et al., 2014; Tamura et al., 2020). Specifically, with 38 dams already constructed or under construction, sediment reaching the delta is expected to decrease by 51%. If all planned dams are completed, the cumulative sediment retention will reach 96% (Kondolf et al., 2014). Moreover, Kummu et al. (2008) indicated that between 1995 and 2000, the Tonle

Sap Lake supplied 0.65 ± 0.6 Mt/yr to the downstream MR. However, since 2001, the Tonle Sap Lake has become a sediment sink, accumulating an average of 1.35 ± 0.7 Mt/yr. The net sediment storage in the lake has further exacerbated the impact of reduced sediment supply due to upstream dam development and sand mining in the river (Sok et al., 2021). The suspended sediment flux in the lower MR is expected to change to 43 Mt/yr by 2020–2029 (Bussi et al., 2021). In addition, Darby et al. (2016) stated that in recent years (1981–2005), the suspended load delivered to the delta has decreased by 52.6 ± 10.2 million tons due to a shift in tropical-cyclone climatology, which may pose further risk to the vulnerable coastal systems. Schmitt et al. (2017) predicted through models that, under the current situation of planned dam construction, sand mining, and groundwater extraction, the sediment supply to the delta will be completely cut off. Coupled with the increased subsidence rate of the delta and rising sea levels, by the end of this century, the MD will almost completely disappear.

Since the 1990s, extensive sand mining in the downstream area of the MR has been attributed to land reclamation and civil industrial activities, making it one of the most severely impacted deltas by riverbed mining (Bravard et al., 2013). Hackney et al., (2020) revealed that the total sand flux entering the MR is about 6.18 ± 2.01 Mt/yr, significantly lower than the current sand extraction rate of 50 Mt/yr. Furthermore, Jordan et al., (2019) demonstrated that sand mining, averaging about 17.8 million m^3/yr, has resulted in extensive fluvial erosion. The large-scale sand extraction has resulted in the creation of numerous pools and mine pits. In Vietnam, the deepest mine pit reaches up to 45 m (Brunier et al., 2014). These deep pits actively trap sediment that should have flowed downstream (Anthony et al., 2015). The depletion of sand sources due to massive extraction has hindered the normal sedimentation process of rivers, leading to the destruction of river and delta ecosystems and negatively impacting delta landform development.

Main cause of erosion along coastlines of the MD is due to sediment depletion associated with increasing number of dams, large-scale sand mining and accelerated subsidence (Fig. 5; Anthony et al., 2015). The sediment is transported by waves and currents and is redistributed along the coast and transported in southwesterly direction. Due to the appearance of sand bars around the river mouths, caused by siltation from river branches, most of the coastline sections about 10–20 km downstream (southern side) of every river mouth are eroded. Study of the past years of Landsat images indicates that the coastal erosion occurs very quickly, with an average erosion rate of about 50 m/y (Li et al., 2017). The MD changes from accreting to eroding around 2005. From 1973 to 2005, the MD's seaward shoreline growth decreased gradually from a mean of 7.8 m/yr to 2.8 m/yr, and after 2005 it became negative, with a retreat rate of −1.4 m/y (Liu et al., 2017b). Strong wave action causes coastal erosion from November to March every year. The erosion rate of the shoreline is increasing, especially in the silty east coast that is influenced by the northeast monsoon while the west coast is mainly affected by the southwest monsoon, which causes coastal erosion every year from June to October (Le Xuan et al., 2022). These erosion trends and their spatiotemporal disparities generate considerable risks to residents, rendering the MR one of the most vulnerable and threatened deltas (Besset et al., 2019).

3.4 Freshwater Resources and Saline Water Intrusion

Freshwater availability during the dry season is bound to decrease in the MD, due to a combination of factors. On the supply side, both dry season local rainfall and MR discharge are likely to decrease under climate change conditions (Anh et al., 2019; Ruiz-Barradas & Nigam, 2018). The ground water resource in the delta is also rather limited, not to mention the fact that groundwater stocks are currently under stress and already show signs of depletion (Erban et al., 2014; Shrestha et al., 2016). On top of potential reduction in rainfall and river discharge, increased evaporation due to rising temperature could further decrease water availability in the delta (Hu & Mo, 2021). On the demand side, water demand for different sectors is likely to increase in the future (MRC, 2018; Tang et al., 2023). Reduced freshwater availability in combination with salt water intrusion and higher water demands may cause severe water stress in the dry season. In fact, local drought events seem to occur with higher frequencies in the MD during dry months (Nguyen, 2017; Nghia et al., 2022). In the dry season, the average discharge of the MR is less than 2,500 m^3/s, and even as low as 1,700 m^3/s, with the groundwater table lowering by 0.01–0.55 m/y in some places (Duy et al., 2021; Lu & Chua, 2021; Tuan & Wyseure, 2006). This will affect nearly 15 billion m^2 of cultivable land and may affect domestic and industrial water supply (Nguyen et al., 2021). The current strains on the water resources base of the delta are threefold: seasonal fresh water flooding during the wet season and seasonal fresh water scarcity during the dry season (Nguyen et al., 2021; MRC, 2022); water quality degradation, in terms of salinity, acidity, and water pollution (Eslami et al., 2021; MRC, 2021; Phong et al., 2014; Smajgl et al., 2015); coastal flooding (Phung et al., 2016).

Groundwater is a primary freshwater source for various domestic, industrial and agricultural purposes, especially in the MD where there is a lack of surface water supply (Tran et al., 2020). Groundwater exploitation in the MD has increased dramatically over the past 25 years. The exploitation in the southern MD in 2019 was 567,364 m^3/d. The most exploited aquifers are the upper-middle Pleistocene and the middle Pliocene, accounting for 63.7% and 24.6%, respectively. The least exploited aquifers are the upper Pleistocene and the upper Miocene, accounting for 0.35% and 0.02%, respectively. In the deeper aquifers, the change in storage is negative due to the high exploitation rate, leading to a decline of the stocks in these aquifers (Hoan et al., 2022). The delta has transformed from an almost undisturbed hydrogeological state to a situation with increasing aquifer depletion (Minderhoud et al., 2017). As groundwater levels drop, the consequent compaction of sediment layers causes land subsidence, at these locations at an average rate of 1.6 cm/yr. If pumping continues at

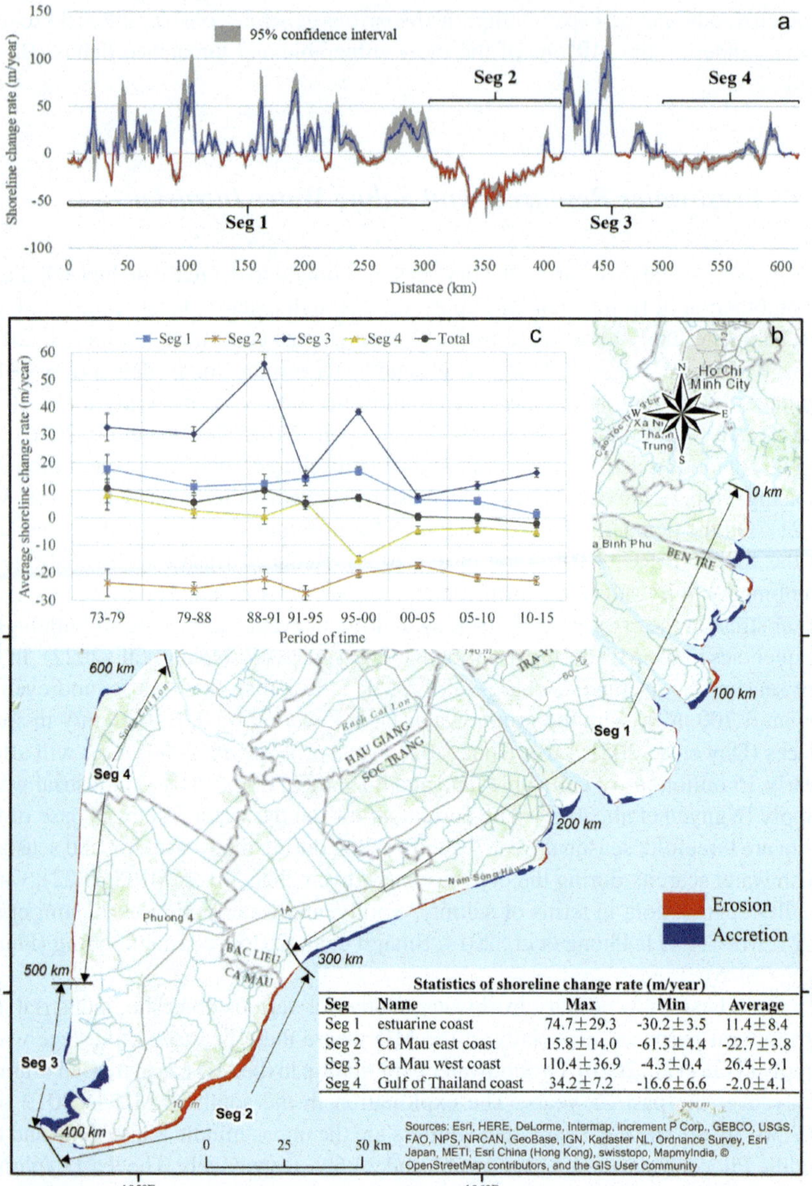

Statistics of shoreline change rate (m/year)

Seg	Name	Max	Min	Average
Seg 1	estuarine coast	74.7±29.3	-30.2±3.5	11.4±8.4
Seg 2	Ca Mau east coast	15.8±14.0	-61.5±4.4	-22.7±3.8
Seg 3	Ca Mau west coast	110.4±36.9	-4.3±0.4	26.4±9.1
Seg 4	Gulf of Thailand coast	34.2±7.2	-16.6±6.6	-2.0±4.1

Fig. 5 Graph of shoreline changes for the Mekong Delta between 1973 and 2015. According to **a** the shoreline change rates along each transect, **b** the delta's shoreline can be divided into four segments: estuarine coast (Seg 1), Ca Mau East Coast (Seg 2), Ca Mau West Coast (Seg 3), and the Gulf of Thailand (GoT) coast (Seg 4). (c)The line graph shows the average shoreline change rates for the four segments, and the table (lower right) displays the maximum, minimum, and average shoreline change rates of the four segments (from Li et al., 2017)

present rates, about 0.88 m of land subsidence is expected by 2050. Hence, groundwater extraction is suggested as the main driver of land subsidence, which in coastal areas poses a flood inundation hazard that is compounded by continued sea level rise (Erban et al., 2014). Additionally, excessive groundwater extraction also amplifies saltwater intrusion into coastal freshwater aquifers (Klessens et al., 2022). Failure to mitigate groundwater extraction-induced subsidence and saltwater intrusion may strongly increases the delta's vulnerability to flooding, salinization, coastal erosion and, ultimately, threatens its nearly 18 million in habitants with permanent inundation (Minderhoud et al., 2020).

The MD, as one of the world's regions most susceptible to saline water intrusion, suffers significantly from salinity (Fig. 6; Loc et al., 2021). For instance, in the 2015–2016 dry season, saltwater intruded over 90 km inland, causing substantial crop damage in 11 out of the 13 provinces in the MD, including nine coastal provinces (Thanh et al., 2023). Given the projected climate change characterized by reduced dry season river discharge, rising sea levels, and increased evaporation, saline water intrusion in the MD is expected to increase significantly. The anticipated sea-level rise by 2030 could affect more than 240 km^2 of agricultural land through salinity intrusion, significantly impacting local water resources that are vital to life and production (Phuong et al., 2020).

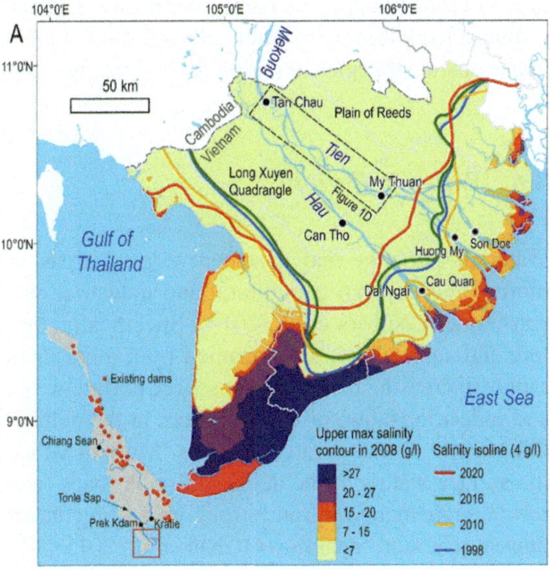

Fig. 6 Saltwater intrusion in the Mekong Delta over 1998, 2010, 2016 and 2020 (from Loc et al., 2021)

3.5 Land Use Changes

The predominant transformation of land use in the MRB has been characterized by the extensive conversion of forested regions into cultivated lands (Fig. 6; Liu et al., 2020; Tang et al., 2021). Over the period of 2003–2014, the estimated area converted to plantations was approximately 33,617 $\pm$ 7342 km^2, while agricultural expansion covered about 14,915 $\pm$ 4682 km^2. Notably, 82% of the deforested areas were transformed into tree plantations. Sam and Khoi (2022) indicated that between 1997 and 2010, the MD experienced significant land use changes, witnessing a 2.35% decline in forested areas and a corresponding 2.29% rise in agricultural land. It is also clear that aquaculture began to develop in coastal areas from the 1980s and expanded rapidly from coastal to inland areas (Fig. 7).

Land use changes in the MR Basin (MRB) have raised significant attention due to their potential and substantial impacts on hydrological processes, ecological systems, and the livelihoods of residents (Sam & Khoi, 2022; Shrestha et al., 2018; Zhai et al., 2016). Zhai et al. (2016) demonstrated that the significant expansion of forest, woodland, and grassland areas in the specified region of MRB from the mid-1980s to 2000 contributed to a notable reduction in runoff. This expansion of green cover likely had a positive impact on water retention and infiltration, leading to fluvial discharge decrease. Shrestha et al. (2018) indicated that the presence of uncertainty in land use demand gives rise to the most significant variations in streamflow and sediment load, leading to potentially substantial implications for water and sediment management strategies in regions experiencing rapid development.

3.6 Water Pollution

Water in estuarine and nearshore coastal areas in the MR is facing more and more pollution from agricultural and industrial chemicals and domestic untreated wastewater (Wilbers et al., 2014). Specifically, the intensification of rice agriculture and aquaculture, accompanied by a more intensive use of agro-chemicals, is rapidly increasing the non-point source pollution. In addition, the rapid urbanization and industrialization of urban centers in the delta is accompanied by an exponential growth of waste water effluents. Most of which are discharged untreated into the surface waters of the delta. Pollution loads (e.g., heavy metals, agro-chemical residues) are increasing, and strongly affecting the dilution and flushing capacities of present river and canal flows (Chau et al., 2015). Meanwhile, groundwater contamination by heavy metals and arsenic was also documented in the delta (Buschmann et al., 2008).

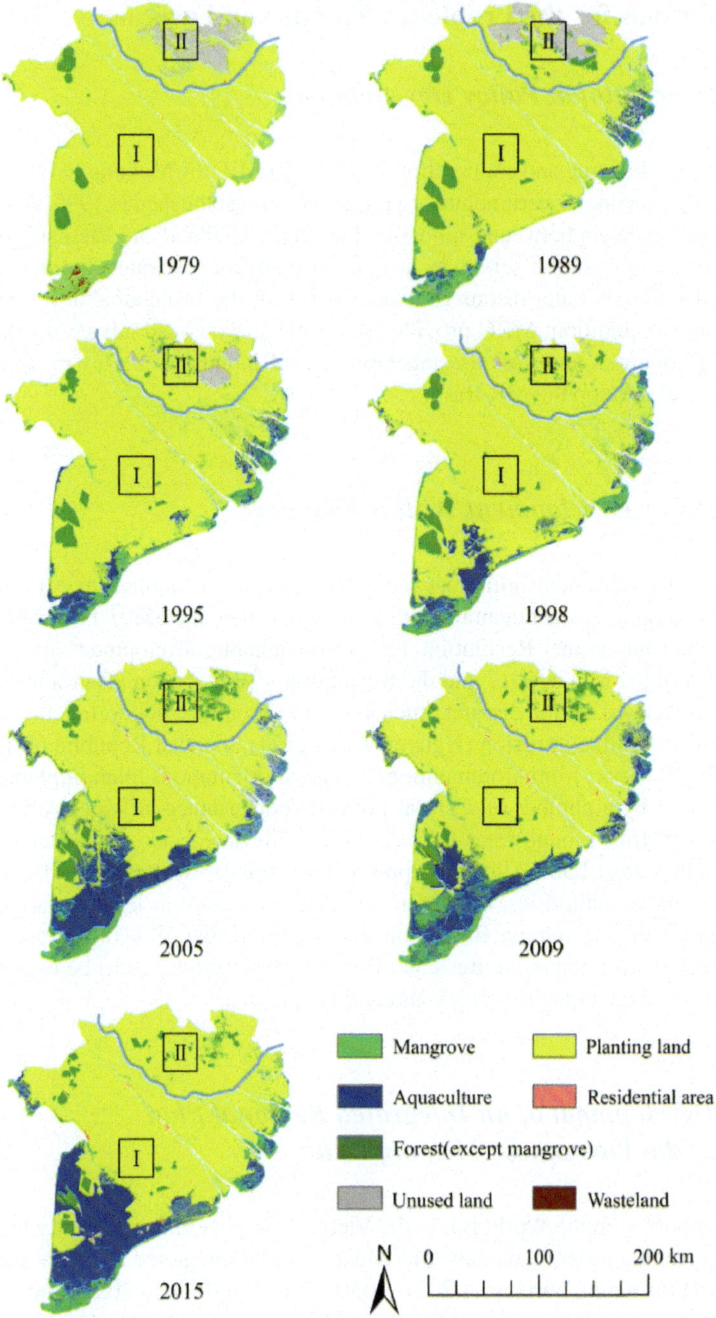

Fig. 7 LULC classification maps of the Mekong Delta in the year 1979, 1989, 1995, 1998, 2005, 2009, and 2015 based on Landsat and HJ1B-CCD1 images (from Liu et al., 2020)

4 Solutions for the Problems, Trends and Forecast

4.1 International Policy Development

For regional dialogue and cooperation in the lower MR, the Mekong River Commission (MRC), an intergovernmental organization, was established in 1995 based on the Mekong Agreement between Cambodia, Lao PDR, Thailand and Vietnam (Nguyen, 2007). The organization serves as a regional platform for water diplomacy and a knowledge hub of water resources management for the sustainable development of the region. In addition, MRC provides scientific analysis and advice for the lower Mekong countries and various stakeholders and interested parties to sustainably manage and develop the MR Basin.

4.2 Policy Development Within Vietnam

A series of government initiatives have been issued by Vietnam to cope with the rapidly changing environmental conditions (Nguyen et al., 2020). For example, the government has issued Resolution 120 on sustainable development and climate–resiliency of the MD in 2017 and the Resolution's 2019 Action Programme to cope with water related natural disasters such as storms, droughts, floods, together with sea level rise and saline intrusion (Nguyen et al., 2021; Socialist Republic of Vietnam, 2017, 2019). Apart from aforementioned general orientations, plan implementation is supported by a number of sectoral policies. For instance, a Water Act has been issued to restrict groundwater exploitation to alleviate land subsidence (HCMC, 2007). The Forest Land Allocation policy, issued in 1991 by the Ca Mau and Bac Lieu provincial authorities, has given forest land for 20-year lease to farmers with a Green Certificate (Green Book), on the condition that 70% of the area should be replanted with mangrove trees and the rest (30% of total area) be exploited for agriculture, aquaculture, livestock and residential area.

4.3 Development of an Integrated Regional Plan of the Vietnamese Mekong Delta

With support from the World Bank, the Vietnamese government recently has developed a planning program namely "the Mekong Delta Integrated Regional Planning" (MDIRP) for a time horizon of 2021–2030 with an outlook to 2050. This has been built on the earlier Mekong Delta Plan established with Dutch donor support in 2013. The MDIRP combines investment, economic development planning and a planning governance structure for the Vietnamese Mekong Delta (13 provinces/cities) based on hydrological, spatial, agricultural and climate change analysis. This planning program is expected to be the key reference for the on-going and future developments of natural disaster risk reduction strategies, climate-change resilient strategies, land use planning, infrastructure investments and socio-economic policies, taking due account of the adaptability and sustainability of future changes. The strategy is designed to both "manage challenges" and "create value". It aims to protect people, improve livelihoods, promote a more balanced regional development and improve the environment. The overall vision is to have prosperous and thriving communities, together turning challenges into opportunities, utilizing and protecting Mekong Delta resources. The strategy is based on a simple and widely accepted premise that agriculture is the most advantageous sector in the region, and that much productive land is currently being underused relative to its potential suitability for higher-value crops. Zones are defined regarding water management and flood protection: protection of the core freshwater zone and coastal areas, improved water quality (especially in relation to agriculture and aquaculture), climate change adaptation and management of natural disaster risks.

4.4 Coastal Restoration and Protection Forecast

Coastal restoration and protection have not adequately taken off and without focused interventions it will not take off in the future. Continued dependence on groundwater exploitation for rural and urban water supply, mono-culture based aquaculture and agriculture in the coastal zones will accelerate land subsidence and must be addressed. They will make coastal areas more vulnerable to coastal erosion, tidal inundation and storm surge induced coastal flooding. Initial response will be to strengthen the sea dyke along the coast, with a refurbished road and sluice gates to close the systems at times of storm surges. Coastal mangrove regeneration, however, is expected to remain limited without key interventions, as it causes sub-optimal water management conditions for the aquaculture sector. By 2030 and beyond, environmental degradation of aquaculture and farmlands may cause accelerating land abandonment in the coastal zones. This would provide opportunities for restructuring of the coastal areas, including re-greening of coastal zones to forests that may be sustainably exploited for diverse sea (food) products. As sea level rise (severely aggravated by land subsidence in the coastal zone) is set to accelerate, active sedimentation of the peninsula should become a priority goal of mangrove regeneration.

4.5 Hydrological Measures

Although several solutions were executed and planned, structural solutions remain predominant. Comprehensive infrastructural measures are implemented by the government of Vietnam for a safe delta and to enhance sustained economic development. Specifically,

(1) Canals, high and low dykes have been built extensively on the delta plain to prevent river flooding and facilitate crop production;
(2) Sea dikes have been built to protect low-lying coastal regions. But the sea dikes are rather low and their main function is to prevent inundation during spring tides;
(3) At some place a setback line to a distance of 3–5 km from the coastlines has been set. The outer land is then accepted to be inundated by sea water. This actually works as a buffer zone where both aquaculture has been developed in combination with mangrove forests;
(4) A surface water supply plant was constructed, and larger facilities for the entire VMD are currently being planned to reduce pressures on groundwater resources;
(5) Tidal barriers and/or sluices with control gates at small sea water entrances/cannels to address salinity intrusion; At the same time careful studies on impacts and effectiveness large scale storm-surge and tidal barriers at main (large) river entrances are implemented;
(6) Natural defenses by the remained mangrove systems (about 15% of the coastlines, mainly in far south provinces i.e. in Bac Lieu and Ca Mau) as an alternative of hard engineering approach.

5 Future Challenges

The resilience of the MD is threatened by higher temperature, accelerated land subsidence, reduced fluvial sediment supply and a rising sea level (Dang et al., 2022). The task to prepare the MD for an uncertain future is complex. It is important to note that the most severe impacts will be caused by a combination of these uncertain and rapidly changing threats. For example, the MD may experience a much higher flood risk due to a combination of sea level rise, higher fluvial peak flows, and possibly more frequent and severe typhoons. Reduced water availability in combination with salt water intrusion and higher water demands could potentially cause severe water stress in the dry season. Such low-probability, high-impact events will significantly affect agricultural production, especially rice and fruit crops and can potentially have an impact on domestic and industrial water supply over MD.

Given the foreseen future change in the delta, while understanding the changes and complexity of the various social-ecological systems in the MD, it is equally important to implement solutions and relevant policies (Nguyen et al., 2020). An ensemble of hard and soft policies is likely to provide the most effective results for people's livelihoods in the MD (Smajgl et al., 2015). Furthermore, integrated planning at basin and delta scale can provide critical information to make delta management robust to future uncertainty and help in establishing delta resilience as a crucial objective in river basin management (Schmitt et al., 2021).

References

Anh, L. T., Takagi, H., & Thao, N. D. (2019a) Storm surge and high waves due to 1997 Typhoon Linda: Uninvestigated worst storm event in Southern Vietnam. *Journal of Japan Society of Civil Engineers. Ser. B3 (Ocean Engineering), 75*(2), 73–78.

Anh, D. T., Hoang, L. P., Bui, M. D., & Rutschmann, P. (2019b). Modelling seasonal flows alteration in the Vietnamese Mekong Delta under upstream discharge changes, rainfall changes and sea level rise. *International Journal of River Basin Management, 17*(4), 435–449.

Anh, L. T., Takagi, H., Thao, N. D., & Esteban, M. (2022). Investigation of awareness of typhoon and Storm Surge in the Mekong Delta-Recollection of 1997 typhoon Linda. *Journal of Japan Society of Civil Engineers, 73*(2), 168–173.

Anthony, E. J., Brunier, G., Besset, M., Goichot, M., Dussouillez, P., & Nguyen, V. L. (2015). Linking rapid erosion of the Mekong River delta to human activities. *Scientific Reports, 5*(1), 14745.

Berchoux, T., Hutton, C., Hensengerth, O., Voepel, H., Tri, V., Vu, P., Hung, N. N., Parsons, D. R., & Darby, S. (2023). Effect of planning policies on land use dynamics and livelihood opportunities under global environmental change: Evidence from the Mekong Delta. *Land Use Policy, 131*, 106752.

Buschmann, J., Berg, M., Stengel, C., Winkel, L., Sampson, M. L., Trang, P. T. K., & Viet, P. H. (2008). Contamination of drinking water resources in the Mekong Delta floodplains: Arsenic and other trace metals pose serious health risks to population. *Environment International, 34*, 756–764.

Besset, M., Gratiot, N., Anthony, E., Bouchette, F., Goichot, M., & Marchesiello, P. (2019). Mangroves and shoreline erosion in the Mekong River delta, Viet Nam. *Estuarine, Coastal and Shelf Science, 226*, 106263.

Bravard, J. P., Goichot, M., & Gaillot, S. (2013). Geography of sand and gravel mining in the Lower Mekong River: First survey and impact assessment. *EchoGéo, 26*, 2–18.

Brunier, G., Anthony, E. J., Goichot, M., Provansal, M., & Dussouillez, P. (2014). Recent morphological changes in the Mekong and Bassac river channels, Mekong Delta: The marked impact of river-bed mining and implications for delta destabilisation. *Geomorphology, 224*, 177–191.

Bussi, G., Darby, S. E., Whitehead, P. G., Jin, L., Dadson, S. J., Voepel, H. E., Vasilopoulos, G., Hackney, C. R., Hutton, C., Berchoux, T., Parsons, D. R., & Nicholas, A. (2021). Impact of dams and climate change on suspended sediment flux to the Mekong delta. *Science of the Total Environment, 755*, 142468.

Chau, N. D. G., Sebesvari, Z., Amelung, W., & Renaud, F. G. (2015). Pesticide pollution of multiple drinking water sources in the Mekong Delta, Vietnam: Evidence from two provinces. *Environmental Science Pollution Research, 22*, 9042–9058.

Dang, A. T. N., Kumar, L., Reid, M., & Nguyen, H. (2021). Remote sensing approach for monitoring coastal wetland in the Mekong Delta, Vietnam: Change trends and their driving forces. *Remote Sensing, 13*, 3359.

Dang, A. T. N., Reid, M., & Kumar, L. (2022). Assessing potential impacts of sea level rise on mangrove ecosystems in the Mekong Delta, Vietnam. *Regional Environmental Change, 22*, 70.

Darby, S. E., Hackney, C. R., Leyland, J., Kummu, M., Lauri, H., Parsons, D. R., & Aalto, R. (2016). Fluvial sediment supply to a mega-delta reduced by shifting tropical-cyclone activity. *Nature, 539*(7628), 276–279.

Duy, N. L., Nguyen, T. V. K., Nguyen, D. V., Tran, A. T., Nguyen, H. T., Heidbüchel, I., Merz, B., & Apel, H. (2021). Groundwater dynamics in the Vietnamese Mekong Delta: Trends, memory effects, and response times. *Journal of Hydrology: Regional Studies, 33*, 100746.

Erban, L. E., Gorelick, S. M., & Zebker, H. A. (2014). Groundwater extraction, land subsidence, and sea-level rise in the Mekong Delta. *Vietnam. Environmental Research Letters, 9*(8), 084010.

Eslami, S., Hoekstra, P., Minderhoud, P. S. J., Trung, N. N., Hoch, J. M., Sutanudjaja, E. H., Dung, D. D., Tho, T. Q., Voepel, H. E., Woillez, M. N., & van der Vegt, M. (2021). Projections of salt intrusion in a mega-delta under climatic and anthropogenic stressors. *Communications Earth and Environment, 2*(1), 142.

FAO. (2016). Fishery and Aquaculture Statistics. Food and Agriculture Organization of the United Nations, 2016.

GSO (General Statistics Office). (2011). Statistical data on Vietnam. Database of the GSO website.

GSO (General Statistics Office). (2020). Data on Population of Mekong Delta, Vietnam in the year 2018. General Statistics Office of Vietnam. Database of the GSO website.

Gupta, A. D., Babel, M. S., & Ngoc, P. (2003). Flood damage assessment in the Mekong Delta, Vietnam. In *Proceedings of the 2nd International Conference of Asia-Pacific Association of Hydrology and Water Resources* (Vol. I, pp. 109–117).

Gupta, H., Kao, S. J., & Dai, M. (2012). The role of mega dams in reducing sediment fluxes: A case study of large Asian Rivers. *Journal of Hydrology, 464*, 447–458.

Hackney, C. R., Darby, S. E., Parsons, D. R., Leyland, J., Best, J. L., Aalto, R., Nicholas, A. P., & Houseago, R. C. (2020). River bank instability from unsustainable sand mining in the lower Mekong River. *Nature Sustainability, 3*, 217–225.

HCMC. (2007). Ho Chi Minh City Water Act 2007. Retrieved 23, 2016.

Heiko, A., Nguyen, N. H., Trinh, T. L., & Vo, K. T. (2012). Flood hydraulics and suspended sediment transport in the plain of reeds, Mekong Delta. In F. Renaud & C. Kuenzer (Eds.), *The Mekong Delta system—Interdisciplinary analyses of a River Delta* (pp. 221–232). Springer.

Hoa, L. T. V. (2008). Infrastructure effects on floods in the Mekong River Delta in Vietnam. *Hydrological Processes, 22*, 1359–1372.

Hoan, T. V., Richter, K. G., Börsig, N., Bauer, J., Ha, N. T., & Norra, S. (2022). An improved groundwater model framework for aquifer structures of the quaternary-formed sediment body in the Southernmost parts of the Mekong Delta, Vietnam. *Hydrology, 9*, 61.

Hoang, L. P., Lauri, H., Kummu, M., Koponen, J., van Vliet, M. T. H., Supit, I., Leemans, R., Kabat, P., & Ludwig, F. (2016). Mekong River flow and hydrological extremes under climate change. *Hydrology and Earth System Sciences, 20*(7), 3027–3041.

Hu, S., & Mo, X. (2021). Attribution of long-term evapotranspiration trends in the Mekong River basin with a remote sensing-based process model. *Remote Sensing, 13*(2), 303.

Jordan, C., Tiede, J., Lojek, O., Visscher, J., Apel, H., Nguyen, H. Q., Quang, C. N. X., & Schlurmann, T. (2019). Sand mining in the Mekong Delta revisited—Current scales of local sediment deficits. *Scientific Reports, 9*(1), 17823.

Katzfey, J., Mcgregor, J., & Ramasamy, S. (2014). High-resolution Climate Projections for Vietnam: Technical Report.

Klessens, T. M. A., Daniel, D., Jiang, Y., Van Breukelen, B. M., Scholten, L., & Pande, S. (2022). Combining water resources, socioenvironmental, and psychological factors in assessing willingness to conserve groundwater in the Vietnamese Mekong Delta. *Journal of Water Resources Planning and Management, 148*(3), 05021034.

Kondolf, G. M., Rubin, Z. K., & Minear, J. T. (2014). Dams on the Mekong: Cumulative sediment starvation. *Water Resource Research, 50*(6), 5158–5169.

Kummu, M., Penny, D., Sarkkula, J., & Koponen, J. (2008). Sediment: Curse or blessing for Tonle Sap Lake?. *AMBIO: A Journal of the Human Environment, 37*(3), 158–163.

Le Xuan, T., Ba, H. T., Thanh, V. Q., Wright, D. P., Tanim, A. H., & Anh, D. T. (2022). Evaluation of coastal protection strategies and proposing multiple lines of defense under climate change in the Mekong Delta for sustainable shoreline protection. *Ocean and Coastal Management, 228*, 106301.

Leinenkugel, P., Esch, T., & Kuenzer, C. (2011). Settlement detection and impervious surface estimation in the Mekong Delta using optical and SAR remote sensing data. *Remote Sensing of Environment, 115*, 3007–3019.

Li, X., Liu, J. P., Saito, Y., & Nguyen, V. L. (2017). Recent evolution of the Mekong Delta and the impacts of dams. *Earth-Science Reviews, 175*, 1–17.

Liu, J. P., DeMaster, D. J., Nittrouer, C. A., Eidam, E. F., & Nguyen, T. T. (2017a). A seismic study of the Mekong subaqueous delta: Proximal versus distal sediment accumulation. *Continental Shelf Research, 147*, 197–212.

Liu, J. P., DeMaster, D. J., Nguyen, T. T., Saito, Y., Nguyen, V. L., Ta, T. K. O., & Li, X. (2017b). Stratigraphic formation of the Mekong River Delta and its recent shoreline changes. *Oceanography, 30*(3), 72–83.

Liu, S., Li, X., Chen, D., Duan, Y., Ji, H., Zhang, L., Chai, Q., & Hu, X. (2020). Understanding the land use/land cover dynamics and impacts of human activities in the Mekong Delta over the last 40 years. *Global Ecology and Conservation, 22*, e00991.

Loc, H. H., Van Binh, D., Park, E., Shrestha, S., Dung, T. D., Son, V. H., Truc, N. H. T., Mai, N. P., & Seijger, C. (2021). Intensifying saline water intrusion and drought in the Mekong Delta: From physical evidence to policy outlooks. *Science of the Total Environment, 757*, 143919.

Lu, X. X., & Chua, S. D. X. (2021). River discharge and water level changes in the Mekong River: Droughts in an era of mega-dams. *Hydrological Processes, 35*(7), e14265.

Mekong River Commission (MRC). (2015). Annual Mekong Flood Report 2013. Mekong River Commission.

Mekong River Commission (MRC). (2018). Irrigation database improvement for the lower Mekong Basin. MRC Secretariat.

Mekong River Commission (MRC). (2021). 2018 Lower Mekong water quality monitoring report. MRC Secretariat.

Mekong River Commission (MRC). (2022). Mekong low flow and drought conditions in 2019–2021: Hydrological conditions in the Lower Mekong River Basin. MRC Secretariat.

Minderhoud, P. S. J., Erkens, G., Pham, V. H., Bui, V. T., & Stouthamer, E. (2017). Impacts of 25 years of groundwater extraction on subsidence in the Mekong Delta, Vietnam. *Environmental Research Letters, 12*(6), 064006.

Minderhoud, P. S. J., Middelkoop, H., Erkens, G., & Stouthamer, E. (2020). Groundwater extraction may drown mega-delta: Projections of extraction-induced subsidence and elevation of the Mekong delta for the 21st century. *Environmental Research Communications, 2*, 011005.

Nghia, B. P. Q., Pal, I., Pramanik, M., & Dasgupta, R. (2022). The impact of climate change on drought and its adaptation strategies: Findings from general circulation models and households in Tien Giang Province, Vietnam. *Climate Change, 175*, 18.

Nguyen, N. A. (2017). Historic drought and salinity intrusion in the Mekong Delta in 2016: Lessons learned and response solutions. *Vietnam Journal of Science, Technology and Engineering, 59*(1), 93–96.

Nguyen, V. L., Ta, T. K. O., & Tateishi, M. (2000). Late Holocene depositional environments and coastal evolution of the Mekong River Delta. Southern Vietnam. *Journal of Asian Earth Sciences, 18*(4), 427–439.

Nguyen, H. N. (2007). Flooding in Mekong River Delta, Viet Nam. Human Development Occasional Papers (1992–2007).

Nguyen, Q. H., Tran, D. D., Dang, K. K., Korbee, D., Pham, L. D. M. H., Vu, L. T., Luu, T. T., Ho, L. H., Nguyen, P. T., Ngo, T. T. T., Nguyen, D. T. K., Wyatt, A., van Aalst, M., Tran, T. A., & Sea, W. B. (2020). Land-use dynamics in the Mekong delta: From national policy to livelihood sustainability. *Sustainable Development, 28*, 448–467.

Nguyen, M. N., Nguyen, P. B., Van, T. D., Phan, V. T., Nguyen, B. T., Pham, V. T., & Nguyen, T. H. (2021). An understanding of water governance systems in responding to extreme droughts in the Vietnamese Mekong Delta. *International Journal of Water Resources Development, 37*, 256–277.

Pengel, B., Malone, T., Tes, S., Katry, P., Pich, S., & Hartman, M. (2007). Towards a new flood forecasting system for the Lower Mekong River Basin (pp. 1–10). Third South-East Asia Water Forum.

Phong, N. D., Hoanh, C. T., Tuong, T. P., & Malano, H. (2014). Effective management for acidic pollution in the canal network of the Mekong Delta of Vietnam: A modeling approach. *Journal of Environmental Management, 140*, 14–25.

Phung, D., Rutherford, S., Dwirahmadi, F., Chu, C., Do, C. M., Nguyen, T., & Duong, N. C. (2016). The spatial distribution of vulnerability to the health impacts of flooding in the Mekong Delta. *Vietnam. International Journal of Biometeorology, 60*, 857–865.

Phuong, T. H., Hung, H. P., Hung, N. T., Man, D. B., Mon, D., Vinh, D. H., Thuc, P. T. B., Thuong, T. V., Ninh, N. T., Phuong, T. A., & Long, N. H. D. (2020). Assessment of changes in the structure land use in Tra Vinh Province under the scenarios of climate change and sea level rise. *Vietnam Journal of Science and Technology, 58*, 70–83.

Ruiz-Barradas, A., & Nigam, S. (2018). Hydroclimate variability and change over the Mekong River Basin: Modeling and predictability and policy implications. *Journal of Hydrometeorology, 19*, 849–869.

Sam, T. T., & Khoi, D. N. (2022). The responses of river discharge and sediment load to historical land-use/land-cover change in the Mekong River Basin. *Environmental Monitoring and Assessment, 194*(10), 700.

Schmitt, R. J., Rubin, Z., & Kondolf, G. M. (2017). Losing ground-scenarios of land loss as consequence of shifting sediment budgets in the Mekong Delta. *Geomorphology, 294*, 58–69.

Schmitt, R. J., Bizzi, S., Castelletti, A., Opperman, J. J., & Kondolf, G. M. (2019). Planning dam portfolios for low sediment trapping shows limits for sustainable hydropower in the Mekong. *Science Advances, 5*(10), eaaw2175.

Schmitt, R. J., Giuliani, M., Bizzi, S., Kondolf, G. M., Daily, G. C., & Castelletti, A. (2021). Strategic basin and delta planning increases the resilience of the Mekong Delta under future uncertainty. *Proceedings of the National Academy of Sciences, 118*(36), e2026127118.

Shrestha, S., Bach, T. V., & Pandey, V. P. (2016). Climate change impacts on groundwater resources in Mekong Delta under representative concentration pathways (RCPs) scenarios. *Environmental Science and Technology, 61*, 1–13.

Shrestha, B., Cochrane, T. A., Caruso, B. S., & Arias, M. E. (2018). Land use change uncertainty impacts on streamflow and sediment projections in areas undergoing rapid development: A case study in the Mekong Basin. *Land Degradation and Development, 29*, 835–848.

Smajgl, A., Toan, T. Q., Nhan, D. K., Ward, J., Trung, N. H., Tri, L. Q., Tri, V. P. D., & Vu, P. T. (2015). Responding to rising sea levels in the Mekong Delta. *Nature Climate Change, 5*, 167–174.

Socialist Republic of Vietnam. (2017). Resolution 120/NQ-CP on Sustainable and Climate Resilient Development in the Mekong Delta.

Socialist Republic of Vietnam. (2019). Decision 417/QĐ-TTg: Action Program on Sustainable Development of the Mekong Delta in Response to Climate Change.

Sok, T., Oeurng, C., Kaing, V., Sauvage, S., Kondolf, G. M., & Sánchez-Pérez, J. M. (2021). Assessment of suspended sediment load variability in the Tonle Sap and Lower Mekong Rivers, Cambodia. *Catena, 202*, 105291.

Syvitski, J. P. M., Kettner, A. J., Overeem, I., Hutton, E. W. H., Hannon, M. T., Brakenridge, G. R., Day, J., Vörösmarty, C., Saito, Y., Giosan, L., & Nicholls, R. J. (2009). Sinking deltas due to human activities. *Nature Geoscience, 2*(10), 292–300.

Szczuciński, W., Jagodzinski, R., Hanebuth, T. J. J., Stattegger, K., Wetzel, A., Mitrega, M., Unverricht, D., & Phung, P. V. (2013). Modern sedimentation and sediment dispersal pattern on the continental shelf off the Mekong River delta, South China Sea. *Global and Planetary Change, 110*, 195–213.

Ta, T. K. O., Nguyen, V. L., Tateishi, M., Kobayashi, I., Tanabe, S., & Saito, Y. (2002). Holocene delta evolution and sediment discharge of the Mekong River, southern Vietnam. *Quaternary Science Reviews, 21*(16), 1807–1819.

Tamura, T., Saito, Y., Nguyen, V. L., Ta, T. K. O., Bateman, M. D., Matsumoto, D., & Yamashita, S. (2012). Origin and evolution of interdistributary delta plains; insights from Mekong River delta. *Geology, 40*, 303–306.

Tamura, T., Nguyen, V. L., Ta, T. K. O., Bateman, M. D., Gugliotta, M., Anthony, E. J., & Saito, Y. (2020). Long-term sediment decline causes ongoing shrinkage of the Mekong megadelta, Vietnam. *Scientific Reports, 10*(1), 8085.

Tang, X. J., Woodcock, C. E., Olofsson, P., & Hutyra, L. R. (2021). Spatiotemporal assessment of land use/land cover change and associated carbon emissions and uptake in the Mekong River Basin. *Remote Sensing of Environment, 256*, 112336.

Tang, R. N., Dai, Z. J., Mei, X. F., Zhou, X. Y., Long, C. Q., & Van, C. M. (2023). Secular trend in water discharge transport in the Lower Mekong River-delta: Effects of multiple anthropogenic stressors, rainfall, and tropical cyclones. *Estuarine, Coastal and Shelf Science, 281*, 108217.

Thanh, T. N., Huynh Van, H., Vo Minh, H., & Tri, V. P. D. (2023). Salinity intrusion trends under the impacts of upstream discharge and sea level rise along the Co Chien River and Hau River in the Vietnamese Mekong Delta. *Climate, 11*(3), 66.

Thi Ha, D., Ouillon, S., & Van Vinh, G. (2018). Water and suspended sediment budgets in the lower Mekong from high-frequency measurements (2009–2016). *Water, 10*, 846.

Tran, D. A., Tsujimura, M., Vo, L. P., Nguyen, V. T., Kambuku, D., & Dang, T. D. (2020). Hydrogeochemical characteristics of a multi-layered coastal aquifer system in the Mekong Delta, Vietnam. *Environmental Geochemistry and Health, 42*, 661–680.

Tuan, L. A., & Wyseure, G. (2006). Water environmental governance in the Mekong River Delta, Vietnam. In *The 2nd International Symposium on Water Environment Partnership in Asia.*

Tung, T. T., Chien, N. Q., Le, H. T., & Hai, L. T. (2020). Modelling storm surge hazard to Mekong Delta. In N. Trung Viet, D. Xiping & T. T. Tung (Eds.), *APAC 2019*. Springer.

Vo, K. T. (2012). Hydrology and hydraulic infrastructure systems in the Mekong Delta, Vietnam. In F. Renaud & C. Kuenzer (Eds.), *The Mekong Delta system—Interdisciplinary analyses of a River Delta, The Mekong Delta System* (pp. 49–82). Springer.

Wassmann, R., Hien, N. X., Hoanh, C. T., & Tuong, T. P. (2004). Sea level rise affecting the Vietnamese Mekong Delta: Water elevation in the flood season and implications for rice production. *Climatic Change, 66*, 89–107.

Weichselgartner, J. (2005). From the field: flood disaster mitigation in the Mekong Delta. In *Proceedings for the 7th European Sociological Association Conference, 'rethinking inequalities', Torun, Disaster and Social Crisis Research Network* (pp. 9–12).

Wilbers, G. J., Becker, M., Nga, L. T., Sebesvari, Z., & Renaud, F. G. (2014). Spatial and temporal variability of surface water pollution in the Mekong Delta, Vietnam. *Science of the Total Environment, 485–486*, 653–665.

Wild, T. B., & Loucks, D. P. (2014). Managing flow, sediment, and hydropower regimes in the Sre Pok, Se San, and Se Kong Rivers of the Mekong Basin. *Water Resource Research, 50*(6), 5141–5157.

Zhai, H. J., Hu, B., Luo, X. Y., Qiu, L., Tang, W. J., & Jiang, M. (2016). Spatial and temporal changes in runoff and sediment loads of the Lancang River over the last 50 years. *Agricultural Water Management, 174*, 74–81.

Mississippi River Delta

Y. Jun Xu, Nina S. N. Lam, Kam-biu Liu, and Kehui Xu

1 Physical Setting

The modern Mississippi River Delta (MRD) is formed where the river empties into the Northern Gulf of Mexico in South Louisiana, United States. Over the past 7,000 years, the Mississippi River (MR) made six major course changes and deposited sediment over 400 km of the previous shoreline from approximately latitude 88°57″ to 91°50″ W (Fig. 1). Draining a total land surface of 3.2 million km^2, the MR created a massive deltaic plain of nearly 30,000 km^2 since the global sea level became stable. The deltaic region is low-lying and much of the area is less than 2 m above sea level, predominantly covered by various types of wetlands (Zhong & Xu, 2011).

The Mississippi River Delta region is characterized by a humid subtropical climate with long hot summers and short mild winters. Long-term annual temperature in the Baton Rouge area, the capital of the State of Louisiana, was reported to be 20 °C, with monthly averages ranging from 11 °C in the coldest month (January) to 28 °C in the warmest month (July) (Xu & Xu, 2016). Long-term annual precipitation in the area was reported to be about 1,477 mm, ranging from 159 mm in July to 81 mm in October.

Y. J. Xu (✉)
School of Renewable Natural Resources, Louisiana State University, Baton Rouge 70803, USA
e-mail: yjxu@lsu.edu

N. S. N. Lam
Department of Environmental Sciences, Louisiana State University, Baton Rouge 70803, USA

K. Liu · K. Xu
Department of Oceanography & Coastal Sciences, Louisiana State University, Baton Rouge 70803, USA

Y. J. Xu · K. Liu · K. Xu
Coastal Studies Institute, Louisiana State University, Baton Rouge 70803, USA

W. Zhang and H. de Vriend (eds.), *Delta Sustainability*,
https://doi.org/10.1007/978-981-97-7259-9_11

Fig. 1 The modern Mississippi River Delta formed over the past 7,000 years with two major outlets: the Mississippi River mainstem (MR mainstem) and the Atchafalaya River

1.1 Riverine Systems, Discharge, Sediment Load

The Mississippi River from its headwaters to its mouth at the Northern Gulf of Mexico is a 10th-order stream system. The river discharges runoff from several large tributary basins, including the Missouri River, Upper Mississippi River, Ohio River, Arkansas River, and Lower Mississippi River Basins (Fig. 2). The Missouri River Basin has the largest land mass (44.6%), followed by the Upper MR Basin (16.5%), the Arkansas River Basin (15.4%), the Ohio River Basin (15.2%), and the Lower MR Basin (8.3%) (Meade & Moody, 2010; Rossi et al., 2009).

The Mississippi River currently has two outlets, the river's mainstem and its distributary—the Atchafalaya River, which carries about 24% of the total Mississippi River flow (Wang & Xu, 2020). Over the past several decades, the MR discharged on average a combined volume of 680 km^3 freshwater annually from the two rivers into the Gulf of Mexico (Joshi & Xu, 2015; Rosen & Xu, 2015). A clear increasing trend of discharge occurred from 1931 to 2017 (Reiman & Xu, 2019), which is due

Fig. 2 Network of major streams and rivers in the Mississippi-Atchafalaya River System. Much of the river flow is contributed by the Ohio, and the Upper and Lower Mississippi Rivers, while much of the sediment is contributed by the Lower Mississippi and Missouri Rivers

mainly to an increase in precipitation in the river basin (Yin et al., 2023). The Ohio River contributes most (47%) to the river total discharge, followed by the Upper Mississippi (19%), Lower Mississippi (14%), Missouri (13%), and Arkansas (7%) (Yin et al., 2023).

From 1950 to 2017, the Mississippi River discharged on average 192 Mt sediment annually, to which the Lower Mississippi River contributed most (33%) followed by the Missouri (32%), Ohio (19%), Upper Mississippi (13%), and Arkansas (2%) (Yin et al., 2023). There is a clear decline in sediment load due mainly to the sharp reduction of sediment load in the Missouri River.

1.2 Wetlands and Ecosystems

The Mississippi River Delta (MRD) contains more than 11,000 km^2 of coastal and estuarine wetlands, which account for nearly 40% of the wetlands in the conterminous United States (Couvillion et al., 2017). These wetlands exhibit a distinct vegetational zonation, which is primarily controlled by a gradient of decreasing salinity landward from the coast. Locally, the salinity along this gradient is controlled by three main factors: proximity to the sea, elevation, and freshwater input (Ryu et al., 2022a). Thus, vegetation communities vary regionally and locally along this environmental gradient. Five wetland vegetation zones or ecosystems are recognized from the coast moving landward: (1) mangroves, (2) salt marshes, (3) brackish/intermediate marshes, (4) freshwater marshes, and (5) bottomland and swamp forests (Fig. 3).

The mangrove ecosystem occupies a narrow zone along the seaward edge of the land fringing the Gulf of Mexico. Only one species of mangrove—*Avicennia germinans* (black mangrove)—is present, and the trees are short-statured, generally less than 2 m tall. Palynological studies indicate that black mangroves migrated to southern Louisiana only about a century ago (Ryu et al., 2022b). Their population expanded significantly during the last few centuries, especially since the beginning of the 21st century, as a result of global warming (Yao et al., 2022a, 2022b), even though their distribution and abundance have been episodically impacted by winter freezes, hurricanes, and human activities (Cohen et al., 2021a, 2021b). Landward from the mangrove zone is a zone of salt marshes dominated by plants tolerant of high salinity

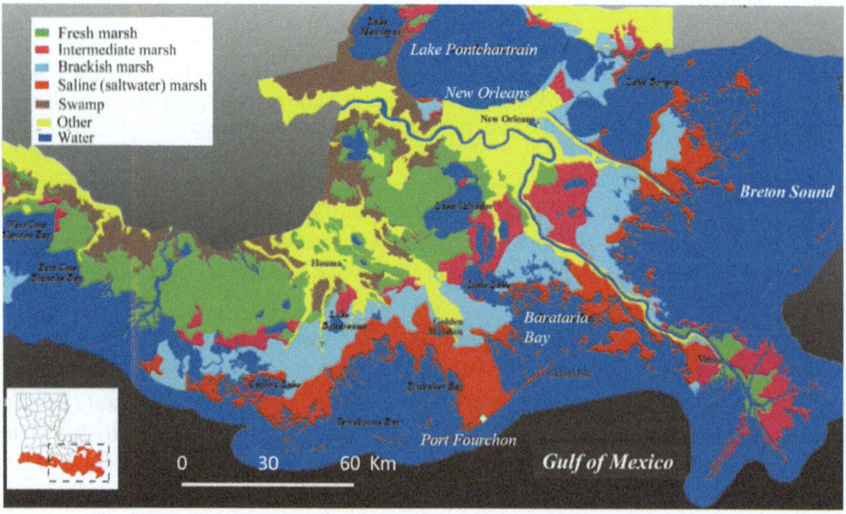

Fig. 3 Ecological zonation in the Mississippi River Delta. A south-to-north transect from Port Fourchon to Lake Pontchartrain encompasses five vegetation zones from a narrow belt of mangrove swamp to salt marshes, brackish marshes, intermediate marshes, fresh marshes, to bottomland and swamp forests. Modified from Sasser et al. (2014)

(10–35 ppt) such as *Spartina alterniflora* (smooth cordgrass), *Juncus roemerianus* (black rush), and some succulents like *Batis maritima* (saltwort) and *Salicornia spp* (pickleweed). This zone is subjected to tidal influences and often storm surges during hurricane events. Behind the salt marshes is a transitional zone of brackish and intermediate marshes, where salinity ranges from 2 to 18 ppt as a function of the interplay between tidal and fluvial influences. Common plants include *Spartina patens* (marshhay cordgrass), *Distichlis spicata* (saltgrass), and *Schoenoplectus americanus* (bullrush), which are often mixed with some broadleaved emergent aquatic plants depending on local hydrological and salinity conditions (Visser et al., 1998). The zone of freshwater marshes lies further landward from the intermediate marshes. Here the salinity is lower (0.5–5.0 ppt), ranging from oligohaline to freshwater as a result of the predominance of fluvial and pluvial input. Species diversity is much higher here than in the previous zones, with plant communities consisting of various mixtures of *Panicum hemitomon* (maidencane), *Zizaniopsis miliacea* (giant cutgrass), *Sagittaria lancifolia* (arrowhead), *Bacopa monnieri* (water hyssop), *Typha* spp. (cattails), and *Phragmites australis* (common weed), among others. The innermost zone on the Mississippi River deltaic plain is a forest ecosystem dominated by the wetland conifer *Taxodium distichum* (baldcypress) and various bottomland hardwood trees such as *Nyssa aquatica* (water tupelo), *Acer rubrum* (red maple), *Salix nigra* (black willow), *Liquidambar styraciflua* (sweetgum), *Fraxinus pennsylvanica* (green ash), and *Quercus* spp. (oaks). However, old-growth swamp forests dominated by baldcypress have been depleted due to human activities, especially logging, drainage alteration, and habitat destruction (Ryu et al., 2018, 2021a, 2021b).

Hitherto, little had been known about the Holocene history of these wetland ecosystems and how they evolved in response to global and regional environmental changes. This dearth of information was due to the total absence of palynological records from the Mississippi River Delta. Recently, six high-resolution multi-proxy paleoecological records along a south-to-north transect from the mangrove zone to the bottomland/swamp forest zone have been published that reveal the Holocene evolutionary history of Louisiana's coastal wetlands (Rodrigues et al., 2021; Ryu et al., 2021a, 2021b, 2022a, 2022b; Yao et al., 2022a, 2022b). It has been well established that the Mississippi River mouth has switched locations several times during the last 6,000 years, resulting in the formation of six delta lobes along the coast (Flocks et al., 2006) (Fig. 4). These processes of delta lobe switching drastically changed the delivery of freshwater and sediments at different times during the Late Holocene to different parts of the Mississippi River deltaic plain and coastal Louisiana. This became the primary external forcing agent causing ecological changes in Louisiana's coastal and deltaic wetlands during the Late Holocene.

However, the trends, sequence, and timing of ecological changes varied geographically along this environmental gradient from coast to inland. At a freshwater wetland site near Lake Pontchartrain along the innermost part of the transect, the environment was an oyster reef in an embayment along the Gulf of Mexico coast about 5000–4500 years ago. The progradation of the St. Bernard delta lobe south of the study site and the subsequent formation of Lake Pontchartrain caused the estuarine/marine environment to transition to a lake after 4000 years ago, and finally to a cypress

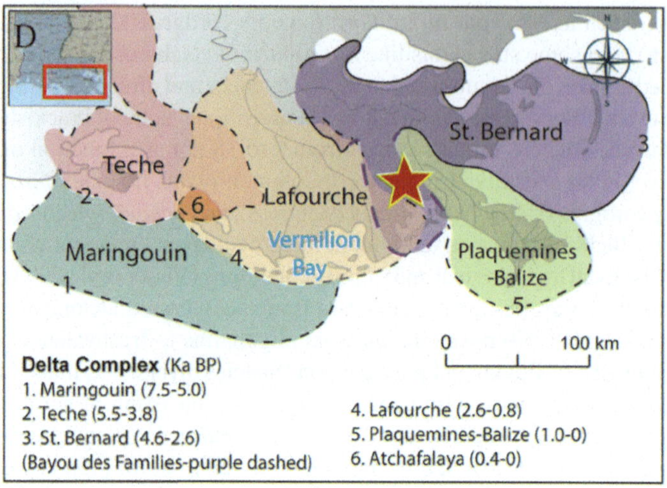

Fig. 4 The Mississippi River Delta complex and age ranges (Ka BP) of the six subdelta lobes (Figure adapted from Ryu et al., 2022b; age ranges based on Flocks et al., 2006). The red star denotes the location of Little Lake, a brackish marsh site along a transect of paleoecological study sites from Port Fourchon to Lake Pontchartrain discussed in this section

swamp after 2000 years ago (Ryu et al., 2021a, 2021b). Two sites from an intermediate marsh and a brackish marsh near Lake Salvador and Little Lake, respectively, elucidated the predominant control of the St. Bernard deltaic cycle on the Holocene ecosystem changes in the middle part of the transect in the Barataria Basin (Fig. 4). Both sites occupied an interdistributary bay with a mixture of marine and freshwater inputs around 3000 years ago. The progradation of the St. Bernard delta lobe, which was active about 4600–2600 years ago (Flocks et al., 2006; Frazier, 1967; Törnqvist et al., 1996), delivered abundant freshwater and sediments to the sites, changing the environments to a freshwater lake, swamp, or marsh after about 2600 years ago. Increasing marine influence during the last 300 years due to relative sea level rise changed these sites into an intermediate marsh and a brackish marsh, respectively (Ryu et al., 2022a, 2022c) (Fig. 5). Three palynology-based multi-proxy records from a backbarrier lagoon (Bay Champagne) near Port Fourchon (Fig. 4) at the seaward end of the transect provide an 8000 year history of coastal ecosystem changes in the mangrove zone. From 8000 to 6500 yr BP the site was a freshwater lake at some distance inland from the coast (Rodrigues et al., 2021). With increasing sediment supply from the St. Bernard (4600–2600 yr BP) and Lafourche (2600–800 yr BP) subdelta lobes, the lake gradually changed to a *Taxodium-Nyssa* freshwater swamp and then a *Typha-Bacopa* freshwater marsh from about 4000 to 700 yr BP (Ryu et al., 2022b). Continued relative sea level rise and shoreline retreat during the past 700 years have rendered the site closer and closer to the coast, allowing more frequent occurrence of seawater intrusion and storm surge impacts to the site. Increasing salinity transformed the freshwater wetland into an intermediate/brackish marsh after 700 yr BP and, eventually, a salt marsh after 200 yr BP. (Ryu et al.,

2022b). The arrival of *Avicennia germinans* about a century ago led to the formation of the modern black mangrove-dominated ecosystem along the Mississippi River deltaic coast in Louisiana (Ryu et al., 2022b; Yao et al., 2022a, 2022b) (Fig. 5).

The Holocene history of wetland ecosystem development is driven by a number of external controlling agents. Geological factors, particularly the Mississippi River delta cycle, played an overriding role in wetland ecological changes during the early part of the Late Holocene (~5000–2000 yr BP). However, other environmental factors such as sea level rise, climate variability, shoreline retreat, hurricane impacts, and anthropogenic effects played an increasingly important role during the past 1000–2000 years, especially the past several centuries. The last factor, human activity, is especially important during the past few centuries. The environmental impacts of human activities have been evidenced by peaks of microscopic charcoal as early as 3000–2000 years ago, suggesting the use of fire for hunting or cooking by the Tchefuncte people around Lake Pontchartrain. Human activities and their impacts on wetland ecosystems intensified during the past few centuries, especially in the form of tree logging, forest clearance, levee building, canal and pipeline construction, drainage alteration, and habitat destruction, resulting in accelerated coastal land loss, increasing salinity, and ecological succession in Louisiana's wetland ecosystems (Ryu et al., 2018, 2021a, 2021b).

The combined effects of these external driving agents, especially sea level rise, land subsidence, shoreline retreat, enhanced hurricane activity, and human impacts, have led to a serious, long-term environmental problem for the estuarine and deltaic wetlands of Louisiana: coastal land loss. From 1932 to 2016, coastal Louisiana has lost a total of about 5,000 km^2 of wetlands, whereby coastal wetlands have turned into open water areas. This is equivalent to a loss of approximately 25% of the 1932 land area, at an average rate of about 60 km^2 per year (Couvillion et al., 2017). This is the most important environmental challenge faced by human societies living in the Mississippi River Delta region.

2 Socio-economic Characteristics

2.1 Population

Unlike most river deltas in the world, the Mississippi River Delta has been experiencing persistent population loss along the coastal, low-lying areas in the past two decades. In contrast, considerable population growth has been observed in the adjacent inland areas. This coastal-inland migration trend is likely due to increasing flood risks caused by slow-moving sea-level rise and land subsidence as well as frequent fast-moving hurricanes experienced in the region. From 2000 to 2021, Louisiana has experienced direct hit from major hurricanes including Katrina (2005), Rita (2005), Gustav (2008), Isaac (2012), Laura (2020), and the latest, Ida (2021). An estimate produced by the National Weather Service suggests that a tropical cyclone makes

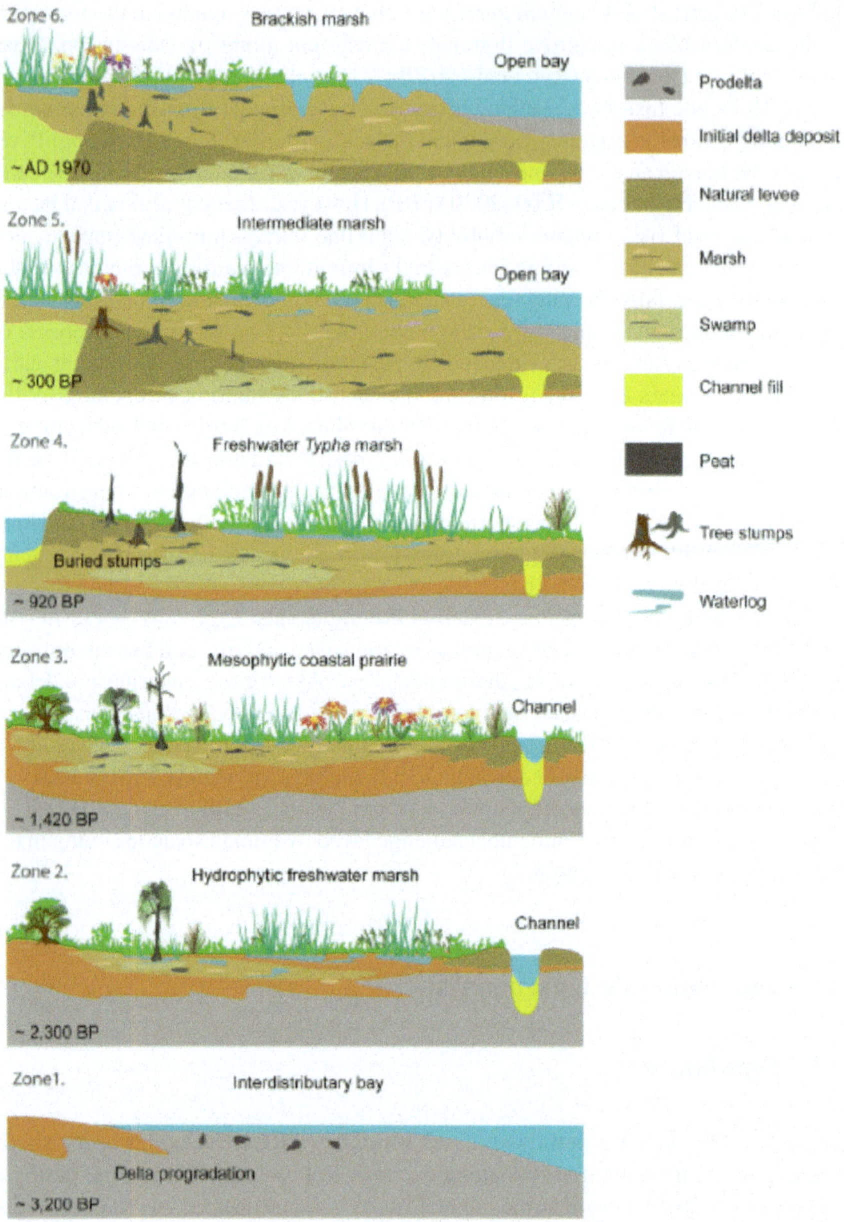

Fig. 5 Concept model showing stages of ecosystem evolution at Little Lake over the last 3200 years (after Ryu et al., 2022b). The location of Little Lake is denoted by the red star in Fig. 4

landfall along the Louisiana coastline about two times every three years, and a hurricane makes landfall once every 2.8 years (Roth, 2023). With global warming, these estimates could be even higher.

Population growth in a region is considered an indicator of recovery and resilience against natural and man-made disasters (Lam et al., 2016). We can examine the trend of population growth or decline in the Lower Mississippi River Basin (LMRB) to understand how population changes in the region and their interactions with the natural environment change over time (Cai et al., 2016; Lam et al., 2018a, 2018b, 2018c). The Lower Mississippi River Basin includes the watersheds of the Mississippi River and its main tributary Atchafalaya River, and it encompasses 26 parishes (in Louisiana, counties are called parishes, see Fig. 6). The Basin includes three major cities including New Orleans, Baton Rouge, and Lafayette. If we consider Lake Pontchartrain as an approximate boundary where parishes north of the Lake are considered the North, or inland, and parishes south of Lake Pontchartrain are called the South, or coastal, we can observe from the Census a distinct trend of population migrating from the coastal South to the inland North.

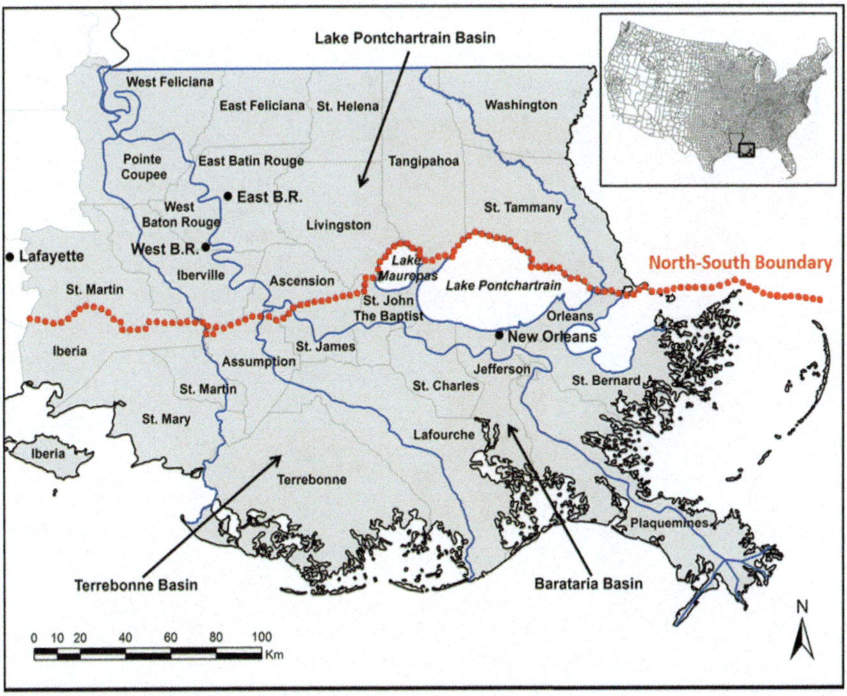

Fig. 6 The Lower Mississippi River Basin (LMRB)—parishes north of the N–S boundary experienced high population growth whereas parishes south of the boundary suffered significant population decline between 2000 and 2020. *Map source* Lam et al. (2018a)

According to the U.S. 2010 Census statistics, the total population in the Basin area was close to 2.7 million. During 2000–2010, two parishes in the North experienced very high population increase, with a rate close to 40% each (Ascension and Livingston parishes). In contrast, two parishes in the South had high population decrease, with St. Bernard Parish leading with a rate of −47%, followed by Orleans Parish (where the City of New Orleans is located) with a rate of −29%. Cumulatively for the last 20 years (2000–2020), Louisiana as a state had a population increase of only 4%, compared with the national average of 19% (authors computed the percentages using the statistics from United Nations-World Population Prospects; https://www.macrotrends.net). Figure 7 maps the cumulative population change rate between 2000 and 2020 for all parishes in Louisiana, which shows that the same trend of population declines in the coastal areas continued.

There are many reasons for population decline in a region. Assuming other things being constant, out-migration is the major contributor to population decline. Migration decisions are complex, and abundant literature has addressed the issue. In the MRD region, we recently analyzed the migration consideration factors based on a 2017 telephone survey of 1,125 residents in the region (Correll et al., 2020). Our study found that at least 21.5% of respondents considered moving away from the

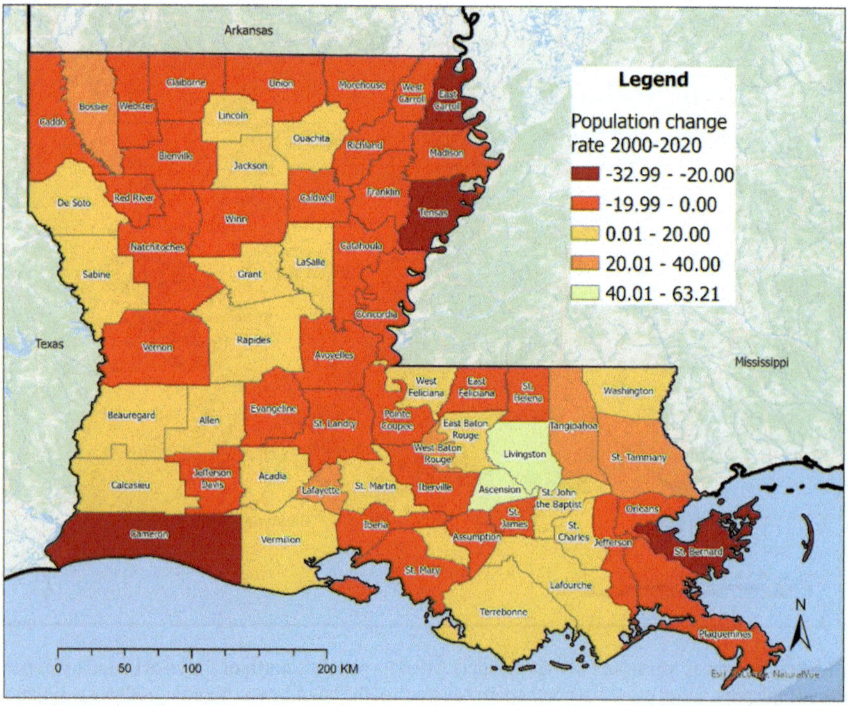

Fig. 7 Population change rate by parish in Louisiana between 2000 and 2010. *Data source* U.S. Census, map produced by Wang (2022)

flood-prone deltaic areas, and they were mostly renters and younger, had experienced flooding before, and were less satisfied with their current living condition. As expected, flood risk was found to be an important deciding factor. However, it was not considered as the most important factor. Instead, economic opportunities had a greater effect on respondents' desire to move or stay. The findings provide useful information and insights into the planning and management of a flood-prone region and whether managed retreat or other mitigation measures would be effective.

Most recently in 2021, Hurricane Ida made landfall as a Category 4 hurricane near Port Fourchon in Lafourche Parish on August 29, the same exact date that Hurricane Katrina hit New Orleans 16 years ago (2005). In addition to causing at least 30 deaths in Louisiana and unpredictably, 50 more deaths in the northeastern states due to torrential rain and flooding (PBS, 2021), the storm knocked down the power grid of the entire City of New Orleans and its surrounding region, with some areas without electricity for over two weeks. This disastrous event has led to extensive outmigration within a year, losing nearly 40,000 people between 2021 and 2022, and they were mostly from the hard-hit New Orleans and coastal areas (The Advocate, 2023). These outmigration responses to the aftermath of Hurricane Ida by the affected communities are very similar to that of Hurricane Katrina 16 years ago (Lam et al., 2009, 2012; LeSage et al., 2011a, 2011b). While this new extensive outmigration could be short-term, the reasons that made people leave are revealing. First, the collapse of the homeowners insurance industry after Ida and the skyrocketing cost of homeowners insurance and flood insurance (e.g., annual premium raised from \$5,000 to \$11,000) are the major push factors that have driven many middle-class homeowners out of the region (The Advocate, 2023). Second, the slow payout from the insurance company for those who had been insured is another reason. Third, in the coastal area (Terrebonne Parish), the fluctuations of the oil and gas industry, in other words, the uncertain economic opportunities in the future, remain to be the biggest consideration of whether to relocate or stay.

2.2 *Major Industries (Energy/Petrochemical Industry, Transportation, Agriculture, Fisheries)*

The Mississippi River Delta is a vital economic region, supporting numerous economic functions not only for the region but also for the rest of the United States. In brief, Louisiana supports the U.S. economy by providing \$9.3 billion per year in tourism. Louisiana has nearly 30% of the commercial fishing landings of the continental United States, with an estimate of \$1.8 billion annual impact from recreational fishing. Louisiana also has five of the nation's 12 largest shipping ports by cargo volume, carrying a fifth of all waterborne commerce in the United States (CPRA, 2012). The delta region has been estimated to contribute a total GDP of 130 billion annually (Haq & Milliman, 2023).

Oil and gas industries have been a dominating economic force in the region for a long time. Recent statistics from the U.S. Energy Information Administration (2022) shows that as of May 2022, Louisiana ranks third in the nation in natural gas production and proven reserves among the states. Louisiana accounts for about 9% of U.S. total marketed natural gas production and holds about 8% of the nation's natural gas reserves. Louisiana's 14 oil refineries account for nearly one-fifth of the nation's refining capacity and can process about 3.2 million barrels of crude oil per day. In 2021, Louisiana shipped 52% of the nation's liquefied natural gas exports and about 13% of its coal exports. New Orleans is the third-largest U.S. coal exporting port. Ports in Louisiana are a major passage connecting the U.S. with the world.

However, supporting the industry has a cost to the fragile wetlands in the MRD. In particular, critical energy infrastructure such as pipelines and canals built for transporting materials in and out of the Delta has led to a more fragmented landscape, hastening the land loss and increasing the threat of flooding in inland areas as water could rush through the canals to inland urban areas during storm surges and hurricanes (Lam et al., 2018c). Figure 8 shows the major pipelines built to support the oil and gas industries crisscrossing the fragile wetlands in coastal Louisiana. At the same time, much of this critical energy infrastructure, such as refineries, petrochemical plants, pipelines, and other support facilities, lie directly in the coastal zone and are themselves at risk to coastal land loss and sea-level rise (Dismukes & Narra, 2015). Given the increasing awareness and support of clean energy production in the coming years, the uncertain future of the oil and gas industry in the MRD remains a great concern to the communities. These complex interactions between economic and environmental factors affect the sustainable future of the MRD. There is an urgent need for a creative sustainability plan so that the MRD can continue supporting the economic functions that link the U.S. with the world while minimizing its loss of land and ecosystem services.

3 Current Issues

3.1 Sea Level Rise and Land Loss

Local sea level can be controlled by both global sea level and local subsidence. Hamlington et al. (2022) performed detailed observation-based trajectory analyses and found that the Gulf coast has experienced a high rate of sea level rise and acceleration during the satellite-altimeter era that leads to a median estimate of increasing sea level from 2020 to 2050 that exceeds 30 cm. Sweet et al. (2022) predicated both mean and extreme water levels along U.S. coastlines and found that the relative sea level rise will be 1.3 m, 2.8 m, and 4.5 m in low, intermediate, and high scenarios, respectively, in the year 2150. Moreover, Dangendorf et al. (2023) recently reported a sea level acceleration in tide gauge records along the Southeast and Gulf coasts of the U.S. that has led to rates (>10 mm/yr since 2010) that are unprecedented in last 120 years.

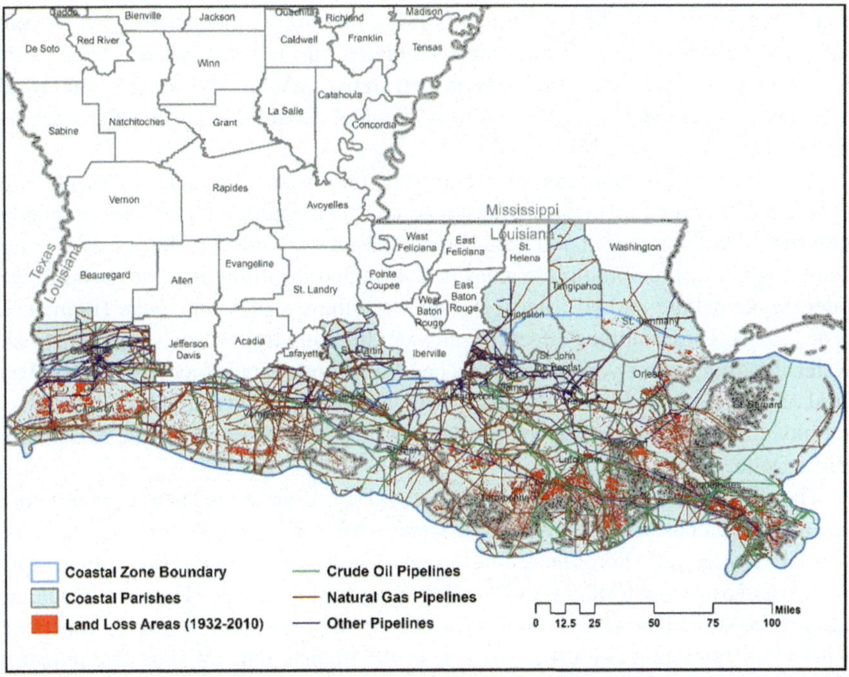

Fig. 8 Pipelines supporting oil and gas industries crisscrossing fragile wetlands in coastal Louisiana, creating fragmentation and hastening land loss. *Map source* Lam et al. (2018a)

High subsidence rates and increasing global sea level rate together have led to the highest land loss rate of Louisiana coast among all U.S. coastlines. Couvillion et al. (2011) did a trend analysis and reported a wetland loss rate of ~16 mi^2 per year from 1985 to 2010. Such extremely high land loss has greatly impacted coastal sediment transport, morphology evolution, ecology and human population and this complicated issue has become a grand challenge for the State of Louisiana.

3.2 Hurricanes and Storm Surges

The Mississippi River Delta is highly susceptible to destructive impacts by tropical cyclones (hurricanes). According to NOAA, the Mississippi River Delta has an average annual probability of 14.3–12.5% of being directly hit by a hurricane of category 1–5 intensity, or an average return period of 7–8 years, which is the highest on the northern Gulf of Mexico coast (excluding southernmost Florida). For category 3–5 hurricanes (the most destructive according to the Saffir-Simpson hurricane intensity scale), the average annual probability is 5.0% and the corresponding average return period is 20 years, also the highest on the northern Gulf of Mexico coast. According to predictions from a dynamic probability model (Elsner & Bossak, 2004), these landfall probabilities are even higher during years characterized by the co-occurrence of La Nina conditions and a negative NAO index, which would result in a much higher risk for the Mississippi River Delta region than years characterized by El Nino and positive NAO.

Intense hurricane landfalls are accompanied by high storm surges (Needham & Keim, 2012), which can cause widespread coastal flooding with extensive inundation and saltwater intrusion into Louisiana's coastal wetlands. In addition, hurricanes can bring heavy precipitation to areas further inland, resulting in widespread fluvial flooding of deltaic and riverine wetlands from the overflowing rivers (Bianchette et al., 2016). The combined effects of saltwater flooding from storm surges and freshwater flooding from rivers and bayous can cause compound flooding events (Wahl et al., 2015; Zscheischler et al., 2018), which can profoundly affect the hydrology, salinity, and biogeochemistry of coastal and estuarine wetlands in the Mississippi River Delta.

The effects of hurricanes and their associated storm surges and heavy precipitation can affect the coastal wetlands in Louisiana in different ways. Hurricanes can cause coastal erosion and shoreline retreat on one hand, and sediment deposition in the coastal and estuarine wetlands on the other. Several studies have documented that the Caminada-Moreau headland on the west side of the Mississippi River bird-foot delta, where Port Fourchon is located, has one of the highest rates of shoreline retreat in North America—an average of 12 m per year over a 35-year period from 1983 to 2018 (Dietz et al., 2018, 2022; Penland & Ritchie 1979, Penland & Suter 1988, Penland et al. 2003, 2005) (Fig. 9). Most significantly, on a decadal timescale, the average shoreline retreat rate (~21 m/yr) was found to be nearly five times higher during the hurricane-active period (1998–2013) than the average rate (~4.7 m/yr) during the hurricane-quiet periods (1983–1997; 2015–2018), suggesting that hurricanes are a major factor causing accelerated shoreline retreat in Louisiana (Dietz et al., 2018) (Fig. 10).

Hurricanes are also an important agent of sediment deposition in the coastal and alluvial wetlands on the Mississippi River deltaic plain. Field measurements made after the landfalls by Hurricanes Katrina and Rita suggested that significant amounts of inorganic sediment were deposited in the coastal wetlands of Louisiana due to storm surges processes (Turner et al., 2006), although the magnitude of these storm surge deposition estimates was in dispute (Törnqvist et al., 2007). A pair of consecutive hurricanes, Gustav and Ike, were shown to have collectively deposited up to 21 cm of sandy sediments on the bottom of a backbarrier lake (Bay Champagne) near Port Fourchon, Louisiana (Dietz et al., 2021, 2022; Liu et al., 2011). However, Dietz et al. (2021) have shown that the lower part of this storm deposit was primarily caused by Hurricane Gustav's high storm surge from the sea, whereas the upper part of this storm deposit was primarily caused by fluvial processes with finer sediments carried by freshwater moving seaward from the wetlands behind the lake due to Hurricane Ike's heavy precipitation. This study underscores the importance of hurricane-induced fluvial flooding as an important agent of sediment accretion not

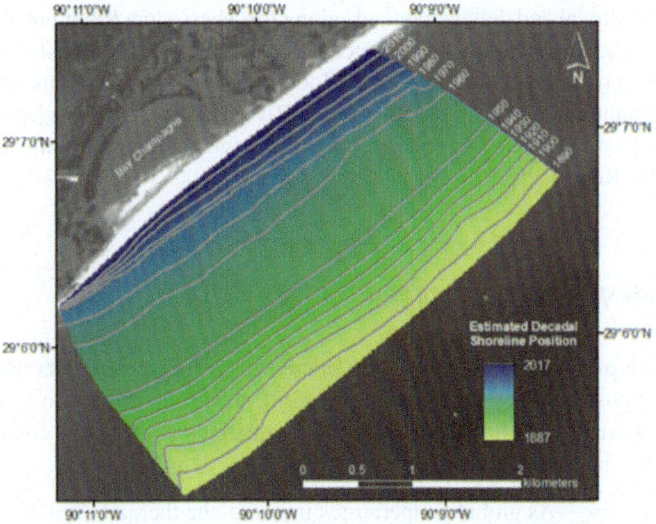

Fig. 9 Reconstructed decadal shoreline positions at Bay Champagne near Port Fourchon, Louisiana, on the west side of the Mississippi River bird-foot delta, from 1887 to 2017 showing rapid shoreline recession. These shoreline positions were determined through the use of a combination of historic maps, photographs, and Landsat imagery (From Dietz et al., 2022)

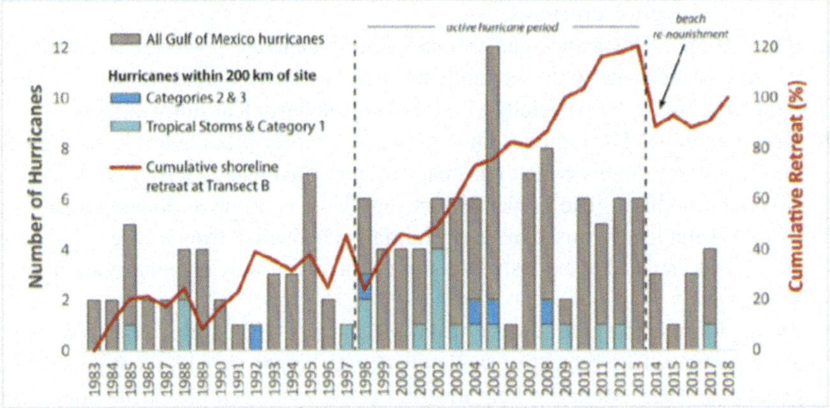

Fig. 10 Relationship between cumulative shoreline recession rates (red line) and annual number of tropical cyclones in the Gulf of Mexico (vertical bars) at tropical storm and category 1 strength (light blue) and category 2 and 3 strengths (dark blue) at Bay Champagne, Louisiana, during 1983–2018. The shoreline recession rates are higher during a more active hurricane period (1998–2013) than during the more quiet period before and after, suggesting that hurricane activity is a major factor affecting coastal erosion in the Mississippi River Delta (From Dietz et al., 2018)

only in the alluvial and deltaic lowlands along the Mississippi River but also in estuarine and backbarrier coastal environments that are contemporaneously subjected to storm surge processes. Thus, compound flooding has become a serious threat facing coastal communities living in the Mississippi River Delta as hurricanes are becoming stronger and wetter as a consequence of anthropogenic climate change (Bevacqua et al., 2019; Kossin, 2018; Sobel et al., 2016).

3.3 *Widespread and Frequent Floods*

Floods have produced serious disasters in the Mississippi River Delta. Because the region is topographically flat and near the Gulf of Mexico, flash floods, backwater floods, and storm surge floods are frequent. Some of the main factors contributing to floods in this region include:

- Sea-level rise—As global temperatures increase, the thermal expansion of ocean waters and melting of polar ice caps cause sea levels to rise. This leads to higher tides and more frequent and severe flooding along the Louisiana coast.
- Subsidence—Much if the MRD naturally subsides over time, but is now starving from sediment replenishment. Subsidence is further exacerbated by human activities, such as oil and groundwater extraction, and the construction of levees and other flood control structures.
- Storm surge—Hurricanes and tropical storms generate powerful storm surges, which push seawater onto the land and inundate coastal areas. Louisiana's low-lying coast makes it especially vulnerable to such tropical storm surges.
- Heavy rainfall—The region is known for its high precipitation levels, and intense rainfall events can overwhelm drainage systems and lead to flash floods.
- Riverine flooding—The Lower Mississippi River is prone to flooding as a result of high-water levels caused by heavy rainfall or snowmelt from upstream drainage areas. This can lead to extensive flooding in nearby low-lying areas without levee protections.
- Wetland loss—The coastal area of the Mississippi River Delta has lost a significant portion of its protective wetlands due to human development and natural processes. These wetlands help absorb excess water and buffer the coast from storm surges, so their loss makes the region more susceptible to flooding.

The Mississippi River Delta is a national energy hub densely distributed with petrochemical plants. There is concern over flood-triggered chemical hazards.

3.4 Nutrient Enrichment and Eutrophication

The Mississippi River discharges a gigantic quantity of nutrients into the Northern Gulf of Mexico, strongly affecting chemical and biological domains in the Mississippi River Delta. From 1970 to 2017, the river delivered annually 1.15 million tons of total nitrogen (TN) and 113,300 tons of total phosphorus (TP), most of which (97%) was contributed by the Upper Mississippi and Ohio Rivers (Yin et al., 2023). The riverine TN load showed a marginal decline, but there was a significant increase in P load. The large river inflow of nutrients has caused eutrophication of the coastal water offshore of the Louisiana coast. In addition to the nutrient delivery, the Mississippi River discharged, on average, 3.95 million tons of dissolved organic carbon and 12.25 million tons of dissolved inorganic carbon annually to the Northern Gulf of Mexico in recent years (Reiman & Xu, 2019). The large quantity of carbon transport can play an important role in aquatic food webs and coastal ecosystem functioning. This is especially the case in flood years, for instance, the 2019 Mississippi River mega flood (Xu & Xu, 2022).

Eutrophication and hypoxic zone occurrence have been observed in the Northern Gulf of Mexico during each summer since the mid-1980s. The large-scale, chronic hypoxic zone was not present prior to the extensive application of chemical fertilizer in the 1950s (Rabalais et al., 2007), highlighting the significance of agricultural management in the Mississippi River Basin on the coastal ecosystem in the GOM (Turner & Rabalais, 2003). In 1997, US EPA established the Mississippi River/Gulf of Mexico Hypoxia Task Force to develop effective strategies for addressing nitrogen and phosphorus pollution problems in the Mississippi River Basin. In 2015, this task force established a goal to reduce nitrate exports by 20% over the next 10 years and the hypoxic zone below 5,000 km^2. Despite the federal and state efforts over the past two decades, hypoxia continues occurring in the Northern Gulf of Mexico (Fig. 11) and the eutrophication issue is far from fixing.

4 Coastal Master Plans to Solve the Problems

The Louisiana's Coastal Protection and Restoration Authority (CPRA) published the first integrated restoration and flood protection plan for coastal Louisiana in 2007, *Louisiana's Comprehensive Master Plan for a Sustainable Coast* (here defined as "Coastal Master Plan"—CMP) (CPRA, 2007). This plan has been mandated by the Louisiana Legislature to be updated regularly (every five years initially, every six years since 2017). In 2012 CPRA issued the second iteration of Louisiana's Comprehensive Master Plan (CPRA, 2012). An updated version was then issued in 2017 (CPRA, 2012), and the draft of the newest version of 2023 was released in early 2023.

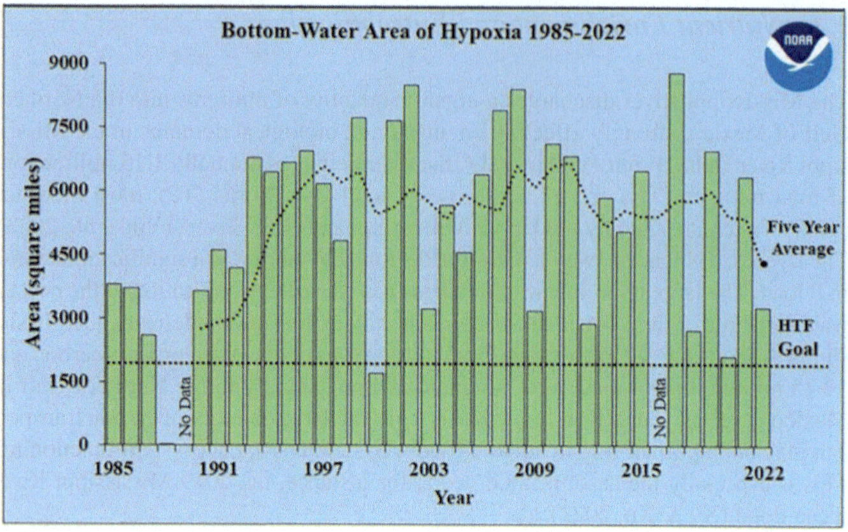

Fig. 11 Long-term size of the hypoxic zone (green bars) measured during the ship surveys since 1985, including the target goal established by the Mississippi River/Gulf of Mexico Watershed Nutrient Task Force and the 5-year average measured size (black dashed lines). *Source* LUMCON/LSU/NOAA

4.1 *Efforts Led by Government Agencies*

The main purpose of the CPRAs is to develop a list of projects, prioritized based on scientific studies and after extensive engagement with stakeholders and communities (CPRA, 2012). In the CPRAs before 2017, most of the projects were structural projects, with very few of them focusing on non-structural measures. For example, from 2007 to 2017, the State of Louisiana completed or funded 135 construction projects, which have benefited over 36,000 acres of land, improved 282 miles of levees, and constructed or is in the process of constructing over 60 miles of barrier islands and berms to help ease the land loss problem. However, there are no documented progresses or impacts of the nonstructural projects.

The 2017 CPRA outlines a $50 billion investment designed to build and maintain land, reduce flood risk to communities, and provide habitats to support ecosystems (CPRA, 2012). A significant shift of emphasis in the 2017 CMP from previous plans is the increased focus on people and communities (Lam et al., 2018a). The Flood Risk and Resilience Program within the Plan focuses on proactive investments to make the coastal communities more resilient. Specifically, the program recommends floodproofing more than 1,400 structures, elevating more than 22,400 structures, and acquiring of approximately 2,400 structures in areas that are most at risk. To achieve this goal, the program includes conducting a coastal flood risk vulnerability analysis, defining nonstructural project areas, prioritizing projects, and facilitating the implementation of projects. In summary, the CMP dedicates $19 billion for structural protection and $6 billion for nonstructural risk reduction. It was estimated that these projects will reduce expected annual damage by $8.3 billion by year 50 as compared to the scenario of "Future Without Action", and they are expected to pay for themselves three times over the course of implementing the plan (CPRA, 2012).

However, the impact of these nonstructural projects remains to be seen. The issue of population declines and long-term sustainability in the region has not been adequately addressed. As already mentioned above, a key question remains: are we restoring land that nobody will want to live in? Addressing this question will require new thinking and new social, environmental, and engineering designs to support the coastal-inland migration trend. For example, could some of the resources be diverted to improve transportation between coastal and inland areas to enable continued economic activities on the coast while securing a safer home base in the inland areas, or could the houses be made smaller but more resistant to storms to save cost and minimize the impacts? We need studies to model and project the various mitigation measures while improving the economic opportunities and human conditions. We also need teams of interdisciplinary scholars, engineers, and stakeholders to map out new and creative strategies to address the long-term sustainability problem in the region.

4.2 Restoration Projects

The CPRAs clude a portfolio of projects, including sediment diversions, marsh creation, barrier island restoration, landbridge, structural and non-structural risk reduction projects and others. Diversion of Mississippi River water and sediment into the adjacent basins to mimic natural deltaic land building processes is a central element of the CPRAs. River diversion projects are generally defined as using "*channels and/or structures to divert sediment and fresh water from the Mississippi and Atchafalaya Rivers into adjacent basins*" (CPRA, 2007). Of this $50-billion 50-year CMP, ~$5.1 billion dollars is proposed for sediment diversion projects in 2012 CMP. The Mid-Barataria Diversion, for instance, projected to cost approximately $1 billion, was awarded "fast track" status by the U.S. White House, which will streamline permitting and reduce the time for project implementation (CPRA, 2012). The newest estimate of this project, however, is more than $2 billion in the year 2023.

To maximize the net land building in MRDP, Xu et al. (2019a, 2019b) proposed that strategies should be focused on (a) enhancing river sediment delivery, (b) increasing sediment retention in receiving basins and (c) decreasing sediment erosion. In terms of river sediment delivery to coastal areas, dam removal and associated sediment bypassing in the Mississippi basin are promising, but long-term federal regulations on dam removals and cooperation among many states throughout the Mississippi Basin are needed (Kemp et al., 2016).

Several approaches can be used to increase sediment retention in receiving basins and minimize erosion in coastal bays and estuaries. Multiple earthen dike structures, called Sediment Retention Enhancement Devices (SREDs, Fig. 12), have been used

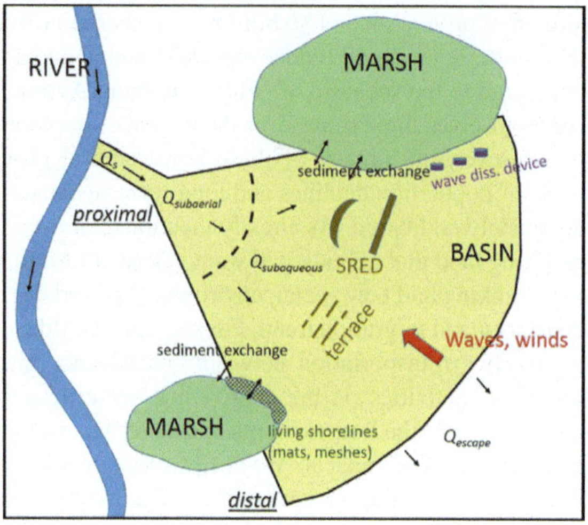

Fig. 12 Processes and sediment budget in a receiving basin of sediment diversion. Q_s is diverted sediment into the basin, $Q_{subaerial}$ is the deposited fluvial sediment as land and $Q_{subaqueous}$ is the deposited fluvial sediment under water. Q_{escape} is the sediment leaving the defined basin area. SRED is Sediment Retention Enhancement Device (From Xu et al., 2019a, 2019b)

in the West Bay diversion on the modern Mississippi bird-foot delta (Balize, Fig. 4) for multiple years and seem to be an effective device. Man-made structures and artificial fragmentation of large receiving basins decrease bay sizes and fetches, thus inhibiting wave development. Terraces are multiple sets of small parallel structures that have been used in Louisiana bays to improve sediment retention (Fig. 12).

Both hard infrastructure (e.g., groins, levees, jetties, breakwaters) and nature-based infrastructure (e.g., oyster reefs, living shorelines) can play a protective role of marsh and bay edges (Xu et al., 2019a, 2019b). Living shorelines can provide protection while also maintaining connectivity between wetlands and adjacent estuaries. The importance of maintaining the connection of the marsh with the estuary is critical as the use of the marsh edge for nursery habitat has been well studied (Baltz et al, 1993). More studies should be conducted on non-rock shoreline protection technologies, such as light-weighted mats and meshes, sediment collection system, wave attenuation devices, and erosion control system. Xu et al. (2019a, 2019b) reviewed multiple diversion studies and found that building diversion projects in more inland areas tend to yield higher sediment retention rates (Fig. 13).

The budget allocations of various coastal protections and restoration projects have been evolving in the past 20 years. Table 1 is used to compare the budget in various CPRAs. Marsh creation has been the highest budget category all the time in three CPRAs but has decreased from 20 billion in 2012 CPRA to 16 billion in 2023 CPRA. Structural and non-structural protections were the next two high budget categories; structural has been always higher than non-structural protection in all three CPRAs. The budget of sediment diversion was 3.8, 5.1 and 2.7 billion in 2012, 2017 and 2023 CPRAs, respectively. It should be noted that a new budget category landbridge of 2.9 billion is listed in 2023 CPRA.

Wiegman et al. (2018) modeled the combined impacts of sea level rise, oil price and management strategy on the cost of coastal protection and restoration projects.

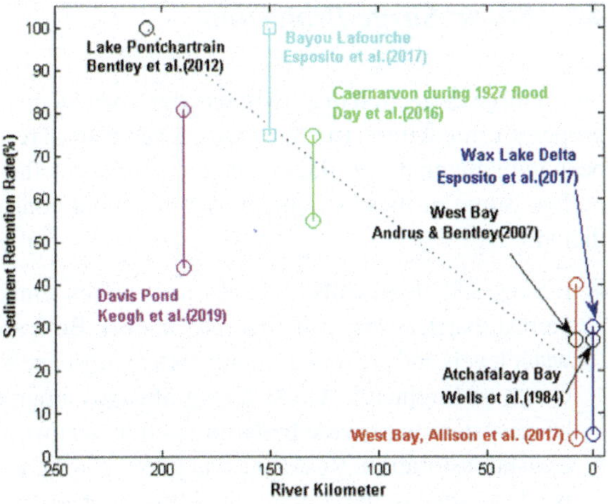

Fig. 13 The semi-quantitative relationship between river kilometer (RK) and sediment retention rate. Circles are for modern Mississippi and Atchafalaya systems whereas the red square is for a crevasse splay of Bay Lafourche system. The dashed line is from rate of Lake Pontchartrain to the middle point of Wax Lake Delta (From Xu et al., 2019a, 2019b; see references therein about diversion sites)

Table 1 Budget of 2012, 2017 and 2023 CPRAs

Projects ($ billions)	2012 CPRA	2017 CPRA	2023 CPRA
Marsh creation	20	17.8	16
Sediment diversion	3.8	5.1	2.7
Barrier Island	1.7	1.5	N/a
Hydrologic restoration	0.7	0.4	0.23
Structural protection	10.9	19.0	14
Non-structural protection	10.2	6.0	11
Landbridge	N/A	N/A	2.9
Ridge restoration	N/A	0.1	0.037
Shoreline protection	N/A	0.9	N/A
Programmatic Restoration	N/A	N/A	2.5
Other/integrated	3.0	N/A	0.64
Total	50	50	50

Because of the convergence of energy and climate trends, sustaining the Louisiana coast with hydraulic dredging will suffer declining returns on investment and marsh creation will likely become unaffordable in the mid to late 21st century, posing a great challenge to future CPRAs. Blum and Roberts (2009) predicted that significant drowning of MRDP is inevitable even if the river sediment loads are restored. Over the past decade, people have been moving from coastal towns and cities to more inland areas (Bailey et al., 2014; Hemmerling, 2018). The CMP considers a 50-year period but planning longer than 50 years should take into consideration population dynamics and the relocation/retreat of hundreds of thousands of Louisiana residents.

4.3 Storm Surge Protection

Storm surge protection in the Mississippi River Delta is crucial for safeguarding lives, property, critical infrastructure, and coastal natural resources. Various strategies have been implemented over the past century to reduce the impacts of storm surges in the region. Some of these strategies that are being considered in the new Master Plan include:

- Levees and floodwalls—Levees are earthen embankments, and floodwalls are vertical barriers made of concrete or steel. Both are designed to hold back rising water levels and prevent flooding. After Hurricane Katrina, the United States Army Corps of Engineers (USACE) significantly upgraded the Greater New Orleans levee system to provide better protection against storm surges.
- Coastal restoration: Restoring and preserving coastal wetlands and barrier islands is a key strategy in reducing the impacts of storm surges. These natural features act as buffers, absorbing wave energy and reducing the height of storm surges. Projects like the Louisiana Coastal Master Plan focus on the restoration of these vital ecosystems.
- Surge barriers—Surge barriers are structures designed to prevent or reduce the flow of storm surge into vulnerable areas. The Inner Harbor Navigation Canal (IHNC) Lake Borgne Surge Barrier, also known as the Great Wall of Louisiana, is an example of a surge barrier constructed to protect parts of New Orleans from storm surges.
- Home elevation: Raising the elevation of homes and other structures in flood-prone areas can help protect them from storm surge damage. This can be done through various methods, such as elevating the structure on pilings or using fill to raise the ground level.
- Flood-resistant building codes: Implementing stricter building codes that require new construction to be more resilient to storm surges can help reduce potential damage. This may include requirements for higher base flood elevation levels, stronger building materials, and better anchoring systems.
- Land-use planning and zoning: By restricting development in high-risk areas and encouraging it in safer locations, land-use planning and zoning can help reduce the exposure of property and infrastructure to storm surges.
- Evacuation planning—Establishing well-planned evacuation routes and procedures can help ensure that residents can safely and efficiently be evacuated the area when a storm surge threatens.
- Public education and awareness: Educating the public about the risks of storm surges and the steps they can take to protect themselves and their property can help build a more resilient community.

If implemented successfully in combination, these strategies can help reduce the impact of storm surges on coastal Louisiana and protect its residents and infrastructure. However, it is important to continue investing in research, infrastructure improvements, and coastal restoration to stay ahead of the increasing risks associated with climate change.

4.4 Maintaining Sustainability

Like many deltaic regions in the world which are faced with increasing threats from sea level rise, natural disasters, and depleting natural resources, a pressing question facing the MRD communities is: will the system be sustainable? Persistent land loss and population decline over the past two decades are the key issues. There have been numerous studies addressing the land loss problem, whereas studies on the population decline in the region are relatively scanty. Many of the previous studies have focused on the natural system, such as developing methods and guidelines for wetland restoration, sediment diversion, and shoreline protection, with little effort paid on examining and quantifying the coupling effect of the human component (Lam et al., 2018a). "Top-down" governmental efforts of coastal restoration, as reflected by the early Louisiana CPRAs (2007, 2012), have seldom addressed the underlying "bottom-up" migration of people from the coast to inland. The decoupling between top-down policies and bottom-up phenomenon could lead to another key question: Are we restoring land that nobody will want to live in? Thus, it is imperative to consider both natural and human systems and their coupling effects in developing policies and measures to increase resilience and sustainability of the MRD region (Cai et al., 2016, 2018; Lam et al., 2016, 2018a; Li & Lam, 2018; Xu et al., 2018; Zhang et al., 2022).

5 Conflicting Perspectives

The Mississippi River Delta is of high complexity in all aspects of natural and human systems. The governmental solutions to solve the current problems have been a subject of controversy due to their potential environmental, social, cultural, and economic impacts. Many government-promoted projects are designed to restore coastal wetlands by diverting sediment and freshwater from the Mississippi River into ambient eroding marshlands. However, there are a number of concerns and debates surrounding these projects. Below briefly summarized are some of these concerns and debates.

Environmental impacts: A number of researchers (e.g., Bargu et al., 2019; Turner et al., 2019) and environmental groups argue that river diversions could harm existing ecosystems by altering the salinity levels of estuaries and marshes, which can negatively affect fish, shellfish, and other wildlife populations. The sudden influx of freshwater into brackish and saltwater environments could disrupt the habitats, impacting the food chain and potentially leading to unintended consequences.

Insufficient sediment: Researchers argue that riverine sediment is insufficient to create new land or protect the sinking land (e.g., Blum & Roberts, 2009; Cortese & Fagherazzi, 2022). A large portion (nearly 40%) of sand sediment, the most important sediment fraction for deltaic building, is trapped in the river channel (Wang & Xu,

2018; Xu et al., 2019a, 2019b). Sea level rise and subsidence in the coming decades will likely override the marginal creation of land from river diversion.

Economic impacts: MRD is an essential part of Louisiana's economy and is of national importance, contributing approximately 25% of the U.S. total fisheries production. Changes in salinity levels near the coast due to river diversions can affect fish and shellfish populations, leading to potential economic losses for the fishing and seafood industries. Oyster beds, in particular, can be significantly impacted by changes in salinity, and fishermen who depend on these resources may face economic hardships (e.g., Lambert et al., 2021).

Costs and benefits: The construction and maintenance of river diversion projects can be extremely expensive, with critics arguing that the costs may outweigh the benefits of restored wetlands. They contend that resources could be better spent on alternative coastal restoration methods or other mitigation measures, e.g., programs to help adaptation to the processes of change in nature. The state government's will in building river diversion overwhelms private concerns (Randolph, 2022).

Landowners' concerns: Some landowners in areas affected by river diversions worry that their properties may be devalued or become uninhabitable due to flooding or erosion caused by the projects. They may also be concerned about the loss of access to their land or having their property expropriated for the construction of diversion structures. The premium of flooding insurance in some Louisiana communities has become cost-prohibitive to home owners.

Effectiveness: There is debate over the effectiveness of river diversions in restoring coastal wetlands. Some critics argue that the rate of wetland loss in Louisiana is so rapid that diversions may not be able to reverse the trend. Others contend that while diversions may help in some areas, they may not be the most effective solution in all cases. A study by Lewis and Ernstson (2019) found that the development of water infrastructures in the region produced cleavages in the region's body politic and eco-hydrology, generating disputes that threaten to slow or obstruct the plan's implementation.

Despite the controversies, proponents of the Mississippi River diversion projects argue that they are necessary for the long-term sustainability of Louisiana's coastal wetlands. They believe that the projects, if well designed and managed, can help protect the state's coast from the effects of sea-level rise, subsidence, and storm surges, while also preserving its unique ecosystems and supporting local industries. However, more evidence-based studies quantifying the long-term impacts of these projects are needed to help settle the controversy and inform the best-practice mitigation and adaptation strategies.

References

Bailey, C., Gramling, R., & Laska, S. B. (2014). Complexities of resilience: Adaptation and change within human communities of coastal Louisiana. In J. W. Day, G. P. Kemp, A. M. Freeman, & D. P. Muth (Eds), *Perspectives on the restoration of the Mississippi Delta* (pp. 125–140). Springer.

Baltz, D. M., Rakocinski, C., & Fleeger, J. W. (1993). Microhabitat use by marsh-edge fishes in a Louisiana estuary. *Environmental Biology of Fishes, 36*(2), 109–126.

Bargu, S., Justic, D., White, J.R., Lane, R., Day, J., Paerl, H., & Raynie, R. (2019). Mississippi River diversions and phytoplankton dynamics in deltaic Gulf of Mexico estuaries: A review. *Estuarine, Coastal and Shelf Science, 221*, 39–52.

Bevacqua, E., Maraun, D., Vousdoukas, M. I., Voukouvalas, E., Vrac, M., Mentaschi, L., & Widmann, M. (2019). Higher probability of compound flooding from precipitation and storm surge in Europe under anthropogenic climate change. *Science Advances, 5*, 5531–5537.

Bianchette, T. A., Liu, K. B., Qiang, Y., & Lam, N. S. N. (2016). Wetland accretion rates along coastal Louisiana: Spatial and temporal variability in light of Hurricane Isaac's impacts. *Water, 8*(1), 1.

Blum, M. D., & Roberts, H. H. (2009). Drowning of the Mississippi Delta due to insufficient sediment supply and global sea-level rise. *Nature Geoscience, 2*, 488–491.

Cai, H., Lam, N. S. N., Zou, L., Qiang, Y., & Li, K. (2016). Assessing community resilience to coastal hazards in the Lower Mississippi River Basin. *Water, 8*(2), 46.

Cai, H., Lam, N. S. N., Zou, L., & Qiang, Y. (2018). Modeling the dynamics of community resilience to coastal hazards using a Bayesian network. *Annals of the American Association of Geographers, 108*(5), 1260–1279.

CPRA. (2007). *I*ntegrated ecosystem restoration and hurricane protection: Louisiana's comprehensive master plan for a sustainable coast. Coastal Protection and Restoration Authority of Louisiana.

CPRA. (2012). Louisiana's comprehensive master plan for a sustainable coast. Coastal Protection and Restoration Authority of Louisiana.

Cohen, M. C. L., de Souza, A. V., Liu, K. B., Rodrigues, E., Yao, Q., Ryu, J., Dietz, M. E., Pessenda, L. C. R., & Rossetti, D. (2021a). Effects of the 2017–2018 winter freeze on the northern limit of the American mangroves. *Mississippi River Delta Plain. Geomorphology, 394*, 107968.

Cohen, M. C. L., de Souza, A. V., Liu, K. B., Rodrigues, E., Yao, Q., Pessenda, L. C. R., Rossetti, D., Ryu, J., & Dietz, M. (2021b). Effects of beach nourishment project on coastal geomorphology and mangrove dynamics in southern Louisiana, USA. *Remote Sensing, 13*, 2688.

Correll, R. M., Lam, N. S. N., Mihunov, V. V., Zou, L., & Cai, H. (2020). Economics over risk: Flooding is not the only driving factor of migration considerations on a vulnerable coast. *Annals of the American Association of Geographers, 111*(2), 1–16.

Cortese, L., & Fagherazzi, S. (2022). Fetch and distance from the bay control accretion and erosion patterns in Terrebonne marshes (Louisiana, USA). *Earth Surface Processes and Landforms, 47*(6), 1455–1465.

Dismukes, D. E., & Narra, S. (2015). Identifying the vulnerabilities of working coasts supporting critical energy infrastructure. *Water, 8*(1), 8.

Couvillion, B. R., Barras, J. A., Steyer, G. D., Sleavin, W., Fischer, M., Beck, H., Trahan, N., Griffin, B., & Heckman, D. (2011). Land area change in coastal Louisiana from 1932 to 2010: U.S. Geological Survey Scientific Investigations. Map 3164 scale 1:265,000, 12 pp. Pamphlet.

Couvillion, B. R., Beck, H., Schoolmaster, D., & Fischer, M. (2017). Land area change in coastal Louisiana (1932–2016). U.S. Department of the Interior, U.S. Geological Survey.

Dangendorf, S., Hendricks, N., Sun, Q. et al. Acceleration of U.S. Southeast and Gulf coast sea-level rise amplified by internal climate variability. *Nature Communications, 14*, 1935.

Dietz, M. E., Liu, K. B., & Bianchette, T. A. (2018). Hurricanes as a major driver of coastal erosion in the Mississippi River Delta: A multi-decadal analysis of shoreline retreat rates at Bay Champagne, Louisiana (USA). *Water, 10*, 1480.

Dietz, M. E., Liu, K. B., Bianchette, T. A., & Smith, D. (2021). Differentiating hurricane deposits in coastal sedimentary records: Two storms, one layer, but different processes. *Environmental Research Communications, 3*, 101001.

Dietz, M. E., Liu, K. B., Bianchette, T. A., Ryu, J., & Yao, Q. (2022). Linking backbarrier lacustrine stratigraphy with spatial dynamics of shoreline retreat in a rapidly subsiding region of the Mississippi River Delta. *Geomorphology, 397*, 108008.

Elsner, J. B., & Bossak, B. H. (2004). Hurricane landfall probability and climate. In R. Murnane, & K. B. Liu (Eds.), *Hurricanes and typhoons: Past, present, and future* (pp. 333–353). Columbia University Press.

Flocks, J. G., Ferina, N. F., Dreher, C., Kindinger, J. L., FitzGerald, D. M., & Kulp, M. A. (2006). High resolution stratigraphy of a Mississippi subdelta-lobe progradation in the Barataria Bight, North-Central Gulf of Mexico. *Journal of Sedimentary Research, 76*(3), 429–443.

Frazier, D. E. (1967). Recent deltaic deposits of the Mississippi River: Their development and chronology. *Gulf Coast Association of Geological Societies: Transactions, 17*, 287–315.

Haq, B. U., & Milliman, J. D. (2023). Perilous future for river deltas. *GSA Today, 33*, 4–12.

Hamlington, B. D., Chambers, D. P., Frederikse, T., Dangendorf, S., Fournier, S., Buzzanga, B., & Nerem, R. S. (2022). Observation-based trajectory of future sea level for the coastal United States tracks near high-end model projections. *Communications Earth and Environment, 3*, 230.

Hemmerling, S. A. (2018). Eroding communities and diverting populations: Historical population dynamics in coastal Louisiana. In J. W. Day, & J. A. Erdman (Eds), *Mississippi Delta restoration* (pp. 201–230). Springer.

Joshi, S., & Xu, Y. J. (2015). Assessment of suspended sand availability under different flow conditions of the Lowermost Mississippi River at Tarbert Landing during 1973–2013. *Water, 7*(12), 7022–7044.

Kemp, G. P., Day, J. W., Rogers, J. D., Giosan, L., & Peyronnin, N. (2016). Enhancing mud supply from the lower Missouri River to the Mississippi river delta USA: Dam bypassing and coastal restoration. *Estuarine Coastal Shelf Science, 183*, 304–313.

Kossin, J. P. (2018). A global slowdown of tropical-cyclone translation speed. *Nature, 558*, 104–107.

Lam, N. S. N., Pace, K., Campanella, R., LeSage, J., & Arenas, H. (2009). Business return in New Orleans: Decision making amid post-Katrina uncertainty. *Plos One, 4*(8), e6765.

Lam, N. S. N., Arenas, H., Pace, R. K., LeSage, J. P., & Campanella, R. (2012). Predictors of business return in New Orleans after Hurricane Katrina. *Plos One, 7*(10), e47935.

Lam, N. S. N., Reams, M., Li, K., Li, C., & Mata, L. (2016). Measuring community resilience to coastal hazards along the Northern Gulf of Mexico. *Natural Hazards Review, 17*(1), 04015013.

Lam, N. S. N., Xu, Y. J., Liu, K. B., Dismukes, D. E., Reams, M., Pace, R. K., Qiang, Y., Narra, S., Li, K., Bianchette, T. A., Cai, H., Zou, L., & Mihunov, V. (2018a). Understanding the Mississippi River Delta as a coupled natural-human system: Research methods, challenges, and prospects. *Water, 10*, 1054.

Lam, N. S. N., Qiang, Y., Li, K., Cai, H., Zou, L., & Mihunov, V. (2018b). Extending resilience assessment to dynamic system modeling: Perspectives on human dynamics and climate change research. *Journal of Coastal Research,* (85), 1401–1405.

Lam, N. S. N., Cheng, W., Zou, L., & Cai, H. (2018c). Effects of landscape fragmentation on land loss. *Remote Sensing of Environment, 209*, 253–262.

Lambert, C. E., Holley, J. R., McComas, K. A., Snider, N. P., & Tucke, G. K. (2021). Eroding land and erasing place: A qualitative study of place attachment, risk perception, and coastal land loss in Southern Louisiana. *Sustainability, 13*(11), 6269.

LeSage, J. P., Pace, R. K., Lam, N. S. N., Campanella, R., & Liu, X. (2011a). New Orleans business recovery in the aftermath of Hurricane Katrina. *Journal of Royal Statistics Society (Series A), 174*(4), 1007–1027.

LeSage, J. P., Pace, R. K., Lam, N. S. N., Campanella, R., & Liu, X. (2011b). Do what the neighbors do: Reopening businesses after Hurricane Katrina. *Significance, 8*(4), 160–163.

Li, K., & Lam, N. S. N. (2018). A spatial dynamic model of population changes in a vulnerable coastal environment. *International Journal of Geographical Information Science, 32*(4), 685–710.

Liu, K. B., Li, C. Y., Bianchette, T. A., McCloskey, T. A., Yao, Q., & Weeks, E. (2011). Storm deposition in a coastal backbarrier lake in Louisiana caused by Hurricanes Gustav and Ike. *Journal of Coastal Research,* (64), 1866–1870.

Lewis, J. A., & Ernstson, H. (2019). Contesting the coast: Ecosystems as infrastructure in the Mississippi River Delta. *Progress in Planning, 129*, 1–30.

Meade, R. H., & Moody, J. A. (2010). Causes for the decline of suspended-sediment discharge in the Mississippi river system, 1940–2007. *Hydrological Processes, 24*(1), 35–49.

Needham, H. F., & Keim, B. D. (2012). A storm surge database for the U.S. Gulf Coast. *International Journal of Climatology, 32*, 2108–2123.

Penland, S., & Ritchie, W. (1979). Short term morphological changes along the Caminada-Moreau coast, Louisiana. *Gulf Coast Association of Geological Societies Transactions, 29*, 342–346.

Penland, S., & Suter, J. R. (1988). Barrier island erosion and protection in Louisiana: A coastal geomorphological perspective. *Gulf Coast Association of Geological Societies Transactions, 38*, 331–342.

Penland, S., Zganjar, C., Westphal, K. A., Connor, P., List, J., & Williams, S. J. (2003). Shoreline changes in the Caminada-Moreau headland and Grand Isle—1887 to 1996 Lafourche and Jefferson Parishes, Louisiana. USGS Open File Report 2003, 11p.

Penland, S., Connor Jr., P. F., Beall Jr., A., Fearnley Jr., S., & Williams Jr., S. J. (2005). Changes in Louisiana's shoreline: 1855–2002. *Journal of Coastal Research,* (44), 7–39.

Public Broadcasting Service (PBS). (2021). Death toll rises after Hurricane Ida's remnants hit Northeast.

Rabalais, N. N., Turner, R. E., Sen Gupta, B. K., Boesch, D. F., Chapman, P., & Murrell, M. C. (2007). Hypoxia in the northern Gulf of Mexico: Does the science support the Plan to Reduce, Mitigate, and Control Hypoxia? *Estuaries and Coasts, 30*(5), 753–772.

Randolph, N. (2022). Modeling authority over a drowning coast. *Environmental Politics, 32*(2), 1–25.

Reiman, J. H., & Xu, Y. J. (2019). Dissolved carbon export and CO2 outgassing from the Lower Mississippi River—Implications of future river carbon fluxes. *Journal of Hydrology, 578*, 124093.

Rosen, T., & Xu, Y. J. (2015). Estimation of sedimentation rates in the distributary basin of the Mississippi River, the Atchafalaya River Basin, USA. *Hydrology Research, 46*, 244–257.

Rossi, A., Massei, N., Laignel, B., Sebag, D., & Copard, Y. (2009). The response of the Mississippi River to climate fluctuations and reservoir construction as indicated by wavelet analysis of streamflow and suspended-sediment load, 1950–1975. *Journal of Hydrology, 377*(3), 237–244.

Roth, D. M. (2023). Louisiana hurricane history. U. S. National Weather Service.

Rodrigues, E., Cohen, M. C. L., Liu, K. B., Pessenda, L. C. R., Yao, Q., Ryu, J. H., Rossetti, D., de Souza, A., & Dietz, M. (2021). The effect of global warming on the establishment of mangroves in coastal Louisiana during the Holocene. *Geomorphology, 381*, 107648.

Ryu, J., Bianchette, T. A., Liu, K. B., Yao, Q., & Maiti, K. (2018). Palynological and geochemical records of environmental changes in a Taxodium swamp near Lake Pontchartrain in southern Louisiana (USA) during the last 150 years. *Journal of Coastal Research,* (85), 381–385.

Ryu, J., Liu, K. B., & Bianchette, T. A. (2021a). Holocene environmental history of a freshwater wetland in southern Louisiana: A sedimentary record of delta development, coastal evolution and human activity. *Journal of Quaternary Science, 36*(6), 980–990.

Ryu, J., Liu, K. B., Bianchette, T. A., & McCloskey, T. A. (2021b). Identifying forcing agents of environmental change and ecological response on the Mississippi River Delta, Southeastern Louisiana. *Science of the Total Environment, 794*(4), 148730.

Ryu, J., Liu, K. B., & McCloskey, T. A. (2022a). Temporal variability in the relative strength of external drivers controlling ecosystem succession in a coastal wetland near Bayou Lafourche, southeast Louisiana, USA. *Quaternary Science Reviews, 276*, 107692.

Ryu, J., Liu, K. B., & McCloskey, T. A. (2022b). The effects of external forcing on the biodiversity and species richness associated with the environmental history of Louisiana's wetlands over the late Holocene. *Quaternary Science Reviews, 298*, 107865.

Sasser, C. E., Visser, J. M., Mouton, E., Linscombe, J., & Hartley, S. B. (2014). Vegetation types in coastal Louisiana in 2013: U.S. Geological Survey Scientific Investigations, Map 3290, 1 sheet, scale 1:550,000.

Sobel, A. H., Camargo, S. J., Hall, T. M., Lee, C.-Y., Tippett, M. K. A., & Wing, A. (2016). Human influence on tropical cyclone intensity. *Science, 353*, 242–246.

Sweet, W. V., et al. (2022). Global and Regional Sea Level Rise Scenarios for the United States: Updated Mean Projections and Extreme Water Level Probabilities along U.S. Coastlines. NOAA Technical Report NOS 01. National Oceanic and Atmospheric Administration, National Ocean Service. Silver Spring.

Törnqvist, T. E., Kidder, T. R., Autin, W. J., van der Borg, K., de Jong, A. F., Klerks, C. J., Snijders, E. M., Storms, J. E., van Dam, R. L., & Wiemann, M. C. (1996). A revised chronology for Mississippi River subdeltas. *Science, 273*(5282), 1693–1696.

Törnqvist, T. E., Paola, C., Parker, G., Liu, K. B., Mohrig, D., Holbrook, J. M., & Twilley, R. R. (2007). Comment on 'Wetland Sedimentation from Hurricanes Katrina and Rita.' *Science, 316*(5822), 201.

Turner, R. E., & Rabalais, N. N. (2003). Linking landscape and water quality in the Mississippi River Basin for 200 Years. *BioScience, 53*, 563–572.

Turner, R. E., Baustian, J. J., Swenson, E. M., & Spicer, J. S. (2006). Wetland sedimentation from Hurricanes Katrina and Rita. *Science, 314*(4798), 449–452.

Turner, R. E., Layne, M., Mo, Y., & Swenson, E. M. (2019). Net land gain or loss for two Mississippi River diversions: Caernarvon and Davis Pond. *Restoration Ecology, 27*(6), 1231–1240.

The Advocate. (2023). Storms, insurance turmoil Fueling population slides. Restore the Mississippi River Delta. Retrieved 12, 2023 from https://mississippiriverdelta.org/whats-at-stake/economy/

U.S. Energy Information Administration. (2022). Louisiana State Energy Profile.

Visser, J. M., Sasser, C. E., Chabreck, R. H., & Linscombe, R. G. (1998). Marsh vegetation types of the Mississippi River Deltaic Plain. *Estuaries, 21*, 818–828.

Wahl, T., Jain, S., Bender, J., Meyers, S. D., & Luther, M. E. (2015). Increasing risk of compound flooding from storm surge and rainfall for major US cities. *Nature Climate Change, 5*, 1093–1097.

Wang, B., & Xu, Y. J. (2018). Decadal-scale riverbed deformation and sand budget of the last 500 km of the Mississippi River: Insights into natural and river engineering effects on large alluvial rivers. *Journal of Geophysical Research: Earth Surface, 123*(5), 874–890.

Wang, B., & Xu, Y. J. (2020). Estimating bed material fluxes upstream and downstream of a controlled large bifurcation - The Mississippi-Atchafalaya river diversion. *Hydrological Processes, 34*(13), 2864–2877.

Wiegman, R., Day, J., D'Elia, C., Rutherford, J., Morris, J., Roy, E., Lane, R., Dismukes, D., & Snyder B. (2018). Modeling impacts of sea-level rise, oil price, and management strategy on the costs of sustaining Mississippi delta marshes with hydraulic dredging. *Science of the Total Environment, 618*, 1547–1559.

Xu, Z., & Xu, Y. J. (2016). A deterministic model for predicting hourly dissolved oxygen change: Development and application to a shallow eutrophic lake. *Water, 8*(2), 41.

Xu, Y. J., Lam, N. S. N., & Liu, K. B. (2018). Assessing resilience and sustainability of the Mississippi River Delta as a coupled natural-human system. *Water, 10*(10), 1317.

Xu, K. H., Bentley, S. J., Day, J. W., & Freeman, A. M. (2019a). A review of sediment diversion in the Mississippi River Deltaic Plain. *Estuarine, Coastal and Shelf Science, 225*, 106241.

Xu, Y. J., Wang, B., & Joshi, S. (2019b). Sediment transport and channel dynamics of the lowermost Mississippi River in recent decades. In *The Proceedings of the 38th IAHR World Congress.*

Xu, Y. J., & Xu, Z. (2022). Carbon dioxide degassing and lateral dissolved carbon export during the unprecedented 2019 Mississippi River mega flood— Implications for large river carbon transport under future climate. *Journal of Hydrology, 614*(B), 128650.

Yao, Q., Cohen, M., Liu, K. B., de Souza, A. V., & Rodrigues, E. (2022a). Nature versus humans in coastal environmental change: Assessing the impacts of Hurricanes Zeta and Ida in the context of beach nourishment projects in the Mississippi River Delta. *Remote Sensing, 14*, 2598.

Yao, Q., Cohen, M., Liu, K. B., Fan, D., Rodrigues, E., Maiti, K., de Souza, A. V., Aragón-Moreno, A. A., Rohli, R. V., Yin, D., & Pessenda, L. C. R. (2022b). Mangrove expansion at poleward range limits in North and South America: Late-Holocene climate variability or Anthropocene global warming? *Catena, 216*, 106413.

Yin, S., Gao, G., Li, Y., Xu, Y. J., Turner, R. E., Ran, L., Wang, X., & Fu, B. (2023). Long term trends of streamflow, sediment load and nutrient fluxes from the Mississippi River Basin: Impacts of climate change and human activities. *Journal of Hydrology, 616*, 128822.

Zhang, W. G., Xu, Y. J., Guo, L. C., Lam, N. S. N., Xu, K., Yang, S. L., Yao, Q., & Liu, K. B. (2022). Comparing the Yangtze and Mississippi Deltas in the light of coupled natural-human dynamics: Lessons learned and implications for management. *Geomorphology, 399*, 108075.

Zhong, B., & Xu, Y. J. (2011). Risk of inundation to coastal wetlands and soil organic carbon and organic nitrogen accounting in Louisiana, USA. *Environmental Science and Technology, 45*, 8241–8246.

Zscheischler, J., Westra, S., van den Hurk, B. J. J. M., Seneviratne, S. I., Ward, P. J., Pitman, A., AghaKouchak, A., Bresch, D. N., Leonard, M., Wahl, T., & Zhang, X. B. (2018). Future climate risk from compound events. *Nature Climate Change, 8*(6), 469–477.

Nile River Delta

Alaa Salem and Yan Liu

1 Physical Setting

1.1 The Nile Catchment

The Nile River is the longest river in the world, extending over more than 6800 km, with a catchment area of about 3.3 million km^2. The present-day Nile has three main tributaries, the Blue Nile, White Nile and Atbara (Fig. 1a). The catchments of the Blue Nile and Atbara, which are approximately 330,000 and 180,000 km^2, respectively, add up to 15% of the total Nile catchment (Woodward et al., 2022). Sediments transported by the Blue Nile (about 61%) and Atbara (approximately 35%) make up the majority of the suspended sediment load of the current Desert Nile in Sudan and Egypt (Padoan et al., 2011). The sediment transported by the White Nile is about 3%, despite of its huge catchment area of more than 1,730,000 km^2. The Blue Nile and Atbara sediments are retrieved from deeply dissected landscapes during summer monsoon. In contrast, the White Nile sediments are retrieved from extremely older, harder rocks, and lower relief, without noticeable seasonal oscillations in runoff (Woodward et al., 2022). The mean annual water discharge and the discharge in times of peak flow are dominated by the Blue Nile, followed by the White Nile and Atbara (Fig. 1b) (Williams & Adamson, 2021).

A. Salem
Faculty of Science, Kafrelsheikh University, Kafrelsheikh 33516, Egypt

Y. Liu (✉)
State Key Laboratory of Estuarine and Coastal Research, East China Normal University, Shanghai 200241, China
e-mail: liuyan@sklec.ecnu.edu.cn

W. Zhang and H. de Vriend (eds.), *Delta Sustainability*,
https://doi.org/10.1007/978-981-97-7259-9_12

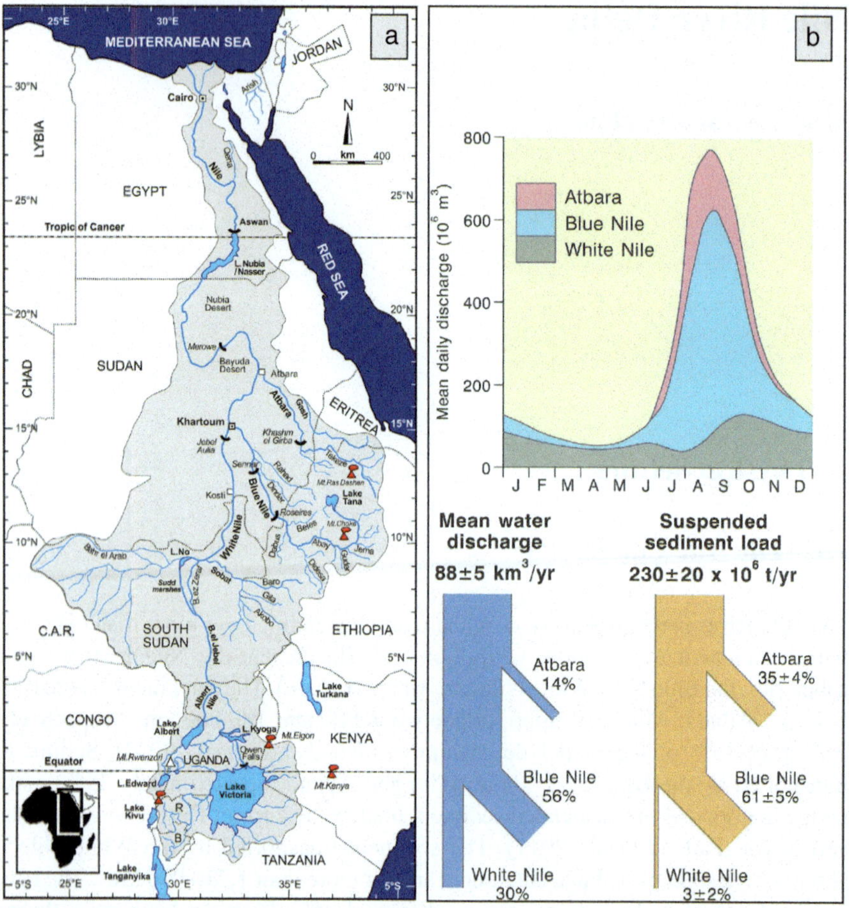

Fig. 1 **a** The Nile basin drainage network (Garzanti et al., 2015); **b** the flow regime of the Nile showing the typical contribution from each of the three main tributary basins (Woodward et al., 2022)

1.2 The Nile Delta

The Nile Delta is one of the oldest deltaic systems in the world (El Banna & Frihy, 2009). It occupies an area of ca. 25,000 km^2 and is bounded by desert to both the east and west. The length of the Nile Delta is approximately 160 km from north to south, and it covers about 240 km along the coastline at the Mediterranean Sea, from Alexandria (west) to Port Said (east) (Fig. 2). The Nile Delta has its apex about 20 km north of Cairo (latitude 30°) where the Nile River splits into two main branches; the Rosetta in the west and the Damietta in the east. The Rosetta branch is about 235 km in length, 200–270 m wide and has an average depth of 2–2.7 m. The Damietta branch is 240 km in length, 170–220 m wide and has a depth ranging from 2.0 to

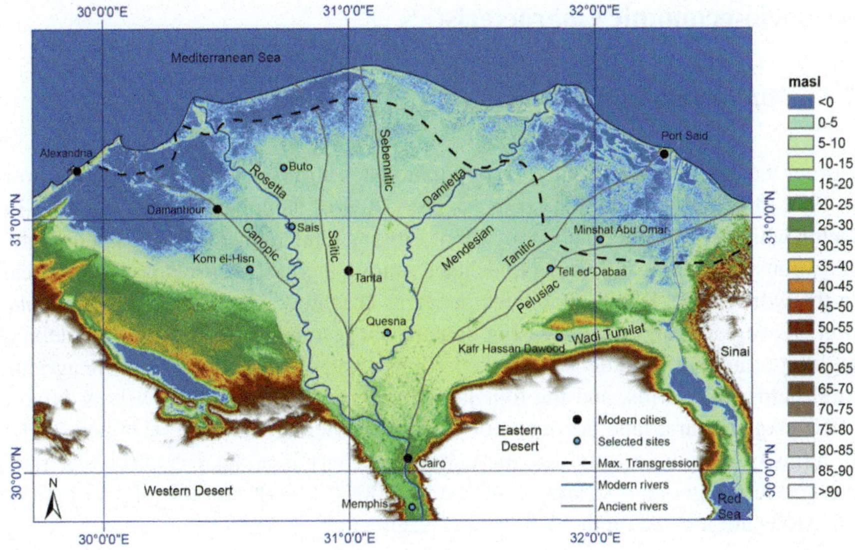

Fig. 2 The Nile Delta and its river system, and the dash line represents the transgression boundary (modern rivers in Blue, and ancient rivers in Grey) (Pennington et al., 2017)

4.5 m. The delta formed by sedimentary processes since the Upper Miocene, and built up by the alluvium brought by the ancient seven active branches of the Nile. Those branches have silted up and replaced by the Damietta and Rosetta branches (Stanley & Warne, 1993). The Holocene sedimentary succession attains a thickness of ca. 5–20 m in most areas of the present delta, declining from south to north, with a maximum thickness of ca. 50 m in the northeast of the delta (Pennington et al., 2017).

The Nile Delta is wave dominated delta. The tide along the Nile Delta coast is semi-diurnal, with a typical average range of 40 cm (Sestini, 1989). The wave action is seasonal, NW and NE are the dominant wind directions (Frihy et al., 2010; Iskander, 2013). By the end of the 18th century, the strong marine current and wave regime along the shoreline and on the shelf had produced a gently arcuate coastline with only a few protuberances, including the Damietta and Rosetta promontories (Stanley & Warne, 1998) and an intermediate relict one at Burullus. These characteristics still apply to the present coastline. The broad beaches of the Delta are backed partially by dunes, cultivated land, urban centers and coastal wetlands forming natural lagoons and reclaimed fish farms. Four brackish coastal lagoons (from west to east, Maryut, Idku, Burullus and Manzala) are separated from the sea by sandy barriers breached by artificial inlets.

The climate is arid and characterized by high evaporation rates (1,500–2,400 mm/year) and very low rainfall (5–200 mm/year). Mean temperatures in the delta are the hottest in July and August, with a maximum average temperature of 34 °C. Winter temperatures are normally in the range from 9 °C at night to 19 °C during the day.

2 Socio-economic Characteristics

2.1 *Population and Urbanization*

During the last forty years, Egypt's population has almost tripled, from 36.6 million in 1972 to 97.5 million in 2017 (Fig. 3). Over 40% of its population lives in the Nile Delta, which constitutes only about 2% of Egypt's total land area. As it is geographically constrained, the Nile delta finds itself among the most densely populated areas in the world, with 1,360 inhabitants per km^2. It is now experiencing high population pressure and an ever-increasing demand for resources. Fortunately, the delta is highly productive, with densely packed villages and cities set among fertile agricultural fields, fish farms, and the four large brackish lagoons (Oczkowski & Nixon, 2010). Agricultural land accounts for 63% of the delta's land use (Hereher, 2010). In terms of economic activities, the fish productivity is about 185,000 t/year from the northern lagoons (Oczkowski & Nixon, 2008), and about 40% of all Egyptian industries are located on the delta plain (Negm et al., 2016).

The expansion of towns and villages dotted around the delta has resulted in rapid urbanization (Shalaby & Moghanm, 2015), leading to unprecedented pressure on the fertile agricultural land of the Nile delta (Fig. 4). As a result, the available area per capita in the delta has dropped from 0.12 ha in 1950 to 0.04 ha in 2017 (CAPMAS, 2017), and there is concern that urban encroachment on fertile and highly productive soils may threaten Egypt's agriculture sustainability and food security (Nizeyimana et al., 2001). The delta is the main agricultural area of Egypt, comprising around

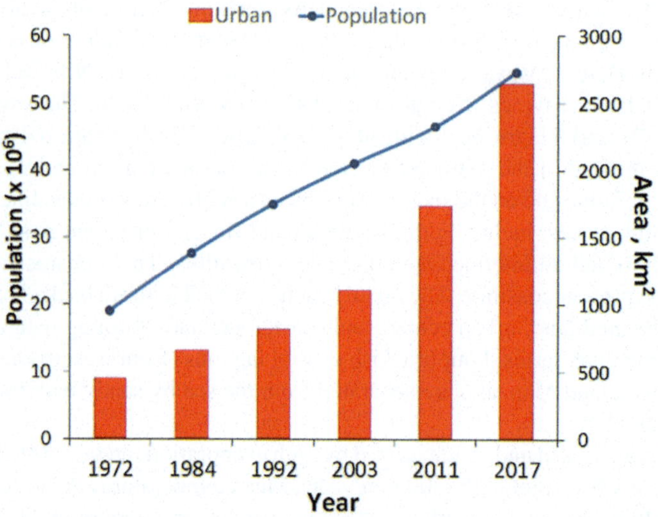

Fig. 3 Change in urban area (km^2) and population over the last 45 years in the Nile delta (Abd-Elmabod et al., 2019)

64% of the total agriculture land area. It yields about two thirds of Egyptian crop production (FAOSTAT, 2018; www.fao.org/faostat/en/), and the fertile soils allow two or three crops to be grown per year (Kassim et al., 2018; Osama et al., 2017).

Abd-Elmabod et al. (2019) found that since 1972, the urban area in the delta has increased more than five-fold, from 452 km^2 in 1972 to over 2600 km^2 in 2017

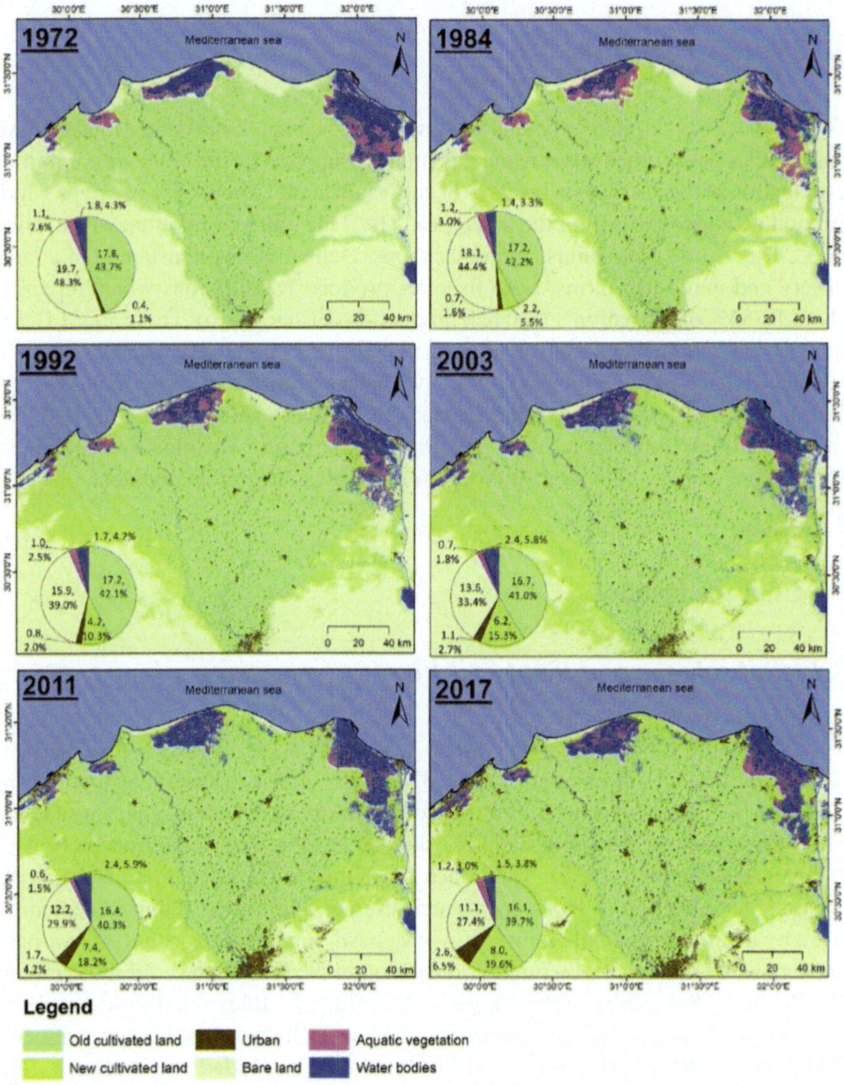

Fig. 4 Urban expansion and change in the major land cover classes from 1972 to 2017. The pie chart illustrates the proportion occupied by each land cover class (×1000 km^2, and %) (Abd-Elmabod et al., 2019)

(Fig. 4). The rate of change increased dramatically after 2003, with the population increasing from 19 to 55 million (Fig. 3). The spatial pattern of expansion (Fig. 4) shows that this is driven by an increase in both small and large settlements, spread fairly uniformly across the delta area. It is not just a result of expansion of Cairo in the south of the delta.

2.2 *Major Industries*

The Nile Delta is suitable for intensive agriculture. Hamza (2006) stated that approximately 85% of Egyptian water resources (mainly supplied by the Nile) are committed to irrigation of the 3.4 million hectares of cultivated lands. About 65% of Egypt's agricultural activities are within the delta area. The delta cultivates cotton, wheat, corn, rice, citrus fruits, groundnuts, and other crops. Delta and lakes sustain both a large fishery and many fish farms. Lake fisheries produce 182,525 tons which represent about 12.54% of the nation's total fish production (Fishar, 2016). There are 21 Egyptian natural and artificial marine fishing ports located on the coasts of the Mediterranean, Red Sea and coastal lakes. Only five marine fishing ports (Port Said, New El-Burullus, Elmaadiya in the Mediterranean; and Attaka and Hurghada in the Red Sea) satisfy the minimum requirements regarding infrastructure, facilities and equipment for handling catches (General Authority for Fish Resources Development, 2015).

3 Current Issues

3.1 *Subsidence and Sea Level Rise*

The Nile delta, only ~1 m above mean sea level on the Mediterranean coast, is subject to uneven rates of submergence. This is a response to several factors leading to increasing land lowering (subsidence) of the northern delta and adjacent seafloor, plus an accelerating rise in eustatic (world) sea level in the Mediterranean (Fig. 5) (Stanley & Clemente, 2017). Three factors leading to subsidence are Neotectonic lowering, compaction of Holocene sequences, and diminished sediment replenishment by much reduced Nile flow to Egypt's coast. Subsidence accounts for variable average land lowering of ~3.7 mm/yr of section in the NW delta, ~ 7.7 mm/yr in the N delta, and ~8.4 mm/yr in the NE delta. Therefore, El-Raey (2010) estimated that even with an SLR of only 0.5 m, nearly all the Nile Delta beaches and approximately 30% of the cities of Alexandria and Port Said will be eroded and damaged. SLR will also hit Egypt's economy, by harming the agricultural sector that supports 15% of Egyptian GDP (Dasgupta et al., 2009; Abu Hatab et al., 2012), and changing the ecology of the coastal lake system (Palmer, 2008).

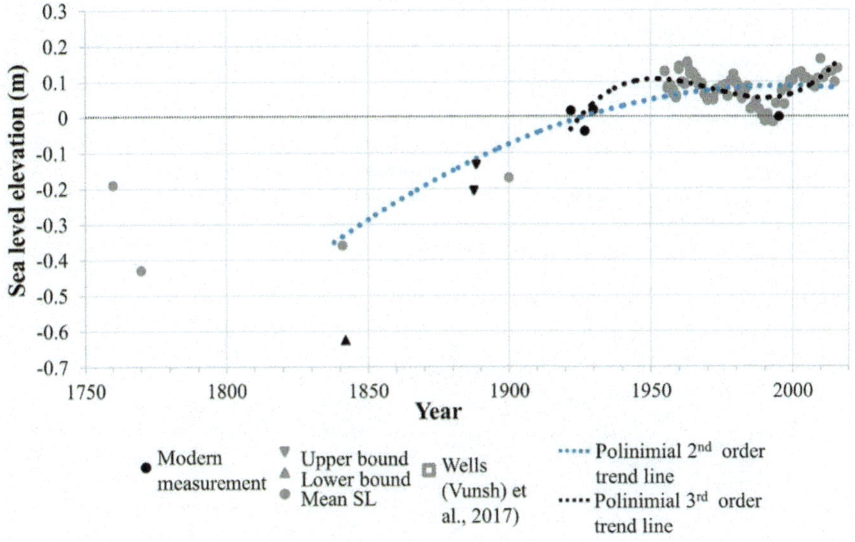

Fig. 5 Sea level rise at eastern Mediterranean sea (Toker et al., 2019)

Nowadays, the Nile Delta coast is partially backed by sand dunes and engineering structures (17.5% and 42.5% respectively), while the remaining coast (40%) is not protected (Fig. 6) (Frihy & El-Sayed, 2013). The assessment of the coastal susceptibility to marine flooding in this sector gives varying results owing to the ranking of the following different zones includes, high vulnerable zones, moderate vulnerable zones and low vulnerable zones (Fig. 6) (Frihy & El-Sayed, 2013).

3.2 *Reduced Sediment Input*

The Nile sediment flux has been drastically modified by human impact in the last century (Garzanti et al., 2015). All of the solid load is trapped presently within artificial reservoirs before reaching Egypt, where the Nile carries only 1% sediment of the pre-dam budget scoured from the catchment (Said, 1993; Stanley & Wingerath, 1996; Abdel-Fattah et al., 2004). The construction of dams to control floods, provide water for irrigation and generate hydroelectricity began with the Aswan Low Dam in southern Egypt (1902), and continued with the Sennar Dam on the Blue Nile (1925) and the Jebel Aulia Dam on the White Nile (1937) (Garzanti et al., 2015). Larger dams were built subsequently at Roseires on the Blue Nile (1966) and at Kashm El Girba on the Atbara (1964) (Fig. 1a). At about the same time the Aswan High Dam was raised, forming the third largest reservoir in the world (named Lake Nubia in Sudan and Lake Nasser in Egypt) with a nominal volume of 157 km^3. In 2009, the Tekeze Dam was built in the Tigrai region of northern Ethiopia, representing Africa's

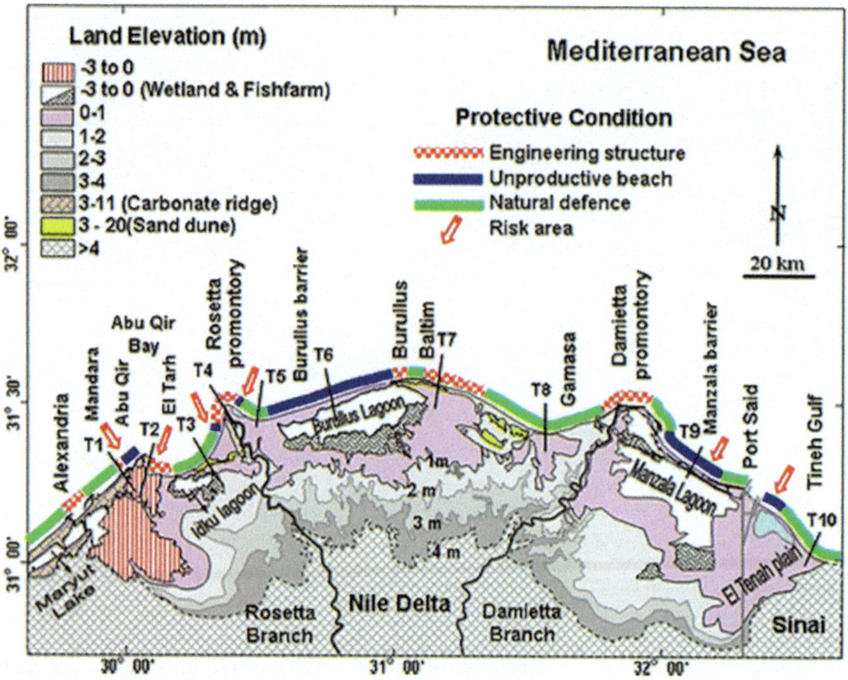

Fig. 6 The lower coastal plain of the Nile Delta showing topographic features, including land elevation relative to mean sea-level, protective state (Frihy & El-Sayed, 2013)

largest arch dam at the time of completion, and the Merowe Dam was built close to the 4th cataract in Nubia, representing the largest contemporary hydropower project in Africa.

After a century of dam construction, detritus that once was flushed during summer floods from Ethiopian highlands to the Mediterranean Sea is now retained before reaching Egypt, where "a once mighty river has now been tamed to a water canal" (Garzanti et al., 2015). In earlier times, annual water discharge in Egypt averaged 84 km^3, fluctuating widely from 900 m^3/s at low flow to 10000 m^3/s in the flood season. Annual sediment fluxes were estimated to vary between 50 and 300 × 10^6 t (Inman & Jenkins, 1984). Before 1902, water discharge was on average 25% greater, with correspondingly greater sediment loads. The delta was thus prograding actively, locally more than 10 m/a (Frihy & Lawrence, 2004; Sestini, 1989). After closure of the Aswan High Dam, Nile sediment has remained trapped within Lake Nubia/Nasser, where it is actively building a new delta ~250 km long and locally >60 m thick (El-Manadely et al., 2002; Moussa, 2013). Annual water discharge was regulated to 55 km^3, maximum velocities were reduced from 1.5–2.0 to 1.0–1.5 m/s, and suspended-sediment concentration during high flow decreased from ~4 to 0.03–0.1 g/l only (Garzanti et al., 2015). The annual sediment load in Egypt, mainly originating from older channel deposits, is now less than 10^6 t (Abdel-Fattah et al., 2004).

3.3 Coastal Erosion

During the past 7500 years, the Nile and other major delta systems in the world have experienced overall growth (Stanley & Warne, 1994). Within this overall growth phase, a series of lesser and more localized accretion and erosion cycles have characterized delta systems (Coleman, 1988; Coleman & Wright, 1975; Penland et al., 1988), including the Nile. During the past centuries, water control measures have seriously curtailed (a) the Nile flow below Aswan and in the delta, (b) flushing of the delta plain during annual floods, (c) influx and deposition of sediment across the delta plain, and (d) bypassing of sediment to and beyond the coast (Stanley & Warne, 1998). At the same time, natural delta destruction processes, particularly coastal currents and subsidence, continue unabated. As a consequence, the natural Holocene delta accretion phase has ended and the Nile's depocenter has entered a human-induced erosion phase. The fluvial-deltaic prism at the Nile mouth is now a relict delta or, in other words, an eroding coastal plain (Stanley & Warne, 1998).

An obvious consequence of reduced Nile flow and sediment discharge in the 20th Century and closure of the Low Dam in 1902 has been accelerated erosion of the promontories at the two Nile mouths (Fig. 7; El Askary & Frihy, 1986; Sestini, 1992). Since 1964, coastal erosion has cut back the north-central delta stretch near the town of Baltim. Coastal areas east of eroded sectors have experienced sand accretion (Frihy et al., 1988; Smith & Abdel-Kader, 1988), which has resulted in an overall straightening of the shoreline (Manohar, 1976; Frihy, 1988; Smith & Abdel-Kader, 1988).

The Nile Delta coast of Egypt is subject to erosion due to the wave action, currents, and relevant coastal processes. This erosion was exacerbated since the construction of the Aswan high dam, where the Nile River's sediments source and discharge were obstructed (El-Raey et al., 1999). Hence, coastal protection structures (such as groins, seawalls, detached and submerged breakwater, jetties, etc.) were constructed to reduce coastal erosion.

Sharaan and Udo (2020) applied the Bruun rule to estimate the projected beach retreat and loss due to sea level rise along the northern coast of Egypt in 2100, including the Nile Delta coastline. The major results are:

(a) The Nile Delta's coastal area are will be subject to severe erosion along the northern coast of Egypt.
(b) The maximal coastal retreat along the Nile Delta coast is expected to occur from Abo Qir to Damietta Promontory, Alexandria, and from El-Manzala to Port Said, respectively.
(c) The Nile Delta's gentle coastal slope increases its vulnerability to SLR. Consequently, the projected erosion rate along the Nile Delta's coasts will be higher than along other Egyptian coasts.

Additionally, Darwish et al., (2017) classified the geomorphological changes along the Nile Delta coast into five relatively homogeneous types. These are: (1) stable (51% of the coast), (2) change of 200–500 m (33% of the coast), (3) change of

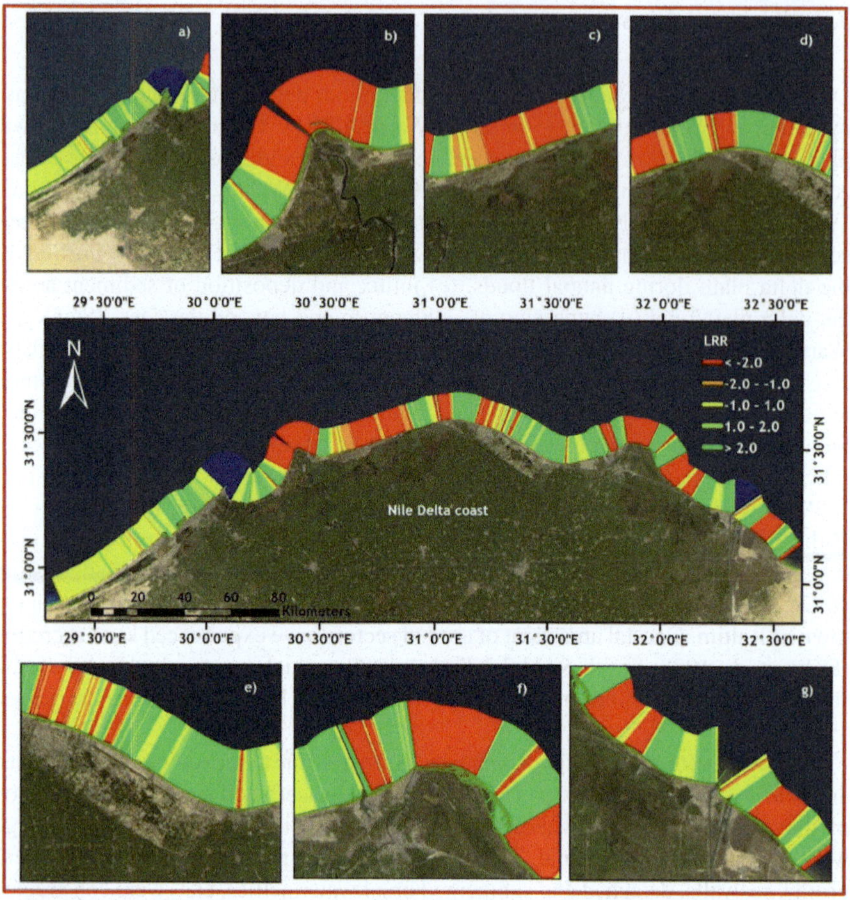

Fig. 7 Shoreline changes "LRR" for the long-term period 1985–2022 at different locations: **a** Alexandria coast and Abo-Qir Bay, **b** Rosetta promontory. **c** Abo-Kashaba Bay, **d** Burullus headland, **e** Gamasa Bay, **f** Damietta promontory, **g** Port Said coast (Elkotby et al., 2023)

500–2000 m (12% of the coast), (4) change of 2000–4000 m (3% of the coast), and, (5) change greater than 4000 m (only on the Rosetta promontory). Between 1945 and 2015, half of the coastline was eroded. The greatest distance of erosion was −5329 m, and the greatest accretion was +2771 m. While several factors are responsible for the coastal change, the Aswan High Dam had the most profound effect on the accelerating coastal erosion rates. This was true especially in front of and around the Rosetta and Damietta promontories. Measurements indicate that, at the Rosetta Promontory, the highest coastline erosion rate increased from −72 m/y between 1945 and 1972 (before the AHD) to −141 m/y between 1972 and 1984 (after the dam) and then to −159 m/y between 1984 and 2001. On the Damietta Promontory, the highest coastline erosion rate was −13 m/y before the dam (between 1945 and 1972) and −45 m/y after the dam (between 1972 and 1984). It was 50 m/y between 1984 and

2001. Beginning in 1991, the Coastal Research Institute built seawalls in severely eroded areas. These included seawalls around Alexandria, the Mohamed Ali Seawall in Abu Quir Bay, the Rosetta Seawall, the Alborg Baltim Seawall, and the Damietta Seawall. Although the seawalls have slowed down erosion, they have caused erosion elsewhere. For example, the eastern side of the Rosetta Promontory has a new coastal erosion zone as a result of construction of the Rosetta Seawall, with an erosion rate of −35 m/y between 2001 and 2015. Also, at the Damietta Promontory, a new coastal erosion zone appeared as a result of construction of the Damietta Seawall.

3.4 Fresh Water Resources

Water resources in Egypt are the Nile discharge, rainfall, deep groundwater, and desalinated water complimented by shallow groundwater in the Nile Delta, as well as both re-used agricultural drainage and treated wastewater as non-conventional water resources (El-Din et al., 2013). Eighty-six percent of that water comes from the Nile and around 11% from underground sources, only 2% is recycled (CAPMAS, 2021). In Egypt, less than 75% of collected wastewater is currently being treated (Elbana et al., 2019). Groundwater plays a vital role in water supply everywhere throughout Egypt. In the desert regions, groundwater is considered as one of the key sources for both rural and agriculture water supplies. Although several aquifer systems exist in Egypt, the most important is the Nile aquifer system, which covers the Nile floodplain and the desert fringes where 90% of Egypt's population lives (Wahaab & Badawy, 2004). In the center of the floodplain, the aquifer is semi-confined (10 m of silt-clay). However, on the edges of the floodplain and the desert fringes, the aquifer becomes phreatic. The thickness of the aquifer ranges from 300 m in the south to 800 m in the Delta (Wahaab & Badawy, 2004).

Of six groundwater aquifers in Egypt (Fig. 8), the Nile aquifer is viewed as a fundamental source of groundwater abstraction as it represents 87% of the total groundwater abstraction in the country (MWRI, 2005). The Nile Delta aquifer is one of the biggest aquifers in Egypt with a total capacity of 500 Bm^3 (Sherif, 1999), around 15 times the amount of water that enters the Delta every year through the river itself. The recharge source of the Nile Delta aquifer is the Nile water leakage from canals, drains, and the irrigated plots themselves (Sefelnasr & Sherif, 2013). The conjunctive utilization of groundwater and surface water plays an essential role during the period of peak irrigation demand. The amount of water abstraction from the Nile aquifer (Delta and Valley) was estimated in 2010 at 6.2 Bm^3, which is below the safe yield (8.4 Bm^3) estimated by MWRI (2012).

The Nile Delta aquifer serves nine governments, and according to FAO (2013) the total irrigated area by groundwater in the Nile Delta would amount to 1740 km^2. Little or no land is being irrigated with groundwater in Damietta, Port Said, and Kafr El Sheikh Provinces, due to the effect of saline water intrusion from the Mediterranean Sea. The highest groundwater use occurs in the El-Beheira governorate. This is due to the inclusion of western new lands (outside the Delta) in the statistics.

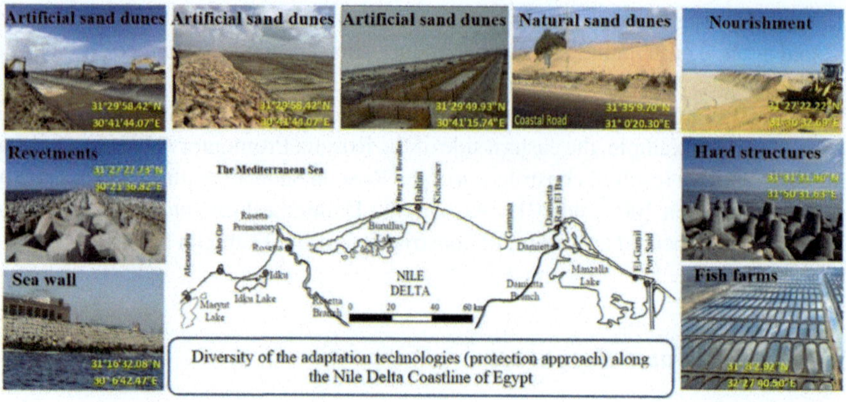

Fig. 8 Locations of selected photographs taken along the Nile Delta region illustrating protection diversity (Sharaan et al., 2022)

3.5 *Environmental Pollution*

During most of Nile delta formation history, natural environments of the Nile delta were capable of sustaining human needs, with minimal impacts on the ecosystem. By the end of the 18th century population increased and technology advanced to such levels that the natural capacity was exceeded and the Nile delta ecosystems no longer functioned as sustainable resources (Stanley & Warne, 1998). In 1964, the construction of the Aswan Dam was intended to control the Nile's waters and provide electric power, which contributed significantly to the development of industry and agriculture in the Delta (Dumont & El-Shabrawy, 2007; Siegel et al., 1994). Since then, sediment (silt rich in nutrients) is trapped behind the Dam and no longer reaches the Nile delta (Oczkowski & Nixon, 2010). Where the suspended-sediment concentration during flow decreased from ~4 to 0.03–0.1 g/l only and the proportion of sediments reaching the coastal area decreased from 100–120 $\times$ 10^6 t/yr to less than 10^6 t/yr (Garzanti et al., 2015). The lack of nutrient-rich silt downstream of the dam leads to seeking alternatives, such as the excessive use of chemical fertilizers, insecticides, and pesticides for agricultural purposes. Usage of agricultural fertilizers increased from 3.4 $\times$ 10^5 t/yr to 13 $\times$ 10^5 t/yr during the period from 1965 to 2002 (FAO, 2008). At the same time, the population in the Nile delta doubled and made the delta the most densely populated area in the world, with 1360 inhabitants/km^2 (FAO, 2008). Moreover, the Egyptian industry has experienced a boom period fueled by low-cost hydroelectricity since the 1960s (Wahaab & Badawy, 2004). The Nile delta houses about 50% of Egyptian industries (Mikhailova, 2001). About 638 M m^3 a^{-1} of water used for Egyptian industry, of which 549 M m^3 a^{-1} is discharged without treatment into the drainage systems, mostly connected to the delta lagoons (Wahaab & Badawy, 2004). Consequently, industrial and agricultural activities associated with population growth in the Nile delta led to increased pollution and infectious disease (Gu et al., 2013; Shetaia et al., 2022; Siegel et al., 1994). Wastewater and sediment discharged along the margins of delta wetlands have increased concentrations of heavy metals, nutrients and other chemicals.

Gu et al. (2013) noticed that heavy metals are particularly concentrated in the post-dam core sediments (above 10–15 cm) from Manzala, Burullus and Edku lagoons (Fig. 11), closely associated with industrial and agricultural development after 1960s (Siegel et al., 1994). Oczkowski et al. (2011) found that the sediments from Lake Manzala clearly reflected the impact of agricultural intensification following completion of the Aswan High Dam and delta barrages in the mid-1960s to early 1970s. Nitrogen and organic matter concentrations more than doubled in three Manzala sediment cores, and the Nitrogen isotope value of $\delta^{15}N$ declined from 5 to <1‰. These changes reflect the increasing use of synthetic fertilizers ($\delta^{15}N \approx 0‰$) from the 1960s to the present.

3.6 Biodiversity Loss Due to Extensive Land Use Change

Lake Burullus is the second largest lake at the northern edge of the Nile Delta, Egypt. It has been recognized as an internationally significant wetland that provides a habitat for migrating birds, fish, herpetofauna, and mammals. However, the lake is experiencing severe human impacts including drainage and conversion to agricultural lands and fish farms. Keshta et al. (2022) used multispectral, moderate-spatial-resolution (30 m^2) Landsat satellite imagery to assess marsh loss in Lake Burullus, Egypt, in the last 35 years (1985–2020). They found a substantial loss (44.8%) of marsh aerial coverage. The drainage and conversion of wetlands into agricultural lands and/or fish farms is concentrated primarily in the western and southern part of the lake, where their surface area doubled (103.2% increase) between 2000 and 2020. The authors recommend that land-use policy makers and environmental government agencies have to raise public awareness among the local communities of Lake Burullus for of economic and environmental consequences of this alarming loss of marshland, which will likely have adverse effects on water quality and cause a decline of the invaluable wetland-ecosystem.

4 Management Measures

4.1 Response to Coastal Erosion

Egypt's Nile Delta with its coastal front on the Mediterranean is considered vulnerable to the impacts of climate change. In addition to the expected rise in sea level, shoreline erosion, stresses on fisheries, and saltwater intrusion into groundwater create major challenges. Therefore, the Egyptian government takes these challenges seriously. Since 1981, the Egyptian General Authority for Shores Protection (SPA) has had the central role and responsibility towards the following tasks: preparing an integrated study for shore protection, coastal conservation, developing proper plans and management strategies for existing and upcoming coastal projects, regular monitoring for the beach morphology, coastal structures maintenance, etc. (Abutaleb et al., 2018). Coastal adaptation structures against SLR in terms of protection technologies include but are not limited to seawalls, beach nourishment, artificial sand dunes, dune rehabilitation, storm surge barriers, and land reclamation (Fig. 8) (Sharaan et al., 2022). The best-known seawall in Egypt is the Mohamed Ali seawall, located at the coast of Abo Qir Bay, meant primarily to protect the low land behind it (1.5–2.5 m below sea level). Other examples are located on the coast of Burg El Burullus and west of Damietta. Also, revetments were implemented at different beaches along the Nile Delta, such as on the Rosetta promontory, east of Burg El-Burullus.

Recently, the Egyptian plans and strategies for promoting the coastal environment considered the climate change impacts (Fig. 8) (Sharaan et al., 2022). The constructed seawalls and revetments are subjected to regular rehabilitation and maintenance. Some implemented projects were reinforced by additional armor units (such as dolomite stones and tetrapod blocks). Also, some of the coastal structures were raised to negotiate the potential wave forces, storms, and overtopping scenarios in the future. Also, in some coastal areas, beaches with seawalls are nourished with sand or provided with sandbags, to further inforce the beach and increase its stability and durability.

4.2 Freshwater

Egypt has proposed a grand plan of sustainable development strategy, named Egypt's vision 2030, that aims to develop the country to a new level in a variety of aspects, including the improvement of the quality of life. Concerning freshwater resources, despite the present severe problems with water shortage, lack of wastewater retreatment, and abuse of recycled water, they have put up new programs on improving the environment.

1. Strengthening the institutional and legislative structure of the water resources management system.
2. Expanding infrastructure for supporting a sustainable water system.
3. Adopting fiscal policy reforms to encourage sustainable consumption patterns of water and natural resources.
4. Raising the awareness to conserve environment and natural resources, providing incentives for more advanced alternatives and technologies for water conservation and natural resources protection.

5 Future Challenges

5.1 Water Issues

Egypt faces various water-related stresses and challenges. Egypt's water is the most important pillar in achieving national wealth and prosperity. With more than 95% of Egypt's renewable freshwater resources originating from outside the country, the most significant water resources challenge is water scarcity. The realization of Egypt's comprehensive "Sustainable Development Strategy 2030" and its related plans concerning social, economic and environmental dimensions depend directy on the state's capability to provide the necessary water resources.

The Cairo Water Week (CWW) www.cairowaterweek.eg is considered as the biggest water-related event in Egypt, organized annually by the Ministry of Water Resources and Irrigation, Egypt in collaboration with national, regional and international partners. It aims to promote water awareness and foster innovative approaches to face the most pressing water-related challenges, stay abreast with global developments to address water challenges and identify modern tools and techniques for managing water resources.

The CWW started its first edition in 2018 with the main theme "Water Conservation for Sustainable Development.", and continued in the following years with different themes. It is becoming a platform for the exchange of expertise, development of appropriate solutions and formulation of recommendations to achieve Egypt's sustainable development goals, as it hosted high-level official meetings, important activities, plenary and technical sessions.

5.2 Potential Impacts of SLR, Coastal Erosion, and Salt Water Intrusion

The Egyptian Water Partnership (EWP) (cedare.org) is registered as an NGO to introduce Integrated Water Resources Management (IWRM). The Egyptian Water Partnership promotes and coordinates water management with the aim of maximizing the social, economic and environmental benefits from water uses. EWP has set a series of adaptive policies against the potential impacts of SLR, coastal erosion, and salt-water intrusion. Among these are:

(1) Changing the cropping patterns in the Nile Delta as a strategic climate change adaptation measure. Crops with high tolerance to salinity should be introduced to agricultural areas close to the Mediterranean.
(2) Develop integrated coastal zone management (ICZM) plans for all coastal governorates and cities concerning natural coastal protection measures such as sand dunes, along with other structures such as seawalls, and concerning the protection of the coastal highway.

(3) Develop coastal protection policies that go in line with ICZM plans; such policies should specify the distance from the coast where damage is to be expected.
(4) In addition, results of Frihy et al. (2021) on the assessment of natural coastal hazards at the Alexandria/Nile Delta interface, could help decision-makers to enact appropriate mitigation measures against natural hazards, particularly at low-lying vulnerable sectors, to alleviate future impacts due to a possible rise in sea level and extreme climatic conditions.

5.3 Wastewater Treatment Station of Bahr El-Bakar Drain—A Huge National Project

Egypt opened the largest wastewater treatment plant in the world. One of the biggest challenges facing Egypt in the coming years is to secure enough water of sufficient quality to sustain its rapidly growing population and expanding economy. With the effect of climate change and threat of the Grand Ethiopian Renaissance Dam (GERD), policymakers have to turn to alternative sources to prevent water stress.

Bahr El-Bakar drain is considered as one of the most polluted drains in Egypt (Fig. 9) (Abu Khatita et al., 2016). Industrial waste and sewage from Cairo and agricultural run-off, pesticides and fertilizers from the Delta all end up flowing into the drain. Unfortunately, at the last decades, great areas on both sides of the drain were using its polluted water for irrigation and fishing causing a high risk to the surrounding environment (Elkiki, 2018). Bahr El Bakar wastewater treatment plant is the largest facility of its kind in the world, treating around 5 million cubic meters of water from the Bahr El Bakar basin every day.

This amount of treated water enables the reclamation and cultivation of 400,000 acres of land, it provides opportunities for agriculture-based industry, cattle growing, etc., and it creates more than 40,000 new job opportunities.

5.4 Black Sands: A Huge National Project Launched on Kafrelsheikh North Coasts

Kafrelsheikh governorate is observing another huge national project along its coasts. It is the black sand factory established in Burullus city in Kafrelsheikh governorate. As a new addition to the series of national projects in Egypt, the black sand plant is the first of its kind in the world with the most advanced mining equipment. It aims to extract 41 radioactive elements and other exploitable minerals from the black sands. A recent aerial survey done by the Egyptian Nuclear Materials Authority proved that there are 11 sites on the Egyptian north coast rich in black sands, extending over 400 km along the coast. These black sands are sources for many economically interesting minerals such as Monazite, Rutile, Ilmenite, Zirconium, Uranium, and

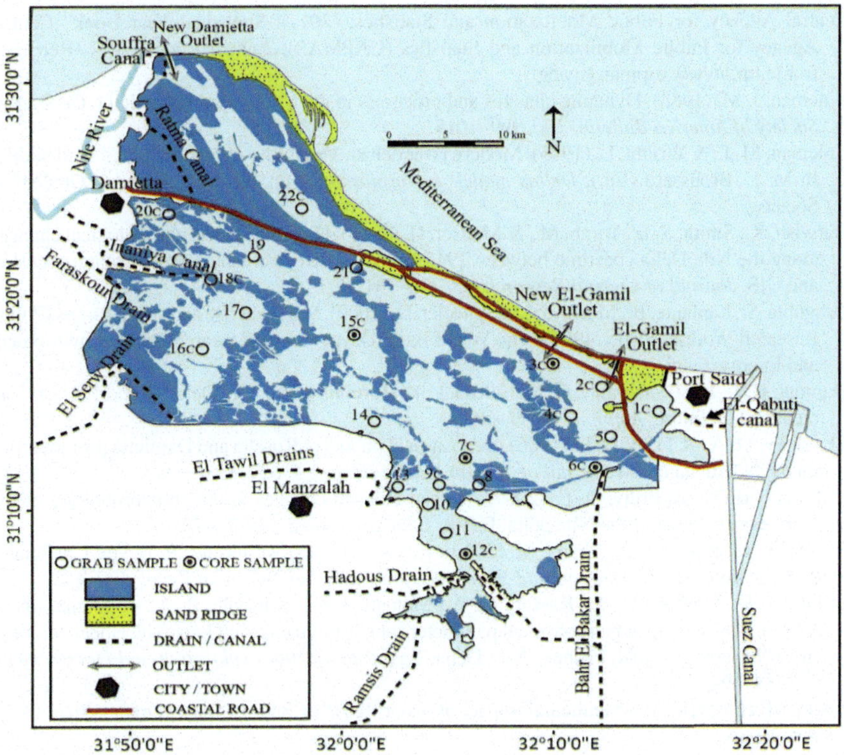

Fig. 9 Bahr El-Bakar drain (Abu Khatita et al., 2016)

other raw materials used in modern industry. The project also contributes to providing more than 5,000 direct and indirect job opportunities.

References

Abdel-Fattah, S., Amin, A., & Van Rijn, L. C. (2004). Sand transport in Nile River, Egypt. *Journal of Hydraulic Engineering, 130*, 488–500.

Abd-Elmabod, S. K., Fitch, A. C., Zhang, Z., Ali, R. R., & Jones, L. (2019). Rapid urbanization threatens fertile agricultural land and soil carbon in the Nile delta. *Journal of Environmental Management, 252*, 109668.

Abu Khatita, A. M., de Wall, H., & Koch, R. (2016). Anthropogenic particle dispersions in topsoils of the Middle Nile Delta: A preliminary study on the contamination around industrial and commercial areas in Egypt. *Environmental Earth Sciences, 75*, 1–19.

Abutaleb, K. A. A., Mohammed, A. H. E. S., & Ahmed, M. H. M. (2018). Climate change impacts, vulnerabilities and adaption measures for Egypt's Nile Delta. *Earth Systems and Environment, 2*, 183–192.

Central Agency for Public Mobilization and Statistics. (2021). Retrieved from http://www.capmas gov.eg/

Central Agency for Public Mobilization and Statistics. (2017). Statistical Year Book" Central Agency for Public Mobilization and Statistics (CAPMAS), Annual Report 2017. Retrieved from http://www.capmas.gov.eg

Coleman, J. M. (1988). Dynamic changes and processes in the Mississippi River delta. *Geological Society of America Bulletin, 100*, 999–1015.

Coleman, M. J., & Wright, L. (1975). Modern river deltas: Variability of processes and sand bodies. In M. L. Broussard (Ed.), *Deltas models for exploration* (pp. 99–149), Houston Geological Society.

Darwish, K., Smith, S. E., Torab, M., & Monsef, H. (2017). Hussein, O Geomorphological changes along the Nile Delta coastline between 1945 and 2015 detected using satellite remote sensing and GIS. *Journal of Coastal Research, 33*, 786–794.

Dasgupta, S., Laplante, B., Murray, S., & Wheeler, D. (2009). Sea-level rise and storm surges [Policy Research Working Paper 4901]. The World Bank-Development Research Group-Environment and Energy Team.

Dumont, H. J., & El-Shabrawy, G. M. (2007). Lake Borullus of the Nile Delta: A short history and an uncertain future. *Ambio, 36*, 677–682.

El Askary, M. A., & Frihy, O. E. (1986). Depositional phases of Rosetta and Damietta promontories on the Nile delta coast. *Journal of African Earth Sciences, 5*, 627–633.

El Banna, M. M., & Frihy, O. E. (2009). Human-induced changes in the geomorphology of the Northeastern coast of the Nile delta, Egypt. *Geomorphology, 107*, 72–78.

Elbana, T. A., Bakr, N., & Elbana, M. (2019). Reuse of treated wastewater in Egypt: Challenges and opportunities. *Unconventional Water Resources and Agriculture in Egypt*, 429–453.

El-Din, E. S., Mesbah, M. A., Kassab, M. A., Mohamed, I. F., Cheadle, B. A., & Teama, M. A. (2013). Assessment of petrophysical parameters of clastics using well logs: The Upper Miocene in El-Wastani gas field, onshore Nile Delta, Egypt. *Petroleum Exploration and Development, 40*, 488–494.

Elkiki, M. H. (2018). Environmental impact of water Reuse of Bahr El Baqar Drain—State of the Art Report.

ElKotby, M. R., Sarhan, T. A., & El-Gamal, M. (2023). Assessment of human interventions presence and their impact on shoreline changes along Nile delta, Egypt. *Oceanologia, 65*, 595–611.

El-Manadely, M. S., Abdel-Bary, R. M., El-Sammany, M. S., & Ahmed, T. A. (2002). Characteristics of the delta formation resulting from sediment deposition in Lake Nasser, Egypt: Approach to tracing lake delta formation. *Lakes and Reservoirs: Research and Management, 7*, 81–86.

El-Raey, M. (2010). Impacts and implications of climate change for the coastal zones of Egypt. *Coastal Zones and Climate Change, 20*, 31–50.

El-Raey, M., Dewidar, K. R., & El-Hattab, M. (1999). Adaptation to the impacts of sea level rise in Egypt. *Mitigation and Adaptation Strategies for Global Change, 4*, 343–361.

Food and Agriculture Organization of the United Nations (FAO). (2008). FAOSTAT Online Statistical Service. FAO. Retrieved from http://faostat.fao.org/default.aspx

Food and Agriculture Organization of the United Nations (FAO). (2013). Monitoring of climate change risk impacts of sea level rise on groundwater and agriculture in the Nile Delta.

Food and Agriculture Organization of the United Nations (FAO). (2018). Statistical data. FAOSTAT.

Fishar, M. R. (2016). Nile Delta (Egypt). In C. Finlayson, G. Milton, R. Prentice, & N. Davidson (Eds.), *The Wetland book* (pp. 1–10). Springer.

Frihy, O. (1988). E Nile Delta shoreline changes: Aerial photographic study of a 28-year period. *Journal of Coastal Research, 10*, 597–606.

Frihy, O. E., & El-Sayed, M. K. (2013). Vulnerability risk assessment and adaptation to climate change induced sea level rise along the Mediterranean coast of Egypt. *Mitigation and Adaptation Strategies for Global Change, 18*, 1215–1237.

Frihy, O. E., Deabes, E. A., & El Gindy, A. A. (2010). Wave climate and nearshore processes on the Mediterranean coast of Egypt. *Journal of Coastal Research, 26*, 103–112.

Frihy, O., & Lawrence, D. (2004). Evolution of the modern Nile delta promontories: Development of accretional features during shoreline retreat. *Environmental Geology, 46*, 914–931.

Frihy, O. E., Mohamed, S., Abdalla, D., & El-Hattab, M. (2021). Assessment of natural coastal hazards at Alexandria/Nile Delta interface, Egypt. *Environmental Earth Sciences, 80*, 1–15.

Garzanti, E., Andò, S., Padoan, M., Vezzoli, G., & El Kammar, A. (2015). The modern Nile sediment system: Processes and products. *Quaternary Science Reviews, 130*, 9–56.

General Authority for Fish Resources Development. (2015). Retrieved from http://www.gafrd.org/

Gu, J., Salem, A., & Chen, Z. (2013). Lagoons of the Nile delta, Egypt, heavy metal sink: With a special reference to the Yangtze estuary of China. *Estuarine, Coastal and Shelf Science, 117*, 282–292.

Hamza, W. (2006). Estuary of the Nile. In P. Wangersky (Ed.), *Estuaries; handbook of environmental chemistry*. Springer.

Hatab, A. R. A., Shoumann, N. A., & Xuexi, H. (2012). Exploring Egypt-China bilateral trade: Dynamics and prospects. *Journal of Economic Studies, 39*, 314–326.

Hereher, M. E. (2010). Vulnerability of the Nile Delta to sea level rise: An assessment using remote sensing. *Geomatics, Natural Hazards and Risk, 1*, 315–321.

Inman, D. L., & Jenkins, S. A. (1984). The Nile littoral cell and man's impact on the coastal zone of the Southeastern Mediterranean. *Coastal Engineering Proceedings, 19*, 109–109.

Iskander, M. M. (2013). Wave climate and coastal structures in the Nile Delta coast of Egypt. *Emirates Journal for Engineering Research, 18*, 43–57.

Kassim, Y., Mahmoud, M., Kurdi, S., & Breisinger, C. (2018). An Agricultural Policy Review of Egypt: First Steps towards a New Strategy, International Food Policy Research Institute.

Keshta, A. E., Alexis Riter, J. C., Kamal H. Shaltout, K. H., Baldwin, A. H., Kearney, M., Sharaf El-Din, A., & Eid, E. M. (2022). Loss of coastal wetlands in Lake Burullus, Egypt: A GIS and remote-sensing study. *Sustainability, 14*, 4980.

Manohar, M. (1976). Dynamic factors affecting the Nile delta coast. In *Proceedings of the Seminar on Nile Delta Sedimentology. Alexandria UNESCO/ASRT/UNDP* (pp. 104–129).

Mikhailova, M. V. (2001). Hydrological regime of the Nile Delta and dynamics of its coastline. *Water Resources, 28*, 477–490.

Moussa, A. M. (2013). A Predicting the deposition in the Aswan High Dam Reservoir using a 2-D model. *Ain Shams Engineering Journal, 4*, 143–153.

MWRI. (2005). National Water Resources Plan for Egypt 2017. Ministry of Water Resources and Irrigation.

MWRI. (2012). Strategy of water resources of Egypt till 2050. Ministry of Water Resources and Irrigation.

Negm, A. M., Saavedra, O., & El-Adawy, A. (2016). Nile Delta biography: Challenges and opportunities. *In the Nile Delta, 55*, 3–18.

Nizeyimana, E. L., Petersen, G. W., Imhoff, M. L., Sinclair, H. R., Jr., Waltman, S. W., Reed-Margetan, D. S., & Russo, J. M. (2001). Assessing the impact of land conversion to urban use on soils with different productivity levels in the USA. *Soil Science Society of America Journal, 65*, 391–402.

Oczkowski, A. J., & Nixon, S. W. (2010). Lagoons of the Nile Delta. In M. J. Kennish & H. W. Paerl (Eds.), *Coastal Lagoons* (pp. 253–282). CRC Press.

Oczkowski, A. J., Flower, R. J., Thompson, J. R., Ayache, F., Ahmed, M. H., Ramdani, M., & Turner, S. (2011). Evidence of North Africa's green revolution preserved in sedimentary organic matter deposited in three coastal lagoons. *Ecological Applications, 21*, 1708–1717.

Oczkowski, A., & Nixon, S. (2008). Increasing nutrient concentrations and the rise and fall of a coastal fishery; a review of data from the Nile Delta, Egypt. *Estuarine, Coastal and Shelf Science, 77*, 309–319.

Osama, S., Elkholy, M., & Kansoh, R. M. (2017). Optimization of the cropping pattern in Egypt. *Alexandria Engineering Journal, 56*, 557–566.

Padoan, M., Garzanti, E., Harlavan, Y., & Villa, I. M. (2011). Tracing nile sediment sources by Sr and Nd isotope signatures (Uganda, Ethiopia, Sudan). *Geochimica Et Cosmochimica Acta, 75*, 3627–3644.

Palmer, A. (2008). Rising sea levels. The World Bank Report. Retrieved from http://www.theworldincrisis.com/artman2/publish/climate/Rising_Sea_Levels.shtml

Penland, S., Boyd, R., & Suter, J. R. (1988). Transgressive depositional systems of the Mississippi Delta plain; a model for barrier shoreline and shelf sand development. *Journal of Sedimentary Research, 58*, 932–949.

Pennington, B. T., Sturt, F., Wilson, P., Rowland, J., & Brown, A. G. (2017). The fluvial evolution of the Holocene Nile Delta. *Quaternary Science Reviews, 170*, 212–231.

Said, R. (1993). The River Nile. Pergamon.

Sefelnasr, A., & Sherif, M. (2013). Impacts of seawater intrusion in the Nile Delta Aquifer, Egypt. *Ground Water, 52*, 264–276.

Sestini, G. (1989). Nile Delta: A review of depositional environments and geological history. *Geological Society, London, Special Publications, 41*, 99–127.

Sestini. (1992). Implications of climatic changes for the Nile delta. In L. Jeftic, J. D. Milliman & G. Sestini (Eds.), *Climatic change and the Mediterranean* (pp. 535–601). Edward Arnold.

Shalaby, A., & Moghanm, F. S. (2015). Assessment of urban sprawl on agricultural soil of northern Nile Delta of Egypt using RS and GIS. *Chinese Geographical Science, 25*, 274–282.

Sharaan, M., & Udo, K. (2020). Projections of future beach loss along the Mediterranean coastline of Egypt due to sea-level rise. *Applied Ocean Research, 94*, 101972.

Sharaan, M., Iskander, M., & Udo, K. (2022). Coastal adaptation to sea level rise: An overview of Egypt's efforts. *Ocean & Coastal Management, 218*, 106024.

Sherif, M. (1999). The Nile Delta aquifer in Egypt. Chapter 17 in seawater intrusion in coastal aquifers, concepts methods and practices. In J. Bear, A. Cheng, S. Sorek, D. Ouazar, & A. Herrera (Eds.), *Theory and application of transport in porous media* (pp. 559–590). Kluwer Academic Publishers.

Shetaia, S. A., Khatita, A. M. A., Abdelhafez, N. A., Shaker, I. M., & El Kafrawy, S. B. (2022). Human-induced sediment degradation of Burullus lagoon, Nile Delta, Egypt: Heavy metals pollution status and potential ecological risk. *Marine Pollution Bulletin, 178*, 113566.

Siegel, F. R., Slaboda, M. L., & Stanley, D. J. (1994). Metal pollution loading, Manzalah lagoon, Nile Delta, Egypt: Implications for aquaculture. *Environmental Geology, 23*, 89–98.

Smith, S. E., & Abdel-Kader, A. (1988). Coastal erosion along the Nile delta. *Journal of Coastal Research, 4*, 245–255.

Stanley, D. J., & Warne, A. G. (1993). Nile Delta: Recent geological evolution and human impact. *Science, 260*, 628–634.

Stanley, D. J., & Warne, A. G. (1994). Worldwide initiation of Holocene marine deltas by deceleration of sea-level rise. *Science, 265*, 228–231.

Stanley, D. J., & Wingerath, J. G. (1996). Clay mineral distributions to interpret Nile cell provenance and dispersal: I. Lower River Nile to delta sector. *Journal of coastal Research, 12*, 911–929.

Stanley, D. J., & Warne, A. G. (1998). Nile Delta in its destruction phase. *Journal of Coastal Research, 14*, 795–825.

Stanley, J. D., & Clemente, P. L. (2017). Increased land subsidence and sea-level rise are submerging Egypt's Nile Delta coastal margin. *GSA Today, 27*, 4–11.

Wahaab, R. A., & Badawy, M. I. (2004). Water quality assessment of the River Nile system: An overview. *Biomedical and Environmental Sciences, 17*, 87–100.

Williams, M. A., & Adamson, D. A. (2021). A land between two Niles: Quaternary geology and biology of the Central Sudan. Routledge.

Woodward, J., Macklin, M. G., Krom, M. D., & Williams, M. A. J. (2022). The River Nile: Evolution and environment. In A. Gupta (Ed.), *Large Rivers: Geomorphology and management* (2nd ed., pp. 959–1000). Wiley.

Pearl River Delta

Haixian Xiong, Jie Ren, Huan Liu, Junjie Deng, Leiping Ye, Benwei Shi, and Jiaxue Wu

1 Physical Setting

1.1 The Pearl River

The Pearl River is one of the world's 25 largest rivers in terms of annual water discharge and sediment load (Eisma, 1998; Zhao et al., 2000). In China, the Pearl River ranks second (after the Yangtze River) in terms of water discharge and third (after the Yellow and Yangtze rivers) in terms of sediment load. The Pearl River originates from the Yunnan Plateau, crossing hill country and even mountainous areas (Wu & Zhou, 2001), and extending 2400 km eastwards to the South China Sea. The river catchment is 450,000 km^2 in area, and experiences a subtropical and tropical monsoon climate straddling the Tropic of Cancer. The dry season extends from October to March, followed by a rainy season from April to September. The multi-year mean annual temperature within the catchment is 14–22 °C and the catchment-wide average precipitation is 1470 mm/yr (Dai et al., 2008). It is a compound river system comprising the West, North and East Rivers and several small rivers that drain into the deltaic basin before entering the South China Sea (Fig. 1a). The average water and sediment discharges from the three main rivers are 280 billion m^3/a and 65 million t/a respectively, according to China Bulletin of Rivers and Sediment (1955–2020). Amongst the three main rivers, the West River is the largest with a total length of 2214 km and a drainage area of 350,000 km^2. It has a mean water discharge of 217

H. Xiong (✉) · J. Ren · H. Liu · J. Deng · L. Ye · J. Wu
School of Marine Sciences, Sun Yat-Sen University, Zhuhai 519082, China
e-mail: xionghx5@mail.sysu.edu.cn

Southern Marine Science and Engineering Guangdong Laboratory, Zhuhai 519082, China

B. Shi
State Key Laboratory of Estuarine and Coastal Research, East China Normal University, Shanghai 200241, China

W. Zhang and H. de Vriend (eds.), *Delta Sustainability*,
https://doi.org/10.1007/978-981-97-7259-9_13

billion m^3/a and an average sediment flux of 58.2 million t/a, which account for 77% and 88% of whole of the Pearl River system, respectively. This makes the West River the most important artery that supplies water and sediment into the deltaic basin. Approximately 80% of sediment influx is deposited within the Pearl River deltaic basin, and the remainder is drained into the South China Sea (Xu et al., 1985). The majority of freshwater and sediment discharged from the Pearl River tends to follow the outgoing tides and flow southwest into the marine area along with the westward coastal current (Gao et al., 2015; Liu et al., 2014).

1.2 The Pearl River Delta

The Pearl River Delta was developed by discharges of the Pearl River system into the Pearl River deltaic basin (Fig. 1). As a result, it is a complex delta that consists of the West-North-Rivers sub-delta, East River sub-delta, Tan River sub-delta and other small river sub-deltas. The geomorphological definition of the delta starts in the west at the eastern entrance of Gaoyao Lingyang Gorge and the Tan River Siqian, and stretches northward to the North River Huangtang, Baoyue, Liuxi River Guangzhou, Shijie, Sui River Huanggang; and eastward to East River Yuanzhou and Zeng River Shatang. It includes the Sanshui Basin east of Gaoyao and south of Shi Jiao, the lower East River delta below Yuanzhou, as well as the lower West-North-Rivers delta and the lower Tan River delta. Administratively, the term Pearl River Delta refers to a much larger area that covers nine prefectures of the province of Guangdong, namely Dongguan, Foshan, Guangzhou, Huizhou, Jiangmen, Shenzhen, Zhaoqing, Zhongshan and Zhuhai and the SARs of Hong Kong and Macau.

The topography of the Pearl River delta is characterized by mountains on three sides and the South China Sea to the south; the deltaic plain accounts for 4/5 of the area, while rocky uplands make up about 1/5. By AD 2020, the total area of the deltaic plain excluding the rocky uplands is estimated at 7500 km^2 (Xiong et al., 2020a, 2020b). The deltaic plain is quite low-lying, and 47% of its area has an elevation lower than 0.4 m above the mean sea level, due to the extensive coastal land reclamation (Huang et al., 1982). On the southern side of the central deltaic basin is an upland landscape, called Wugui mountains (Fig. 1b). This upland landscape separates the Pearl River mouth into two parts. Four deltaic river outlets (i.e. Humen, Jiaomen, Hongqimen, Hengmen) run into the Lingdingyang Bay, or the Pearl River estuary, east of Wugui mountains, whereas four other outlets (i.e. Modaomen, Jitimen, Hutiaomen and Yamen) drain directly into the South China Sea on the west of Wugui mountains. The mouths of these active deltaic outlets are areas that receive the fluvial sediment which forms fast-growing mudflats that have been rapidly reclaimed in past decades. The subaqueous part of the Holocene delta extends as far as 40 km offshore from the outer coastline of Modaomen and as deep as 50 m water depth. The underwater landforms include subtidal flats (water depth <10 m, slope $1'40''$), delta front slopes (10–30 m, $19'25''$ to $1°25'$), estuarine bays (<20 m, $3'25''$) and prodelta (30–50 m, $27'$) (Huang et al., 1995).

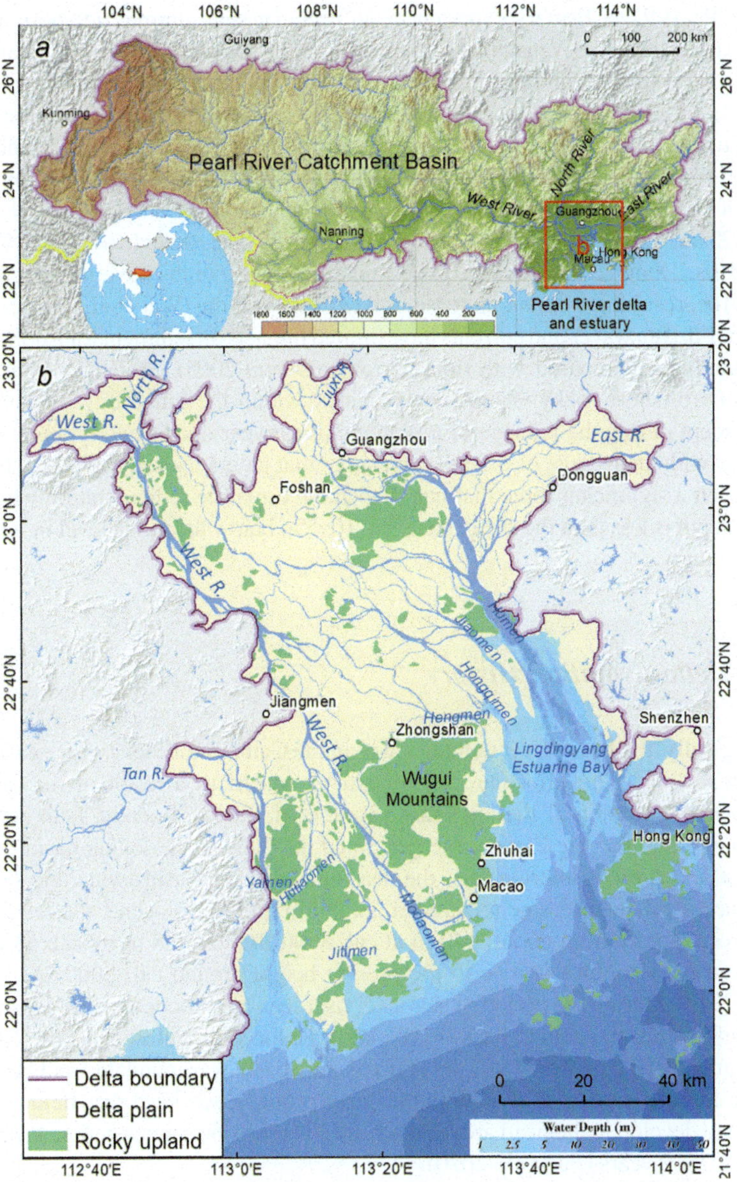

Fig. 1 Landscape maps of the Pearl River catchment basin (**a**) and the Pearl River delta and estuary (**b**)

The Pearl River Estuary experiences an irregular semi-diurnal tide with a small tidal range. Near the Wanshan Archipelago at the mouth of the Lingdingyang, the tidal range is 0.8–0.9 m, reaching 1.7 m near Humen, with an average tidal range of about 1.5 m. Tidal range can increase to 2.29 m and 3.36 m in these two locations, respectively, during the highest astronomical tide. The dominant semi-diurnal tidal constituent in the region is typically M2, followed by K1, O1, and S2. The amplitude ratio of (K1 + O1)/M2 ranges from 0.94 at Humen to 1.77 at Wanshan Islands in PRE (Zhao, 1990). The tidal prism in Lingding Bay decreased from the 1900s to the 2010s (Deng & Bao, 2011). Tidal ranges at river mouths have increased since the 2010s, resulting from an increase in bay water volume (Wei et al., 2021).

Wind-driven waves and currents have limited impacts on sediment transport within the estuary, as compared with tidal currents (Owen, 2005). Protected from storm waves by offshore rocky islands, wave energy within the Lingdingyang estuary is low, except during the passage of a typhoon. Each year one or two typhoons strike the area with an average of 0.5-m storm surge that enters the estuary. The typhoons have been a significant force that has shaped the estuarine sedimentation where the water depth is less than the storm base, i.e. 10–5 m below mean sea level in the Pearl River estuary.

1.3 Delta Evolution History

Geologically, the Pearl River delta lies within the Cathaysia Block, where a regional subsidence since the Miocene dips towards southeast as a result of the ongoing continental extension of the landmass (Zong et al., 2015). Locally, fault activities continuing into the late Pleistocene generated the fault-block deltaic basin. During the last and present interglacials, the sea inundated the basin twice and left two sequences of deltaic sediment (Zong et al., 2009). In the Holocene, the relative sea level rose rapidly from about −40 m at cal. 8000 BC to −0.8 m by cal. 5500 BC, followed by a continuous rise to the present height around cal. 2000 BC (Xiong et al., 2018). During the period of sea-level rise, seawater flooded the palaeo-incised channels within the deltaic basin in the early Holocene to allow the fast sediment accumulation. By cal. 5500 BC, the entire deltaic basin was inundated and became a huge estuarine bay. Coinciding with the deceleration of sea-level rise, deltaic plains started to develop in the head area and advanced across the central deltaic basin over the past 7500 years (Fu et al., 2020).

In the central basin and the bay area, except at locations that were close to former rocky islands where sediment accumulation continued at a higher rate (Wu et al., 2007), the sediment accumulation rate was very slow during the middle Holocene. This implies that the sedimentation in this period was concentrated within the head area; and the central basin and the bay area were starved of sediment supply and received only small amounts of estuarine/prodelta sediments. By about 4000 years

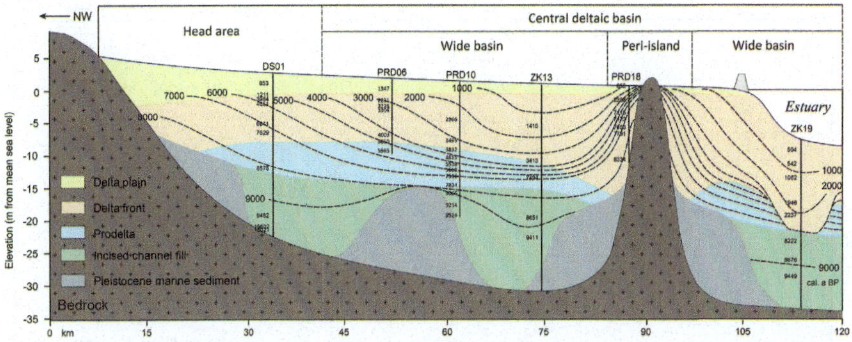

Fig. 2 A summary of the Holocene sediment accumulation history for the Pearl River delta. This integrated model shows a systematic delta progradation operating over paleo-landscapes, resulting in different Holocene sediment accumulation histories between the head area, central wide basin and peri-island area (after Fu et al., 2020)

ago, whilst the head area was almost filled up and changed into a freshwater floodplain, the accumulation of delta front sediment in the central deltaic basin started to accelerate, first in the more up-estuary locations, then in the central area and later in the more down-estuary locations (Fig. 2). During this infilling process, the deposits at these locations changed from delta front facies to delta plain facies around 2000 years ago (Zong et al., 2009). The past 2000 years saw an accelerated delta progradation due to human activities, such as agriculture-induced soil erosion and land reclamation (Xiong et al., 2020a, 2020b).

Palaeo-shorelines of the Pearl River delta can be identified by evidences of intertidal facies from borehole sediments, archaeological relics (i.e. shell mounds, settlement sites and tombs), historical records (county chronicles) on the village development and land reclamation records. The position of the most landward shoreline (5500 BC) is inferred based on sedimentary records from the head areas of the Pearl River delta (Fig. 3). By 5500 BC, the sea reached the maximum flooding extent into the head areas, and marine regression began. The bore-hole sedimentary records from the head areas indicates a sharp decrease in marine and brackish water diatoms and a change from planktonic dominated assemblages to benthic dominated assemblages. This change suggests a switch from subtidal to intertidal conditions, which is dated to around cal. 5500 BC in the head of West-North-Rivers delta plain, and cal. 4800 BC in the East River delta.

The second shoreline (2500 BC) aligns closely with the seaward limit of the Neolithic shell mounds. These shell mounds were found mostly at the foot of rocky uplands (Huang, 1996) which scattered across the head areas of the deltaic basin and along the northern deltaic boundary. These mounds were dated to mostly between 4500 and 2500 BC (Huang et al., 1982; Li et al., 1991). Shell mounds on the East River delta are found mostly along the northern edge of the delta plains, such as the one at Jinlansi. In other words, the 2500 BC shoreline could be delineated along the seaward margin of these shell mounds. Noteworthy is that the mounds along the 2500

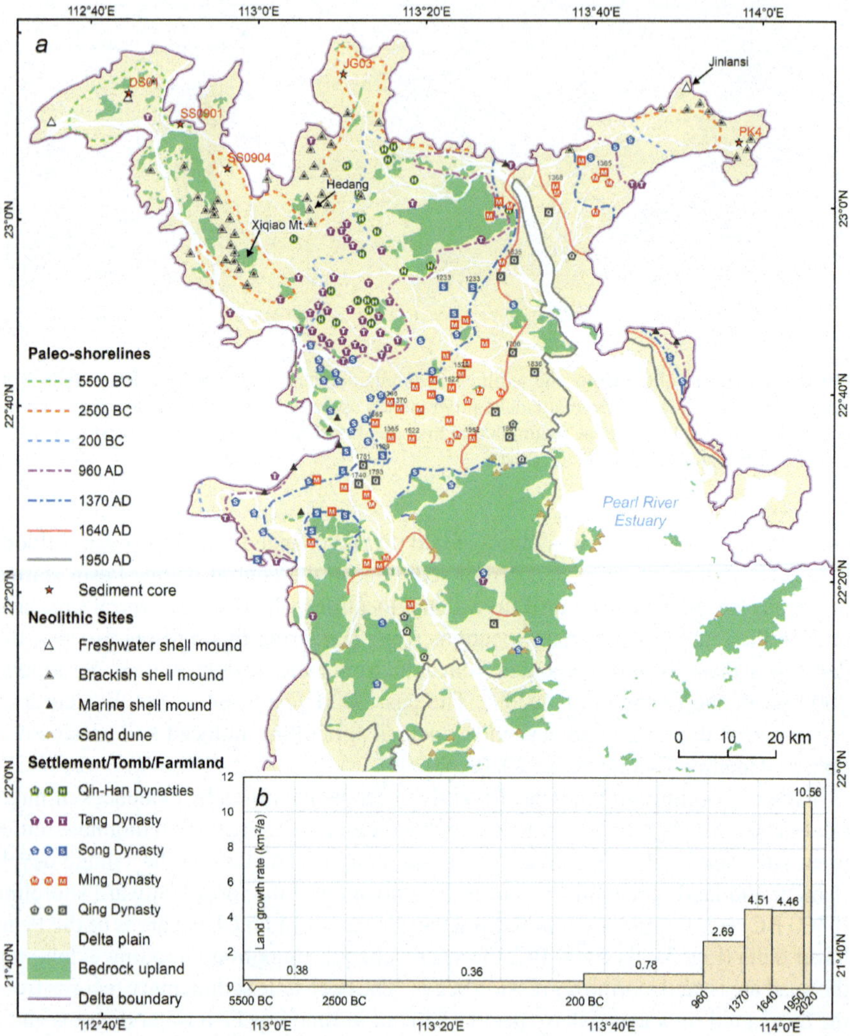

Fig. 3 **a** The deltaic shorelines of 5500 BC, 2500 BC, 200 BC, AD 960, AD 1370, AD 1640 and AD 1950, with locations of sediment cores (DS01, SS0901, SS0904, JG03 and PK4), archaeological sites, and ancient settlements, tombs and farmlands. **b** Land growth rate during seven periods as 0.38 km^2/a in 5500–2500 BC, 0.36 km^2/a in 2500–200 BC, 0.78 km^2/a in 200 BC–AD 960, 2.69 km^2/a in Song-Yuan (AD 960–1370), 4.51 km^2/a in Ming (AD 1370–1640), 4.46 km^2/a in Qing-1950s (AD 1640–1950) and 10.56 km^2/a in AD 1950–2020 (after Xiong et al., 2020a, 2020b)

BC shoreline contain shells dominated by brackish-marine water species, whereas those located further inland are with a mixture of brackish water and freshwater species (Huang, 1996). Such shell mounds are almost absent from the central basin.

By the beginning of the Han Dynasty (206 BC to AD 220), extensive freshwater wetlands emerged in the head area of the West-North-Rivers delta (Zong et al., 2013). Similarly, freshwater wetlands also emerged around the head area of the Liuxi River plains, but the deltaic areas near Guangzhou City was still under saline conditions by this time (cal. 340 BC, Zong et al., 2013). This period of landscape change coincided with the increase of population. As a result, agricultural cultivation in the northern part of the deltaic basin started to intensify. Since the establishment of the Nanyue Kingdom (204–112 BC) at the site of the present Guangzhou City, villages proliferated on the emerged delta plains along the northern edge of the central deltaic basin. Although there are few archaeological sites for the Bronze Age, the earliest sites of the Han settlements are helpful for identifying the 200 BC shoreline that runs along the seaward limit of the Han settlement sites.

As the deltaic shoreline advanced, new immigrants tended to occupy and reclaim the newly emerged tidal flats. The shorelines of AD 960 (end of Tang), AD 1370 (end of Song-Yuan) and AD 1640 (end of Ming) are identified as each runs through the seaward edge of the settlements/farmlands of each of these dynasties respectively. The deltaic shorelines between AD 1950 and 2020 can be accurately obtained from the survey maps and satellite images.

Together, these shorelines show a three-stage developmental history of the deltaic plain (Fig. 3). First, a consistently slow land growth rate at about 0.36–0.38 km^2/a occurred from 5500 to 200 BC, i.e. the Neolithic and Bronze Age. Second, the average growth rate increased substantially to 0.78 km^2/a over a period between 200 BC and AD 960. This rate was accelerated sharply to 2.69 km^2/a in AD 960–1370 and further to about 4.50 km^2/a in AD 1370–1950. In other words, the agricultural stage saw a rising trend of land growth. The third stage (1950–2020 AD) mainly corresponds to the industrial era for the Pearl River delta area, during which the average land growth rate increased sharply to 10.56 km^2/a.

2 Socio-economic Characteristics

2.1 Population and Economic Activities

The Pearl River Delta metropolitan area is composed of Hong Kong, Macao SARs and nine cities of Guandong Province: Guangzhou, Shenzhen, Zhuhai, Foshan, Huizhou, Dongguan, Zhongshan, Jiangmen, and Zhaoqing, covering a land area of 5.6×10^4 km^2 (Fig. 4). This area is much larger than the geomorphological Pearl River delta which has a delta plain of about 7,500 km^2. Nowadays, it is also known as the "Guangdong-Hong Kong-Macao Greater Bay Area".

Fig. 4 The Pearl River Delta metropolitan area, also known as the Guangdong-Hong Kong-Macao Greater Bay Area (Greater Bay Area)

In 2022, the total population of the Pearl River Delta reached 86 million, an increase of 45 million compared to 2000, with an average annual growth rate of 8.7%. The population growth pattern in the Pearl River Delta has shifted from a predominantly natural increase model characterized by high birth rates, low death rates, net emigration, and limited immigration before the era of Reform and Opening-up Policy to a mechanical growth model characterized by low birth rates, low death rates, net immigration, and limited emigration. In terms of spatial distribution, economically developed cities (such as Hong Kong, Macao, Shenzhen, Guangzhou, Dongguan, Foshan, and Zhongshan) have a much higher population density than underdeveloped cities (such as Jiangmen, Zhaoqing, and Huizhou), and there is a clear positive correlation between population distribution and economic development levels. With the urbanization, the population in the Pearl River Delta has further concentrated in urban areas, with the four cities of Guangzhou, Shenzhen, Dongguan, and Foshan hosting over 70% of the region's population. This has resulted in high urban population density and immense pressure on urban ecosystems.

The GDP of the Pearl River Delta has rapidly increased since 2000 and reached 13 trillion RMB by 2022. Its GDP per capita is 154,000 RMB. The urbanization rate is as high as 89%. However, there are significant disparities in economic and social development among the cities in the Pearl River Delta metropolitan area, and the spatial heterogeneity of population distribution poses challenges for regional development planning and resource utilization.

2.2 *Industries*

The Pearl River Delta metropolitan area is one of the most urbanized and industrialized regions in China. Since the start of Reform and Opening-up Policy, it has entered

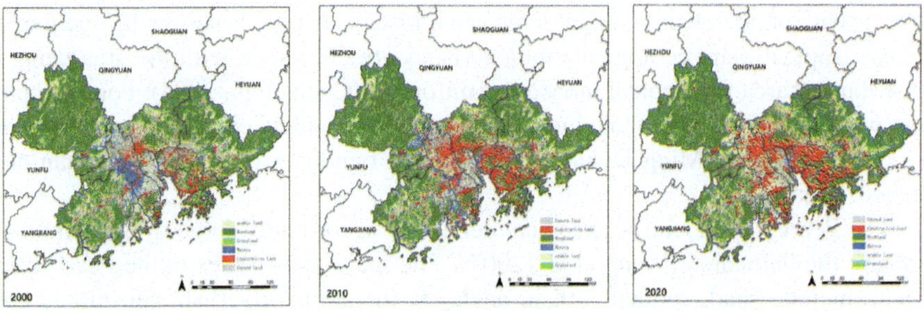

(a) (b) (c)

Fig. 5 Land use interpretation map of Guangdong, Hong Kong and Macao Greater Bay Area in 2000 (**a**), 2010 (**b**) and 2020 (**c**) (after Li et al., 2023)

the fast lane of industrialization, with the industrial economy gradually taking the lead in the national economy, and urban land space rapidly expanding (Fig. 5). In 1996, the three-sector structure (extraction of raw materials (primary), manufacturing (secondary), and service industries (tertiary)) of the PRD was 7.99:49.86:42.15, with the secondary sector dominant, and urban construction land accounting for 4.99% of the total land area. As the industrial structure upgraded and optimized, by 2022, the three-sector structure had changed to 3:32:65, with the tertiary sector taking the lead, and the proportion of urban construction land increasing to 10.05%.

In terms of industrial spatial distribution within the Pearl River Delta, the eastern bank of the Pearl River mainly concentrates on the electronics and information manufacturing industry, primarily following an embedded development model. Hong Kong is an international financial hub whilst Shenzhen serves as the regional finance center of southern China. On the western bank, the primary focus is on private enterprises, employing an endogenous development model.

2.3 *Natural Resources*

The Pearl River Delta has a rich abundance of vegetation and animal resources (Zhuang, 2012). The delta is home to over 500 common species, belonging to more than 130 families and 373 genera, of which 42% are purely tropical and 53% are pantropical. Evergreen monsoon rainforests, mainly composed of fengshui forests and mixed forests, are predominantly secondary in nature and primarily consist of tropical components. They occupy a relatively small area (Wu et al., 2010). In the extensive terrace and low hill areas, evergreen shrubs are more common (Wu et al., 2010). Tropical components dominate the grass slopes, which grow on lateritic soil layers and are prone to soil erosion. Carnivorous plants such as Nepenthes mirabilis are tropical. Tropical plant ecological characteristics are also fully reflected in commonly cultivated trees, such as aerial roots, buttress roots, and stilt roots found in banyan trees, which can form a single forest; old stems flowering in starfruit and pineapple trees; year-round flowering in gardenia and hibiscus; drip-tip leaves in the bodhi tree. These tropical plant ecological characteristics are consistent with the area being a distribution area for tropical monsoon rainforests (Chen et al., 2004). For example, the white champaca (also known as agarwood) was widely cultivated in Dongguan County during the Ming Dynasty, becoming a famous product of Dongguan and called "Guanxiang".

The Pearl River Delta is a tropical delta, and accordingly, large tropical animals inhabit the delta area (Wang et al., 2007). The main fish species in the Pearl River Delta include shad, spotted catfish, anchovy, threadfin, silverfish, eel, moray eel, grass carp, black carp, white carp, red-eyed catfish, Hainan red bream, big-eyed red bream, striped fish, Guangdong catfish, bream, yellow-tailed fusilier, spiny eel, southern white shrimp, dace, and sucker fish, among others. The Pearl River Delta is a water network area, dominated by swamps, depressions, and lowland forests, and as a result, it is home to many typical tropical megafauna, which also reflect the delta's original tropical characteristics. Asian elephants, Malay crocodiles, and peacocks, for example, are mentioned in ancient texts. The "Commentary on Waterways" states that "there are countless species of turtles, crocodiles, rare and strange creatures, and thousands of other species that cannot be recorded". Today, fishermen in Pingshou have also caught crocodile skulls in the river, indicating that crocodiles once lived in Guangzhou. Excavations in Shunde and Xinhui have uncovered entire crocodile skeletons, with radiocarbon dating from Tangxia indicating 3020 ± 80 years, and Leliu dating back to 2540 ± 105 years. This shows that the Pearl River Delta has been a distribution area for tropical animals since ancient times. The typical large benthic animals in the coastal wetlands of the Pearl River Delta are predominantly crustaceans, followed by polychaetes and mollusks, with dominant species mainly found in crustaceans, including shrimp, Chinese eyes shrimp, and Japanese big-eyed crabs (Wang et al., 2007).

3 Current Issues

3.1 Flooding and Waterlogging

The low-lying delta plain has long been vulnerable to seasonal floods and typhoon-induced storm surges. In history, the relatively low and weak dikes of the West and North Rivers were frequently damaged during flood seasons, causing inundation of farmlands and human settlements. For example, the well-known "Yi-Mao-year flood" (July, 1915) was a once-in-two-hundred-year catastrophic flood hitting the Pearl River delta plain. Guangzhou city was flooded for seven days (July 11–17), with over 10,000 people being killed or injured. Furthermore, the frequent typhoons from the Northwest Pacific Ocean often attack the Pearl River delta and generate disastrous storm surges. Historic documents reveal that six strong typhoons landing the delta or vicinity in AD 957, 1245, 1862, 1874, 1906 and 1937 caused death tolls of over 10,000 (Huang & Yim, 2007).

Fig. 6 Waterlogging events happening to Guangzhou in May 22, 2020 (left) and Shenzhen in Jun 17, 2023

In recent years, with the increase in extreme weather events, urban waterlogging has become a serious natural disaster impacting cities in the Pearl River delta. According to incomplete statistics, in 2010 alone, Guangzhou, a mega city in the Pearl River delta, experienced three severe urban waterlogging events, resulting in economic losses exceeding 1 billion RMB. Urban waterlogging in the delta occurs around 20 times per year. In terms of spatial distribution, central cities such as Guangzhou, Shenzhen, and Dongguan experience the most frequent waterlogging (over 60%) and suffer most seriously (Fig. 6). Among them, the old districts in the central cities are the main areas susceptible to urban waterlogging, accounting for over 80% of the total events, while new districts or suburban areas experience fewer. The main causes of urban waterlogging include climate factors, urban drainage systems, geographical locations, and management issues. The severity of urban waterlogging can be reduced through measures such as urban functional zoning, industrial relocation, and cost of living considerations.

3.2 *Sand Mining and Dredging*

Sand mining and dredging of navigation channels in the Pearl River estuary have drastically changed the shoreline and underwater topography, causing abnormalities in the delta's dynamics, sedimentation, and geomorphologic processes. Lingdingyang Bay is the main area of sand mining and dredging. An increase of the average water depth in this sand mining area from 1972 to 2017 is approximately 2 m. The most serious area of sand mining is located in the middle shoal. Sand excavation caused the seabed in this area to be generally 5–10 m deep, and the maximum erosion reached 17 m. Many deep pools are formed in the seabed, showing that sand mining activities have a significant impact on the topography of estuaries. During 1980–2015, channel dredging and sand mining removed nearly 3.6 Mt/year of sediment from upper Lingdingyang, which is equivalent to 30% of the total sediment load during this period (Wu et al., 2016). The accretion rate of the underwater Pearl River Delta has decreased from 3 mm/yr in the early 20th century to 0.5 mm/yr.

Large-scale sand mining also took place in the delta distributary network, and has deepened riverbeds, enhanced tidal dynamics in the river network area, and led to a steady increase in the tidal level of the Pearl River estuary at a rate of 1.4 mm/yr from 1965 to 2012 (Wu et al., 2012). The tidal influence has extended upstream, and the boundaries of the tidal zone, tidal current, and saltwater intrusion have all moved upstream, exacerbating saltwater intrusion in the delta plain. Due to channel dredging, high-salinity water from the continental shelf can more easily invade the Lingdingyang along the deep trough. The deepening of riverbeds also weakens the discharge capacity of the runoff entering the sea, intensifying waterlogging in the delta plain and increasing the risk of flood disasters. There are significant differences in the hydrodynamics between the eastern and western parts of Lingdingyang, with the western part dominated by river action and the eastern part dominated by tidal action. Fine-grained sediments input by rivers mainly accumulate in the western part. However, under the influence of high-intensity human activities, the above-mentioned dynamics and sedimentation patterns may have already changed.

3.3 Land Reclamation

Since AD 1950, large-scale land reclamation projects have taken place in both the east and west outlets of the Pearl River mouth (Fig. 7). Such activity has drastically advanced the shoreline and decreased the tidal prism of the estuarine bays. Within the east outlets, land growth occurred along the edge of the Lingdingyang, whereas within the west outlet, land growth took place in the form of filling up the shallow marine areas between islands. The annual land growth rate is initially constant at about 5.0 km^2/a in the 1950s and 1960s, followed by an increase to 7.4 km^2/a in the 1970s, 18.1 km^2/a in the 1980s, and the highest level of 21.8 km^2/a in the 1990s. Since entering the 21st century, the land growth rate decreases in the first decade to 13.4 km^2/a and drops further down to 3.3 km^2/a in the second decade. The changes in land growth rate happened when the sediment discharge continued to increase slightly from 71.9 million ton/a in the 1950s to 88.1 million ton/a in the 1980s, followed by a period of drastically decreasing discharge towards the present. By 2010s, sediment discharge had decreased to only 23.7 million ton/a. It is noteworthy that the water discharge during the past seven decades fluctuated narrowly between 260 and 300 billion m^3/a, a trend that is clearly different from the one of the sediment discharge. As a function of the land growth rate and sediment discharge, the land formation efficiency (the net land area formed per mass of delivered sediment) displays a general increasing trend from 0.07 to 0.38 m^2/ton until end of the 2010s when it suddenly drops by two third to 0.14 m^2/ton. This seems to suggest that the improvement of engineering techniques in land reclamation is finally insufficient to counter-balance the effects of the reduced sediment supply. Moreover, the government also started to regulate the reclamation and halted many coastal reclamation projects.

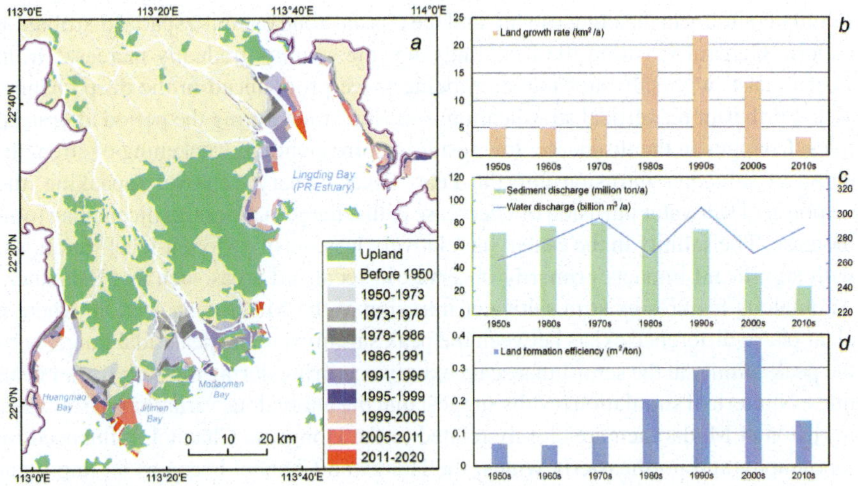

Fig. 7 **a** Areas reclaimed since 1950 AD in both the eastern and western outlets of the Pearl River mouths. **b** Average land growth rate (km^2/a), **c** average sediment (million ton/a) and water (billion m^3/a) discharge and **d** average land formation efficiency (m^2/ton) over the past seven decades (after Xiong et al., 2020a, 2020b)

3.4 Saltwater Intrusion

Saltwater intrusion has greatly affected the residential and agricultural freshwater supplies and industrial production in cities of the Pearl River delta (Liu et al., 2019). In the fourth season of 2021, a rare and long-lasting drought occurred in the Pearl River basin, making saltwater intrude far upstream into Guangzhou City and even the upper East River delta. In October, the waterworks in the central Dongguan City recorded a salinity of 1281 mg/L. Another severe saltwater intrusion happened from the end of 2003 to early 2004, during which more than 5 million people were affected in towns and cities. In October 2003, the salinity in Zhuhai was up to 7000 mg/L, and that of Zhongshan and Guangzhou reached 3000 mg/L. Intense saltwater intrusion has brought some visible influences on environmental elements through biochemical processes, such as inhibiting the nitrification process, changing the concentration and distribution of nutrients as well as the nutrient structure, and resulting in the change of dominant communities in the estuary (Pettit et al., 2016; Yin & Harrison, 2008). The possible factors that can intensify the saltwater intrusion are decreased river discharge, sea-level rise, sand mining and dredging as well as the complex tidal dynamics.

Overall, the salinity in the estuary of the Pearl River (mainly in the Lingdingyang) exhibits irregular temporal variations. During a tidal cycle, the salinity shows overall irregular semi-diurnal variations, but during the period after neap tides, there are more evident irregular diurnal variations. During the dry season and under the influence of the river outflow affected by the Coriolis effect and northeasterly winds, the salinity contours in the outer estuary and the coastal zone exhibit a spatial distribution pattern. Generally, within the Lingdingyang, the salinity gradually increases from west to east, with saltwater intrusion being more pronounced in the deep channel area. Furthermore, vertical stratification is most distinct during the period after neap tides. Changes in the discharge flow result in corresponding weakening or strengthening of saltwater intrusion into the Pearl River estuary; relatively speaking, the response of saltwater intrusion to a decrease in discharge is more significant than to an increase. The salinity in the estuary is relatively less sensitive to sea-level variations, with significant impacts primarily observed in localized areas such as Modaomen. Anomalous features exist in saltwater intrusion at the Modaomen section, where a clear phase difference exists between the peak salinity and the peak tidal range, with the peak salinity at the semi-lunar scale mainly occurring during the period after neap tides. Numerical simulation results suggest that the anomalous variations in saltwater intrusion at Modaomen are closely related to the Hongwan Channel. Influenced by topography and northeasterly winds, the Hongwan Channel becomes an important conduit for transporting salt to the Modaomen Channel. Although the effect of northeasterly winds can significantly enhance saltwater intrusion at Modaomen, it is not the fundamental cause of the anomalous saltwater intrusion. During the period after neap tides, tidal dynamics weaken while the influence of northeasterly winds relatively strengthens. This leads to more severe saltwater intrusion in the Hongwan Channel, ultimately resulting in the peak salinity in the upstream section of the Modaomen Channel occurring during the period after neap tide. The interaction between the Hongwan Channel, wind stress, and tides is the dynamic cause of the anomalous characteristics of saltwater intrusion at the Modaomen Channel.

3.5 *Patchy Hypoxia*

Eutrophication of the Pearl River water and the resulting environmental and ecological disasters, such as water hypoxia and algal blooms, are "bottleneck" issues in the development of the "beautiful bay area" of the Guangdong-Hong Kong-Macao Greater Bay Area and major challenges for comprehensive water environment management and water ecological security. In recent years, the estuarine hypoxia problem is reported to get worsen in extent and intensity (Fig. 8). Studies on the summertime hypoxia of Pearl River Estuary showed that particle-attached bacteria are more abundant than free-living bacteria (Li et al., 2018; Liu et al., 2020). Consequently, it is vital to investigate the role of particle resuspension in the dissolved oxygen (DO) budget, so that we could more clearly identify the degree of coastal hypoxia. Particle resuspension has gained an increasing attention because it is highly associated with oxygen dynamics. An enhancement of oxygen consumption due to particle resuspension was identified in coastal oceans and lakes through observations and numerical models. Moreover, particle resuspension can indirectly affect dissolved oxygen processes by changing biogeochemical cycles, such as the rate of photosynthesis and nitrification. Recently, oxygen consumption within hypoxic

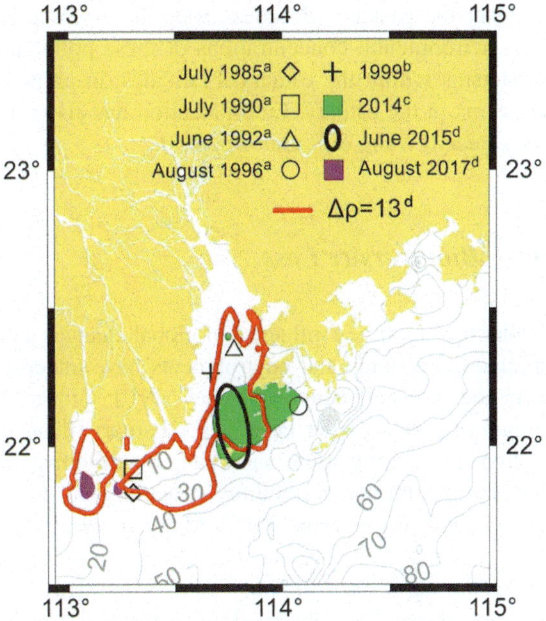

Fig. 8 Reported hypoxia in the Pearl River Estuary and adjacent shelf sea in past three decades. Dataset is from **a** Lin et al. (2001), **b** Yin et al. (2004), **c** unpublished data at https://ocean.ust.hk/home/about-ocean-hk/, and **d** Cui et al. (2019)

waters was found to be mainly generated by particle-attached bacteria in the Pearl River Estuary (Wu et al., 2021). Nevertheless, the impact of particle resuspension on the degree of hypoxia and the DO budget has not yet been straightforward to identify.

3.6 Environmental Pollution

Since the 1980s, population growth and economic development in the Pearl River delta region have led to an increased emission of various pollutants into the Pearl River estuary (Ji et al., 2011). In fact, the estuary was included in the list of hypoxic areas, sometimes referred to as dead zones, by the United Nations Environment Programme (UNEP) in 2006, where nutrients from sewage, animal waste, and fertilizer runoff promoted algal blooms, followed by the subsequent decomposition which consumed dissolved oxygen (Wang & Rainbow, 2020). Pollutants that may directly impact the estuary include toxic metals such as zinc, lead, and cadmium (Yu et al., 2017), as well as ecotoxic organic compounds like polycyclic aromatic hydrocarbons (PAHs) (Wang et al., 2021) from petroleum and its derivatives, certain pesticides, and polychlorinated biphenyls (PCBs) (Xie et al., 2021), among other man-made organochlorines. Relatively recent pollutants include brominated organic compounds such as flame retardants and emerging contaminants like pharmaceuticals and endocrine disruptors (Su et al., 2020). In the 21st century, microplastics (Xu et al., 2015) and synthetic nanoparticles have also been added to the list of contaminants of ecotoxicological concern (Ahmad et al., 2020).

Over the past few decades, extensive research has been conducted to measure the environmental concentrations of these pollutants in the water, sediments, and organisms within the estuary. Pesticides, in particular, have raised environmental concerns in the estuary. Eutrophication has also raised environmental concerns on the estuary (Wang & Rainbow, 2020).

3.7 Biodiversity Loss

Under the combined influence of global changes and human activities, riverine and marine ecosystems and environments have undergone significant transformations over the past half century (Doney, 2010; Giosan et al., 2014). In the ecological monitoring area of the Pearl River Estuary, there has been an increase in the N/P, hypoxia, acidification, and changes in biodiversity. These are characterized by unstable planktonic community structures and frequent harmful algal blooms (Bi et al., 2022; Lu & Gan, 2015). Major biodiversity trends include miniaturization (Wang et al. 2023) and an increase in dinoflagellates (Cheung et al., 2021). By analyzing an 18-year monthly monitoring dataset (2000–2017) in the Pearl River Estuary, researchers observed an increasing trend of the diatom to dinoflagellate ratio (Diatom/Dino) (Cheung et al., 2021).

Furthermore, there is a low density of fish eggs and fry, indicative of a scarcity of fisheries resources (Cheang et al., 2020). Compared with the record of 330 species in the 1980s, the number of fish species in the Pearl River Estuary has declined by 45 (Zhou et al., 2019). The status of the endangered dolphins is a cause for concern (Xu et al., 2015). From 1997 to 2016, there has been an initial increase followed by a subsequent decrease in the distribution range of white dolphins (Wang et al. 2022). The core habitats have shifted towards main shipping lanes and areas near uninhabited islands, possibly due to a reduction in food resources, which may have forced dolphins to make this choice.

3.8 Sea-Level Rise

From 1980 to 2022, sea-level is rising along the Chinese coast (Fig. 9) and the rate of sea-level rise at the South China Sea was recorded at 3.6 mm/yr. Projections indicate that over the next three decades, there will be an estimated increase in sea level rise ranging from approximately 70 to 166 mm (State Ocean Administration, 2022). According to the IPCC report, the future sea level rise will continue to accelerate under various waming scenarios. The potential impact of this phenomenon includes enhanced stratification (Ma et al., 2023) within the Pearl River Estuary and a possible amplification of tidal range (De Dominicis et al., 2020) due to its funnel-shaped morphology. Furthermore, sea-level rise may enhance saltwater intrusion (Yuan et al., 2015). Changes in tides may lead to a reduction in sedimentation rates

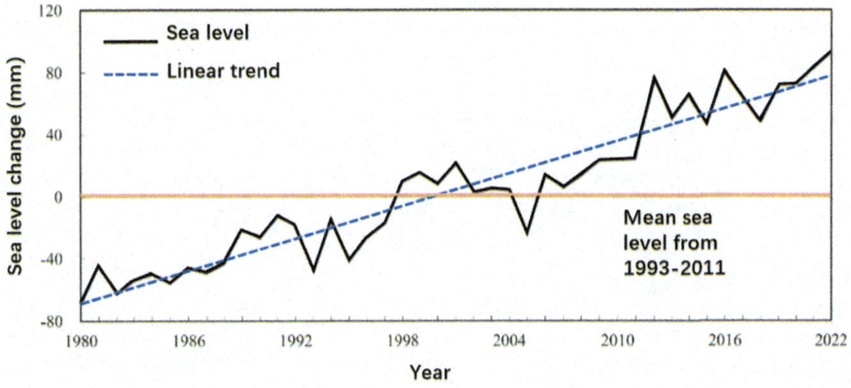

Fig. 9 Sea level anomalies from 1980 to 2022 along the Chinese coast (Ministry of Natural Resources, 2023)

at the delta front (Deng et al., 2020). The impacts of sea-level rise on hydrodynamics and sediment dynamics in the deltaic coastal zone may become more significant than before due to restricted human activities.

3.9 Reduction of River Sediment Supply

From the 1950s to the 1980s, a slight increase of sediment flux from the river occurred (Fig. 10), which can be attributed to deforestation in the catchment. Since 1990s, dam construction became a dominant factor for the reduction of sediment supply. The sediment supply decreased from around 80 to 20 Mt/yr. An obvious deceleration in estuarine infilling has been observed in the comparison of historical bathymetric charts (Wei et al., 2021). However, uncertainties persist regarding future morphodynamic evolution along deltaic coastal areas under simultaneous influences of rising sea levels and reduced riverine sediment input.

4 Solutions for the Problems

4.1 Water Safety Management

As an extremely low-lying delta, water safety management are a matter of primary importance to the Pearl River delta. Local administrations have implemented a range of engineering and administrative measures to resolve issues of floods, storm surges and saltwater intrusion.

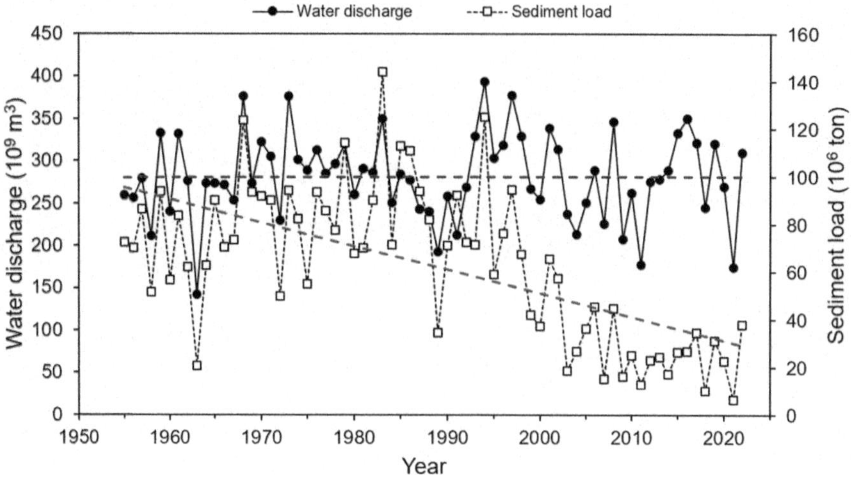

Fig. 10 Changes in water discharge and sediment load of the Pearl River entering the sea from 1955 to 2022

After the Yi-Mao-year Flood, the Minister of the Navy of the Republic of China, Tan Xueheng, and others, such as Ke Weilian, proposed a systematic river management measure that enclosed all the deltaic plains with dykes and sluices to form a series of enclosures. Ke Weilian was responsible for the design and construction of the Lubao Sluice from 1921–1923, which was the largest modern sluice in Guangdong during the Republic of China period. By 1950, several thousands of patchy enclosures were developed in the Pearl River deltaic plain. Since the establishment of the People's Republic of China, the small patchy enclosures were re-organized and merged to larger enclosures. The associated measures included reinforcement of dikes and sluices for the large enclosures, preserving major distributaries and blocking small ones. According to an investigation by the Foshan Administration, the number of enclosures in the West-North-Rivers Delta decreased from 2,950 in AD 1950 to 441 in AD 1961. Nowadays, there are 53 mega enclosures with a protected area of over 50,000 acres (Huang & Zhang, 2004). The management of enclosures and sluices has caused significant changes in the flow and riverbed evolution processes of the Pearl River Delta river network (Li, 1985). In recent years, flooding events caused by river floods have rarely been reported, owing to the enhanced dike-sluice-enclosure system. Only intense typhoon storm surges can generate disastrous inundation, such as typhoon Hato that made landfall in 2017.

Engineering measures (such as sluice construction) and non-engineering measures (such as reservoir operation and administrative regulation) are used for prevention of saltwater intrusion (Liu et al., 2019). The engineering measures like submerged dams and tidal gates are difficult to implement, due to the impact on estuary functions such as flood discharge and navigation. With characteristics like less investment and quick effects, non-engineering measures play an important role in guaranteeing the drinking water safety in regions affected by saltwater intrusion. Integrated water resources management is one of the non-engineering measures to alleviate the water resources problems. According to the Comprehensive Planning of Water Resources in the Pearl River, the joint optimal operation of upstream key reservoirs in the West, North and East Rivers, such as the Tianshengqiao, Longtan and Baise reservoirs, can increase the guaranteed water supply rate to 70%. If the Datengxia dam project in the West River is commissioned and put into operation with the upstream reservoirs, the rate will exceed 90%, which means that water supply in the Pearl River delta can be guaranteed during the drought season. Regulations on Sand Dredging Management have also been implemented in the Rivers of the Guangdong Province since May 2005. The policy strictly plans the sand dredging zone (forbidden mining zone, mineable zone, and reserved zone) to ensure the balance of channel scouring and siltation. It is possible to issue licenses to entities and individuals engaged in sand mining, to approve the river sections and quantities allowed for sand mining each year and to impose higher resource fees for the purpose of limiting sand dredging. Through the combination of administrative enforcement and economic regulations, river sand mining and dredging can be controlled, thus helping to alleviate the saltwater intrusion problem.

4.2 Environmental Protection

The rapid economic development of the Pearl River delta in the past decades has posed great stresses to coastal environments and ecology, causing environmental pollution, water eutrophication, enhanced estuarine hypoxia, wetland loss and biological community disturbance. A range of environmental protection measures are thus introduced in order to achieve an environment-friendly sustainable development situation. Among these, a large-scale policy of industrial transfer and update has been implemented since 2008, which have transferred large number of pollutive low-end industries from the delta metropolitan area to the neighboring prefectures or other inland provinces. This policy has a great positive effect on the improvement of the air and water qualities. Sand mining and dredging in the estuarine bays and rivers are strictly regulated to prevent contaminants being resuspended from sediments. 3–4 months fishing bans have been implemented in the Pearl River and coastal seas to avoid overfishing.

The intense land reclamation and coastal construction projects were the most important driving forces that caused wetland loss and degradation. Under the overall arrangement of the central government of China, laws and regulations for wetlands protection and restoration were implemented recently, with several "red (forbidden) lines" being set up for coastal ecological security. For example, (1) coastal land reclamation is forbidden since 2018 except for national strategic engineering projects, (2) by 2020, the littoral marine reserve area shall not be less than 10%, (3) at least 35% of the coastline shall be kept in a natural state. Mangroves have been planted on mudflats to increase the ecological function. Two mangrove reserves have been developed in the apex of Shenzhen Bay (Mai Po and Futian) and northern Qi'ao Island.

The Chinese white dolphin (Sousa chinensis) is a key and beloved species on the verge of extinction within the Pearl River estuary. The dolphins' habitat spans the Pearl River estuary, and is closely associated with the estuarine mixing zone between the river and the ocean. The population inhabiting the estuary, is believed to number around 2,500 individuals. In recent years, the Chinese white dolphin has been facing a number of threats: habitat loss, overfishing, water pollution and heavy marine traffic, along with coastal development. These threats have had a major and cumulative impact on the population for years. Governments of Guangdong, Hong Kong and Macau have collaborated to protect the remaining population of the species. A local conservation area that includes a core zone, a buffer zone and an experimental zone (Fig. 11) was set up in 1999 and was upgraded to national reserve level in 2003. Within this area, fishing, shipping, pollution discharge and marine construction are strictly regulated to maintain biodiversity and the marine ecosystem. However, according to the latest investigation, their main activity space is largely beyond the conservation area (Fig. 11). That means the previously planned protection strategy needs reexamination and modification.

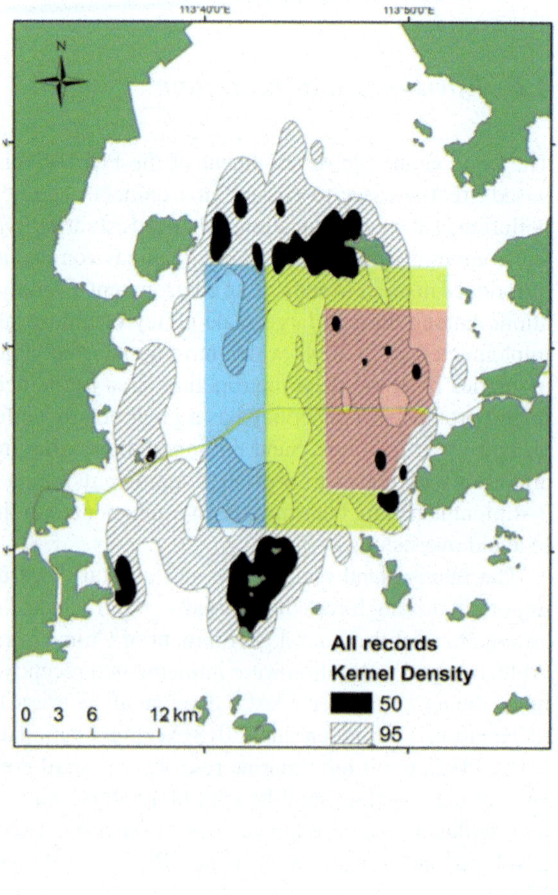

Fig. 11 The present conservation area for Chinese white dolphins within the Pearl River estuary, with the core zone, buffer zone and experimental zone marked in red, yellow and blue colors, respectively. The latest recorded activity space of the dolphins is marked in black and grey patches

5 Paths for the Future

It is estimated that by 2050, the population of the Pearl River Delta will reach 120 million. Climate change and rapid urbanization and economic development will increase the risk of flooding, especially in low-lying areas. Under climate change, the series of problems brought about by urban expansion, such as increasing near-surface temperature, humidity, wind speed, and regional precipitation, have become increasingly prominent, leading to higher average temperatures and an increase in extreme rainfall days. According to statistics, in the past ten years, Guangdong has been directly hit by typhoons 23 times, and with the warming of the climate and the rise of the sea level, urban flooding and typhoon disasters have become more severe.

Interconnected enclosures and sluices, combined with reservoirs, have solved the problem of upstream flood disasters. However, interconnected enclosures and sluices block the natural exchange of water and material between the inner rivers and the outer distributaries or main channels, thus increasing flood risks. This weakens the hydrodynamic conditions of the inner rivers and their self-purification capacity for water pollution, affecting the hatching and growth of young fish as well as fish migration.

Future solutions should be based on the new pattern of changes in the Pearl River Estuary, focusing on key scientific issues related to estuary evolution and management for water security, and advancing in the sequence of "estuary change mechanism and trends—management targets—technology research and development—application demonstration". The focus should be on the key scientific issues of spatiotemporal changes in estuaries driven by intense human activities, the change mechanism and evolution trend of the Pearl River Delta, the potential safety risks of dikes under changing conditions and the vulnerability of water security, and estuary management measures and control theories. Subsequently, technological research and development and demonstration area research should be conducted. Future research content includes: processes, mechanisms, and evolution trends of estuarine and river network changes; process-based multi-dimensional spatiotemporal scale coupling simulation and long-term model prediction; adaptive estuary management and water security risk assessment under complex changing conditions; multi-objective management and water-sand control measures for "network-bay"; dike protection technology and demonstration for dangerous sections of river network dikes; and coastal protection and eco-dike construction technology and demonstration for the Pearl River Estuary.

References

Ahmad, M., Li, J. L., Wang, P. D., Hozzein, W. N., & Li, W. J. (2020). Environmental perspectives of microplastic pollution in the aquatic environment: A review. *Marine Life Science and Technology, 2*, 414–430.

Bi, S., Lai, H., Guo, D., Liu, X., Wang, G., Chen, X., Liu, S., Yi, H., Su, Y., & Li, G. (2022). Spatio-temporal variation of bacterioplankton community structure in the Pearl River: Impacts of artificial fishery habitat and physicochemical factors. *BMC Ecology and Evolution, 22*, 10.

Chen, B., Chen, Y., Yin, G., Ye, S., Ouyang, W., & Lin, M. (2004). Study on the water quality of urban forest ecosystem in the Pearl River Delta. *Forest Research, 17*(4), 453–460. (in Chinese with English abstract).

Cheang, C. C., Lee, B. Y., Ip, B. H., Yiu, W. H., Tsang, L. M., & Ang, P. O., Jr. (2020). Fish and crustacean biodiversity in an outer maritime estuary of the Pearl River Delta revealed by environmental DNA. *Marine Pollution Bulletin, 161*, 111707.

Cheung, Y. Y., Cheung, S., Mak, J., Liu, K., Xia, X., Zhang, X., Yung, Y., & Liu, H. (2021). Distinct interaction effects of warming and anthropogenic input on diatoms and dinoflagellates in an urbanized estuarine ecosystem. *Global Change Biology, 27*, 3463–3473.

Cui, Y., Wu, J., Ren, J., & Xu, J. (2019). Physical dynamics structures and oxygen budget of summer hypoxia in the Pearl River Estuary. *Limnology and Oceanography, 64*(1), 131–148.

Dai, S., Yang, S., & Cai, A. (2008). Impacts of dams on the sediment flux of the Pearl River, southern China. *Catena, 76*, 36–43.

De Dominicis, M., Wolf, J., Jevrejeva, S., Zheng, P., & Hu, Z. (2020). Future interactions between sea level rise, tides, and storm surges in the world's largest urban area. *Geophysical Research Letters, 47*, e2020GL087002.

Deng, J., & Bao, Y. (2011). Morphologic evolution and hydrodynamic variation during the last 30 years in the Lingding Bay, South China Sea. *Journal of Coastal Research, 64*(SI), 1482–1489.

Deng, J., Yao, Q., & Wu, J. (2020). Estuarine morphology and depositional processes in front of lateral river-dominated outlets in a tide-dominated estuary: A case study of the Lingding Bay, South China Sea. *Journal of Asian Earth Sciences, 196*, 104382.

Doney, S. C. (2010). The growing human footprint on coastal and open ocean biogeochemistry. *Science, 328*, 1512–1516.

Eisma, D. (1998). Intertidal deposits: River mouths. tidal flats and coastal lagoons. CRC Press.

Fu, S., Xiong, H., Zong, Y., & Huang, G. (2020). Reasons for the low sedimentation and slow progradation in the Pearl River delta, southern China, during the middle Holocene. *Marine Geology, 423*, 106133.

Gao, S., Liu, Y., Yang, Y., Liu, P. J., Zhang, Y., & Wang, Y. P. (2015). Evolution status of the distal mud deposit associated with the Pearl River, northern South China Sea continental shelf. *Journal of Asian Earth Science, 114*, 562–573.

Giosan, L., Syvitski, J. P. M., Constantinescu, S., & Day, J. (2014). Protect the world's deltas. *Nature, 516*, 31–33.

Huang, G., & Yim, W. W. S. (2007). Reconstruction of an 8,000-year record of typhoons in the Pearl River estuary, China. In *Proceedings International Conference on Climate Change* (pp. 1–12).

Huang, Z., Zhang, W., & Chai, F. (1995). The submerged Zhujiang Delta. *Acta Geographica Sinica, 50*(3), 206–214. (in Chinese with English abstract).

Huang, Z., & Zhang, W. (2004). Impacts of artificial factors on the evolution of geomorphology during recent thirty years in the Zhujiang delta. *Quaternary Research, 24*(4), 394–401.

Huang, G. (1996). Neolithic culture and paleogeographic environment in the Zhujiang Delta. *Acta Geologia Sinica, 51*, 508–517. (in Chinese with English abstract).

Huang, Z., Li, P., Zhang, Z., Li, K., & Qiao, P. (1982). Zhujiang Delta. General Scientific Press. (in Chinese).

Ji, X., Sheng, J., Tang, L., Liu, D., & Yang, X. (2011). Process study of circulation in the Pearl River Estuary and adjacent coastal waters in the wet season using a triply-nested coastal circulation model. *Ocean Modelling, 38*, 138–160.

Li, P., Qiao, P., Zheng, H., Fang, G., & Huang, G. (1991). The environment evolution of the Zhujiang Delta in the last 10000 years. China Ocean Press.

Li, J., Salam, N., Wang, P. D., Chen, L. X., Jiao, J. Y., Li, X., Xie, W. D., Han, M. X., Fang, B. Z., Mou, X. Z., & Li, W. J. (2018). Discordance between resident and active bacterioplankton in free-living and particle-associated communities in estuary ecosystem. *Microbial Ecology, 76*(3), 637–647.

Li, H., Huang, Y., Zhou, Y., Wang, S., Guo, W., Liu, Y., Wang, J., Xu, Q., Zhou, X., & Yi, K. (2023). Spatial and temporal evolution of ecosystem service values and topography-driven effects based on land use change: A case study of the Guangdong-Hong Kong-Macao Greater Bay Area. *Sustainability, 15*(12), 9691.

Li, Z. (1985). The effects of United Embankments on the flow shift and channel changes in the Zhujiang Delta. *Tropical Geography, 5*(2), 99–107. (in Chinese).

Lin, H., Liu, S., & Han, W. (2001). Potential trigger CTB, from seasonal bottom water hypoxia in the Pearl River estuary (in Chinese with English abstracts). *Journal of Zhanjiang Ocean University, 21*, 25–28.

Liu, Y., Gao, S., Wang, Y. P., Yang, Y., Long, J., Zhang, Y., & Wu, X. (2014). Distal mud deposits associated with the Pearl River over the Northwestern continental shelf of the South China Sea. *Marine Geology, 347*, 43–57.

Liu, B., Peng, S., Liao, Y., & Wang, H. (2019). The characteristics and causes of increasingly severe saltwater intrusion in Pearl River Estuary. *Estuarine, Coastal and Shelf Science, 220*, 54–63.

Liu, Y., Lin, Q., Feng, J., Yang, F., Du, H., Hu, Z., & Wang, H. (2020). Differences in metabolic potential between particle-associated and free-living bacteria along Pearl River Estuary. *Science of the Total Environment, 728*, 138856.

Lu, Z., & Gan, J. (2015). Controls of seasonal variability of phytoplankton blooms in the Pearl River Estuary. *Deep Sea Research Part II: Topical Studies in Oceanography, 117*, 86–96.

Ma, M., Zhang, W., Chen, W., Deng, J., & Schrumm, C. (2023). Impacts of morphological change and sea-level rise on stratification in the Pearl River Estuary. *Frontiers in Marine Science, 10*, 1072080.

Ministry of Natural Resources. (2023). China Sea Level Bulletin 2022.

Owen, R. B. (2005). Modern fine-grained sedimentation—Spatial variability and environmental controls on an inner pericontinental shelf, Hong Kong. *Marine Geology, 214*, 1–26.

Pettit, N. E., Bayliss, P., & Bartolo, R. (2016). Dynamics of plant communities and the impact of saltwater intrusion on the floodplains of Kakadu National Park. *Marine and Freshwater Research, 69*(7), 1124–1133.

State Ocean Administration. (2022). *China sea level bulletin.*

Su, C., Cui, Y., Liu, D., Zhang, H., & Baninla, Y. (2020). Endocrine disrupting compounds, pharmaceuticals and personal care products in the aquatic environment of China: Which chemicals are the prioritized ones? *Science of the Total Environment, 720*, 137652.

Wang, H. J., Yang, Z. S., Saito, Y., Liu, J. P., Sun, X. X., & Wang, Y. (2007). Stepwise decreases of the Huanghe (Yellow river) sediment load (1950–2005): Impacts of climate change and human activities. *Global and Planetary Change, 57*, 331–354.

Wang, W. X., & Rainbow, P. S. (2020). Pollution in the Pearl River Estuary. In W. X. Wang & P. S. Rainbow (Eds.), *Environmental pollution of the Pearl River Estuary* (pp. 13–35). Springer.

Wang, Y., Wu, F., Gu, Y., Huang, H., Gong, X., & Liao, X. (2021). Polycyclic Aromatic Hydrocarbons (PAHs) in the intertidal sediments of Pearl River Estuary: Characterization, source diagnostics, and ecological risk assessment. *Marine Pollution Bulletin, 173*, 113140.

Wang, X., Chen, T., Li, M., & Wang, Y. (2022). Long-term changes in habitat use of Indo-Pacific humpback dolphins (Sousa chinensis) in response to anthropogenic coastline shift in Lingding Bay of Pearl River Estuary, China. *Acta Ecologica Sinica, 42*(7), 2962–2973. (in Chinese with English abstract).

Wang, F., Li, X., Tang, X., Sun, X., Zhang, J., Yang, D., Xu, L., Zhang, H., Yuan, H., Wang, Y., Yao, Y., Wang, C., Guo, Y., Ren, Q., Li, Y., Zhang, R., Wang, X., Zhang, B., & Sha, Z. (2023). The seas around China in a warming climate. *Nature Reviews Earth & Environment 4*, 535–551.

Wei, X., Cai, S., & Zhan, W. (2021). Impact of anthropogenic activities on morphological and deposition flux changes in the Pearl River Estuary. *China. Scientific Reports, 11*, 16643.

Wu, C., Ren, J., Bao, Y., Lei, Y., & Shi, H. (2007). A long-term morphological modeling study on the evolution of the Pearl River Delta, network system, and estuarine bays since 6000 yr BP. *Special Papers-Geological Society of America, 426*, 199–210.

Wu, C. S., Yang, S. L., & Lei, Y. P. (2012). Quantifying the anthropogenic and climatic impacts on water discharge and sediment load in the Pearl River (Zhujiang), China (1954–2009). *Journal of Hydrology, 452–453*, 190–204.

Wu, T., Chen, B., Xiao, Y., Pan, Y., Chen, Y., & Xiao, J. (2010). Leaf stoichiometry of trees in three forest types in Pearl River Delta, South China. *Chinese Journal of Plant Ecology, 34*(1), 58–63. (in Chinese with English abstract).

Wu, Z., Saito, Y., Zhao D., Zhou Z., Cao, Z., Li, S., Shang, J., & Liang, Y. (2016). Impact of human activities on subaqueous topographic change in Lingding Bay of the Pearl River estuary, China, during 1955–2013. *Scientific Reports, 6*, 37742.

Wu, Z., Li, Q. P., Ge, Z., Huang, B., & Dong, C. (2021). Impacts of biogenic polyunsaturated aldehydes on metabolism and community composition of particle-attached bacteria in coastal hypoxia. *Biogeosciences, 18*(3), 1049–1065.

Xiong, H., Zong, Y., Qian, P., Huang, G., & Fu, S. (2018). Holocene sea-level history of the Northern coast of South China Sea. *Quaternary Science Reviews, 194*, 12–26.

Xie, Q., Yu, R. Q., Yu, R., Wang, Z., Zhang, X., & Wu, Y. (2021). Historic changes of polychlorinated biphenyls (PCBs) in juvenile and adult cetaceans from the Pearl River estuary from 2003 to 2020. *Science of the Total Environment, 800*, 149512.

Xiong, H., Zong, Y., Huang, G., & Fu, S. (2020a). Human drivers accelerated the advance of Pearl River deltaic shoreline in the past 7500 years. *Quaternary Science Reviews, 246*, 106545.

Xiong, H., Zong, Y., Huang, G., & Fu, S. (2020b). Seabed erosion and deposition related to the typhoon activity of the past millennium on the southeast coast of China. *Earth Surface Processes and Landforms, 45*, 1695–1704.

Xu, X., Song, J., Zhang, Z., Li, P., Yang, G., & Zhou, K. (2015). The world's second largest population of humpback dolphins in the waters of Zhanjiang deserves the highest conservation priority. *Scientific Reports, 5*, 8147.

Xu, J., Li, Y., Can, F., & Chen, Q. (1985). Evolution of channels and shoals in Lingding Sea, Pearl River Estuary. Ocean Press. (in Chinese).

Yin, K., & Harrison, P. J. (2008). Nitrogen over enrichment in subtropical Pearl River estuarine coastal waters: Possible causes and consequences. *Continental Shelf Research, 28*, 1435–1442.

Yin, K., Lin, Z., & Ke, Z. (2004). Temporal and spatial distribution of dissolved oxygen in the Pearl River Estuary and adjacent coastal waters. *Continental Shelf Research, 24*(16), 1935–1948.

Yu, R., He, L., Cai, R., Li, B., Li, Z., & Yang, K. (2017). Heavy metal pollution and health risk in China. *Global Health Journal, 1*, 47–55.

Zhao, H. (1990). The evolution of the Pearl River Estuary. China Ocean Press. (in Chinese).

Zhao, C. H., Zhu, Z. H., & Zhou, D. Z. (2000). Worldwide rivers and dams. China Water Conservancy and Hydroelectric Press. (in Chinese).

Zhuang, X. (2012). Floristic composition and conservation of the Fungshui woods in Guangdong′s Pearl River Delta and Hong Kong. *Forestry Science and Technology of Guangdong Province, 1*, 72–76. (in Chinese).

Zhou, L., Wang, G., Kuang, T., Guo, D., & Li, G. (2019). Fish assemblage in the Pearl River Estuary: Spatial-seasonal variation, environmental influence and trends over the past three decades. *Journal of Applied Ichthyology, 35*(4), 884–895.

Zong, Y., Huang, G., Switzer, A. D., Yu, F., & Yim, W. W. S. (2009). An evolutionary model for the Holocene formation of the Pearl River delta, China. *Holocene, 19*, 129–142.

Zong, Y., Zheng, Z., Huang, K., Sun, Y., Wang, N., Tang, M., & Huang, G. (2013). Changes in sea level, water salinity and wetland habitat linked to the late agricultural development in the Pearl River delta plain of China. *Quaternary Science Reviews, 70*, 145–157.

Zong, Y., Huang, G., Li, X. Y., & Sun, Y. Y. (2015). Late Quaternary tectonics, sea-level change and lithostratigraphy along the northern coast of the South China Sea. *Geological Society London Special Publications, 429*, 123–136.

Rhine-Meuse-Scheldt Delta

Huib de Vriend, Zhengbing Wang, Bas vanMaren, and Zhong Peng

1 Physical Setting

1.1 River Basins

The interlocking Rhine, Meuse and Scheldt basins cover a large part of Northwest Europe, as can be seen in Figs. 1, 2 and 3. The rivers' main characteristics are however quite different, the Rhine being fed by a mixture of snowmelt, rainwater and groundwater whereas the discharge in the Meuse and Scheldt primarily originates from rainwater. With a catchment size of 185000 km^2, the average river discharge of the Rhine is 2310 m^3/s (Frings et al., 2019). The catchments and annual river discharge of the Scheldt (22.000 km^2 and 100 m^3/s; Fettweis et al., 1998) and Meuse (33000 km^2 and 250 m^3/s; Delta-Alliance, 2024) are much lower.

When entering the Netherlands, the river Rhine transports coarse sandy material, the river Meuse gravely material and the river Scheldt fine muddy sediment. Rhine and Meuse exhibit significant downstream fining, such that most of the sediment in the delta proper is fine sandy to muddy. The most complete sediment budget for the Rhine river is probably developed by Frings et al. (2019). On average 2.11 million ton of mud (silt and clay), 0.59 million ton of sand, and 0.11 million ton of gravel are annually supplied to the Rhine delta (defined by Frings et al. as the border between the Netherlands and Germany at Lobith). The amount of sediment annually delivered

H. de Vriend (✉) · Z. Wang · B. vanMaren
Faculty of Civil Engineering and Geosciences, Delft University of Technology, Delft 2628 CN, The Netherlands
e-mail: h.j.devriend@tudelft.nl

Z. Wang · B. vanMaren
Deltares, Delft 2600 MH, The Netherlands

Z. Peng
State Key Laboratory of Estuarine and Coastal Research, East China Normal University, Shanghai 200241, China

W. Zhang and H. de Vriend (eds.), *Delta Sustainability*,
https://doi.org/10.1007/978-981-97-7259-9_14

to the sea is substantially less (0.90 and 0.15 million ton of mud and sand; no gravel), due to floodplain sedimentation, but especially due to dredging. The suspended load (mainly consisting of mud) supplied by the Scheldt river is 0.24–0.29 million ton/year (Van Hoestenberge et al., 2014), calculated at the approximate landward limit of maximum salt intrusion (landward of Antwerp). Earlier estimates by Fettweis et al. (1998) were considerably larger (0.75–2.2 million ton/year). Upon entering the Rhine-Meuse delta, the Meuse transports only 0.2 million ton of sand and 0.4 million ton of mud (Frings et al., 2019), although Cox et al. (2022) mention sediment budgets with a suspended load (sand and mud) of 0.7 million ton/year and a bedload transport (mainly sand) of 0.48 million ton/year.

In the downstream reaches of the Rhine-Meuse delta large amounts of sediment settle in fairways and harbours, which are subsequently dredged and deposited in the North Sea or within the system itself. Additionally, both systems trap fine sediments, which partly originate from the sea (Cox et al., 2021; Frings et al., 2011). The annual dredging volumes are about 10 million m^3/year in both the Rhine-Meuse estuary (Cox et al., 2022) and the Scheldt estuary (van Dijk et al., 2021). All sediment dredged within the Scheldt estuary (mostly sandy, but also more muddy in its upper reaches)

Fig. 1 Basin of the Rhine (From Schulte-Wülwer-Leidig et al., 2018)

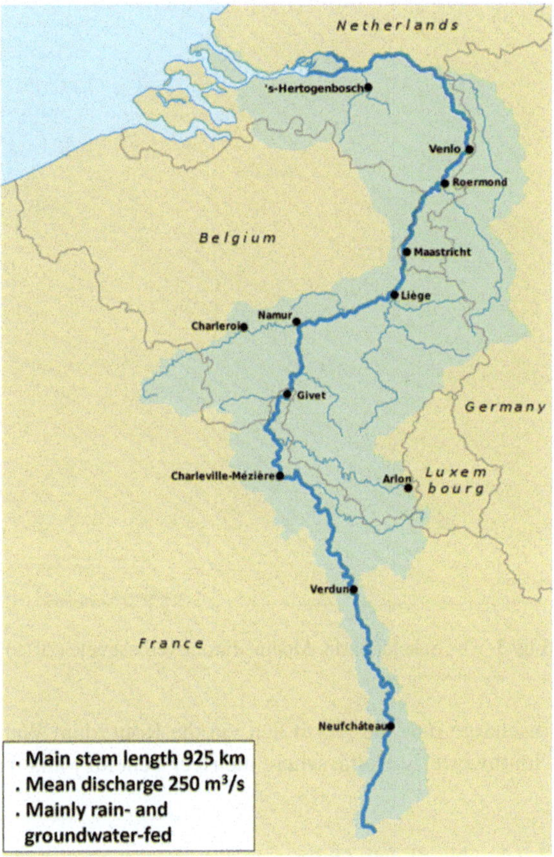

Fig. 2 Basin of the Meuse. Modified from Pseudomas (2022)

are deposited within the Scheldt estuary. Part of the (largely muddy) sediment dredged from the Rhine-Meise estuary and deposited in the North sea is transported back into the Rhine-Meuse estuary by marine processes (van Maren & van Kessel, 2016).

1.2 Delta Systems

Deltaic deposits of the rivers Rhine and Meuse cover a large part of the Netherlands (Fig. 4), a country in northwest Europe, with an area of 41,543 km^2 and a population of 17.8 million people. According to Tockner et al. (2009), the delta defined in this way covers 25,347 km^2, making it the largest in Europe.

After flowing into the Netherlands, the Rhine and Meuse get connected and flow into the sea via various branches. The northernmost branch, the river IJssel, discharges via Lake IJssel into the Wadden Sea and further into the North Sea. The other branches

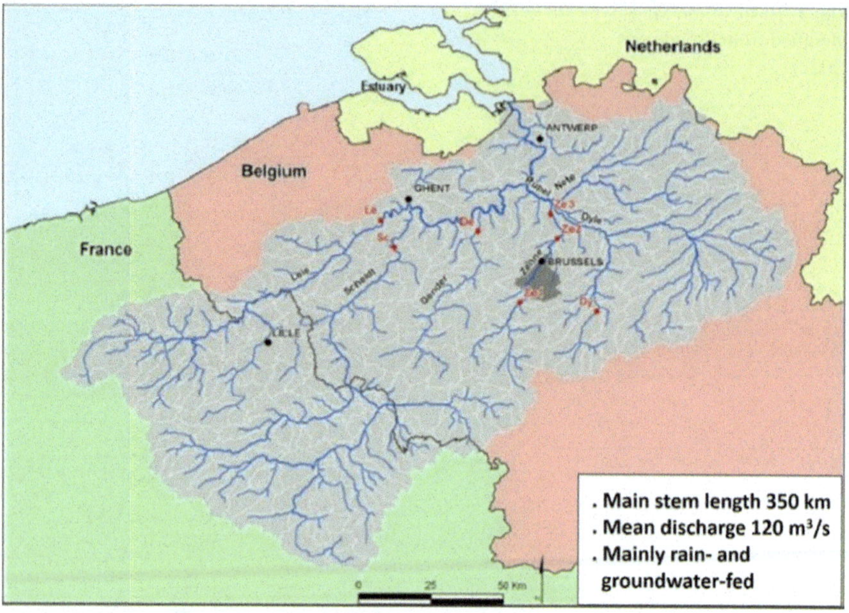

Fig. 3 The Scheldt basin. Modified from Veroustraete (2016)

discharge into the North Sea via the Rotterdam Waterway and the estuaries in the Southwesterly Delta, where the river Scheldt joins from the south.

1.2.1 Northerly Delta

In the Rhine-Meuse-Scheldt delta proper, i.e. the area influenced by tides, it is useful to distinguish between the northerly and the southerly basin. The northerly basin, with Rotterdam as the principal centre of activity, consists of a number of relatively narrow channels and waterways (Fig. 5). This figure also shows the principal human interventions in this system, meant primarily for flood protection and water management. Note the north–south directed channels, where current velocities have strongly increased since the implementation of these measures, because the northernmost channels (Nieuwe Waterweg, Nieuwe Maas) remained tidal, whereas the former estuary in the south (Haringvliet/Hollandsch Diep) has become almost non-tidal.

1.2.2 Southerly Delta

The Southerly basin is of a quite different nature, with large estuaries and sea arms like the Grevelingen (dammed off), the Eastern Scheldt (provided with a movable storm surge barrier) and the Western Scheldt (open). Figure 6 also shows the intense

Fig. 4 Main rivers in the Netherlands. Modified from https://www.rijkswaterstaat.nl/water/waterbeheer/beheer-en-ontwikkeling-rijkswateren/rivieren

urbanization in Belgian Flanders, which is driven to a large extent by port activities along the river Scheldt, the Western Scheldt estuary and the North Sea coast (Zeebrugge).

1.3 Wetlands and Ecosystems

Before 1970, the freshwater tidal marshes in the SW Delta were characterized by distinct vegetation zones closely tied to tidal inundation and exposure to tidal currents. In the lower intertidal regions, rushes prevailed, followed by reed marshes at higher elevations, and alluvial forests at the highest points. Since the estuary's closure, however, the dominant vegetation consists of extensive reeds and rushes, gradually transitioning to a large freshwater micro-tidal region known as the Biesbosch,

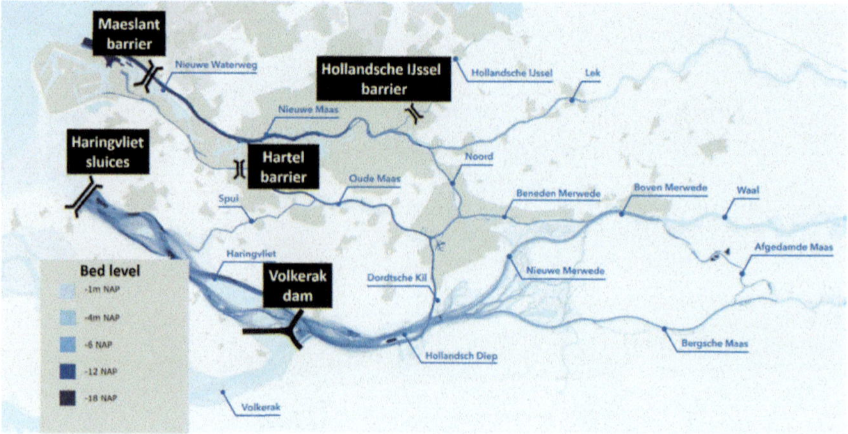

Fig. 5 Northerly basin of the Rhine-Meuse-delta (Rijkswaterstaat, 2019)

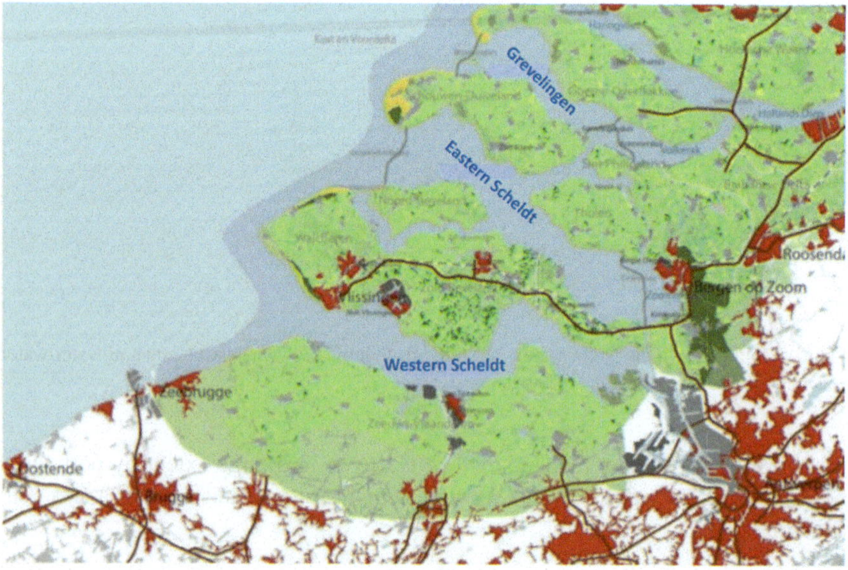

Fig. 6 Southerly basin of the Rhine-Meuse-Scheldt delta (Rijkswaterstaat, 2017)

featuring tidal forests and reed marsh areas (Nienhuis, 2008a). These freshwater tidal marshes exhibit distinct vegetation zones shaped by tidal inundation and currents. In the lower intertidal areas, Scirpus marshes dominate, while higher elevations feature reed marshes, ruderal vegetation, and alluvial forests (Van de Rijt & Coops, 1993). The evolution of vegetation is significantly influenced by the utilization of rushes, reeds, and coppice wood (Smit & Coops, 1991).

The estuarine environment in the RMS-delta encompasses encompasses both freshwater and brackish-water tidal systems, harboring a diverse range of life forms. These include unique plant species, plankton, macroinvertebrates, fish, various bird species, and mammals (Peelen in 1967; Wolff in 1973; Vaas in 1968). Additionally, the invertebrate and fish populations show distinct distribution patterns along an environmental gradient, ranging from marine species in the west to brackish-water and freshwater species in the east. Since the 1960s, a collective effort by the countries along the Rhine River has successfully reversed its decline in three crucial aspects: water quality, biodiversity, and floodplain availability. The most substantial achievements have been made in enhancing water quality. As a result, the Rhine now maintains adequate levels of dissolved oxygen for fish throughout the year.

1.4 Environmental Dynamics

1.4.1 Climate

The Netherlands have a temperate maritime climate with mild winters (min 0.7 °C) and cool summers (max 23.1 °C) according to the decadal mean (1991–2020) in station De Bilt (source KNMI.nl). The mean annual precipitation is 85 cm, with rainfall fairly evenly distributed over the year (only in spring there is slightly less rainfall). The potential evaporation is 56.3 cm (with evaporation rates exceeding rainfall between April and August)—KNMI (2024).

Temperature

Mean air temperatures in the Netherlands have risen significantly in recent decades, as shown in Fig. 7. According to the IPCC-projections (IPCC, 2014), the mean air temperature in Northern Europe must be expected to rise further, especially in winter. Higher air temperatures will lead to changes in snowmelt, which will affect the Rhine discharge regime in particular. They will also translate into higher water temperatures, hence into effects on the ecosystem and indirectly on the economy (cooling water availability for energy production). High temperature extremes (heat waves) are expected to increase markedly.

Precipitation

Precipitation in the Netherlands is about 0.9 m/year, and it has been gradually increasing over the last century (Fig. 7). The IPCC-projections show an increase in mean precipitation in Northern Europe, with more droughts and extreme precipitation events. This will influence the discharge regimes of all three rivers considered

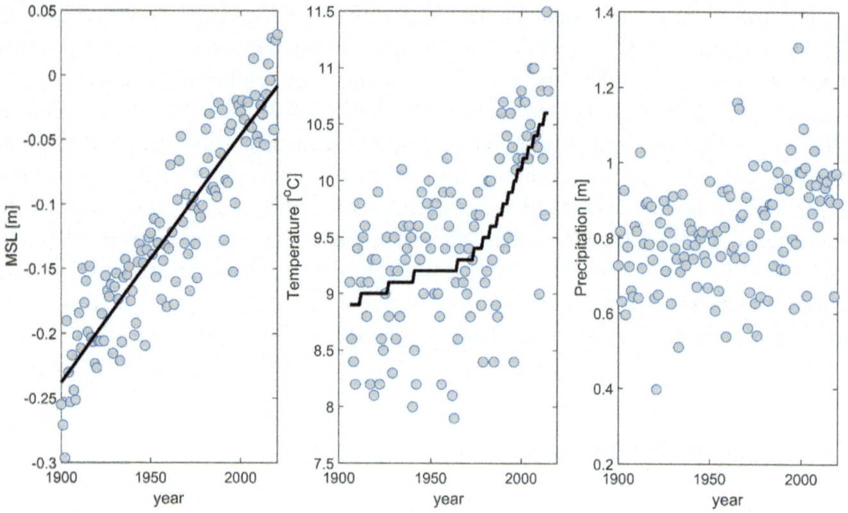

Fig. 7 Development of Mean Sea Level (MSL, left, data from www.psmsl.org/data/), annually average temperature (middle, data from KNMI.nl; trendline from https://www.clo.nl) and annual precipitation (data from KNMI.nl) measured between 1900 and 2020. MSL is defined relative to MSL in the period 2007–2020, and based on an average of measurements of yearly sea level at 7 stations along the Dutch coast (Vlissingen, Hoek van Holland, Maassluis, IJmuiden, Den Helder, Harlingen, Delfzijl). The average rise in MSL over the period 1900–2020 is 0.19 cm/year. Temperature data is measured at the main meteorological station of De Bilt

herein. Furthermore, the probability of extreme flood levels is expected to increase (e.g. Hegnauer et al., 2014).

Storminess

The North Sea and the Northwest European coasts are moderately impacted by storms (Fig. 8) and this does not appear to change in time. Neither do the IPCC climate projections show significant changes of storminess in this region.

For estuaries storm-induced surges are even more important than the storms as such. They can reach significant heights. Figure 9 shows an extrapolation of almost 120 years of data from Hook of Holland for various climate scenarios. Apart from storm surges, Bénard convection over the North Sea may lead to seiches that resonate in the harbour area of Rotterdam (De Jong & Battjes, 2010).

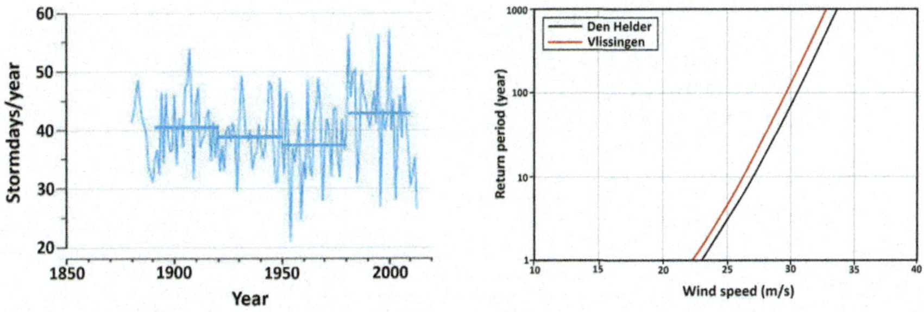

Fig. 8 Annual number of storm days on the North Sea (left) and return period of wind speed (right), both from KNMI.nl

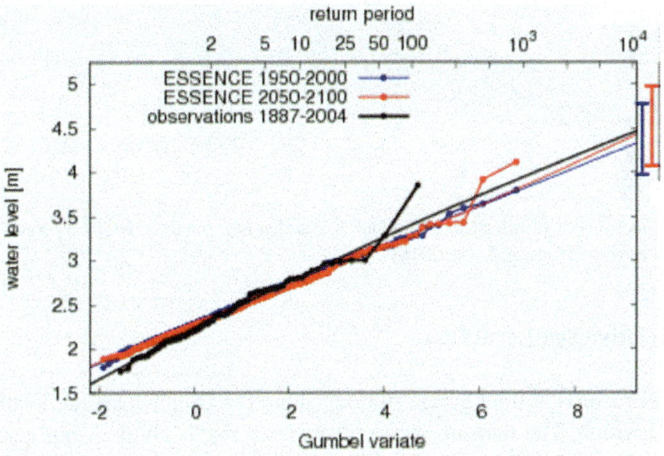

Fig. 9 Water levels at Hook of Holland extrapolated to 2100 (Sterl et al., 2008)

1.4.2 Tides

The tides in the present-day Rhine-Meuse estuary enter through its main outlet (the Nieuwe Waterweg or New Waterway). The mean tidal range is 1.9 m, with an amplitude slowly decreasing in the landward direction (Vellinga et al., 2014). Tidal propagation and circulation have been strongly modified by the various closure works executed in the past 60 years (see sections hereafter). Tides originally entered the estuary through the funnel-shaped Haringvliet, which is much wider and deeper (see Fig. 5). Tides in the Scheldt estuary are larger (3.7 m at the mouth) and increase in the landward direction (up to 5.4 m) (Wang et al., 2019).

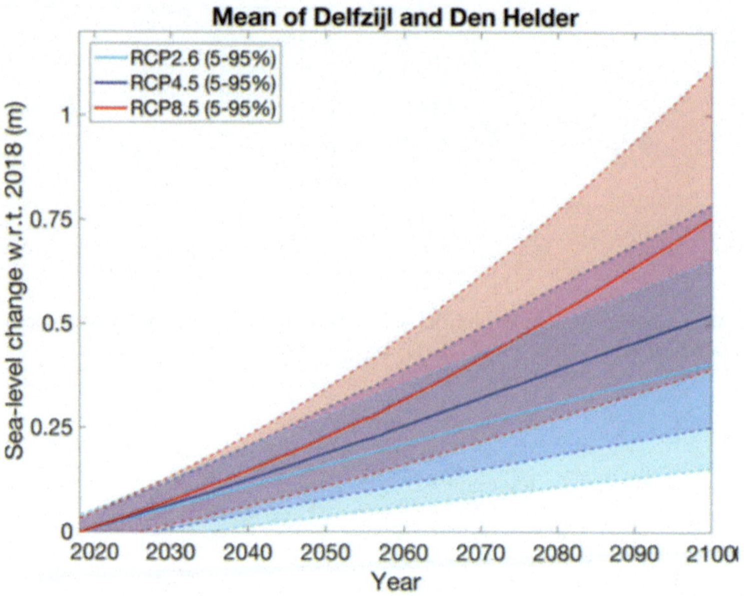

Fig. 10 Projections of future SLR in the Dutch Wadden Sea for three RCPs by Vermeersen et al. (2018), Figure from Wang and Xu (2018)

1.4.3 Relative Sea Level Rise

Water level records at tidal gauge stations along the Dutch coast are available since 1848 (Maassluis). The measured rate of sea level rise is about 2 mm/year (also see Fig. 7).

A recent study concluded that the sea level rise along the Dutch coast is accelerating (Vermeersen et al., 2018; Steffbauer et al., 2022), with a breakpoint in the early 1990s. Analysis suggests that the SLR rate increases from 1.7 $\pm$ 0.3 mm/yr before the breakpoint to 2.7 $\pm$ 0.4 mm/yr after the breakpoint.

Future sea level rise depends on the climate scenario and is very uncertain. For the Dutch Wadden Sea Vermeersen et al. (2018) made projections for the future sea level rise for three Representative Concentration Pathways (RCPs), as defined in IPCC AR5 (Fig. 10).

1.4.4 Subsidence

Subsidence in the Netherlands (see Fig. 11) is mainly due to hydrocarbon mining and peat oxidation, rather than to the compaction of recently deposited delta material. In the past millennia subsidence has been as important a factor in flood protection as sea level rise (see Fig. 12). At present most of the western part of the country lies

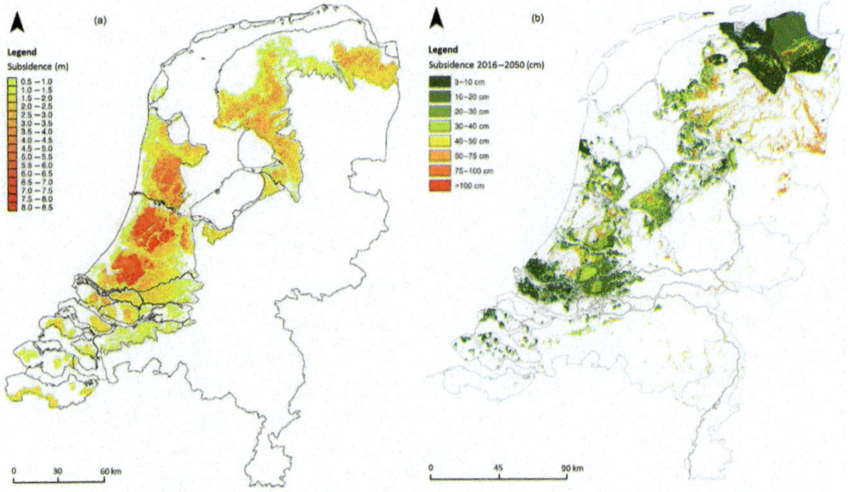

Fig. 11 Historic land subsidence (past 1000 years) due to drainage and peat mining (from Erkens et al., 2016) and expected land subsidence up to 2050 due to continues compaction of soft soils, but also due to gas extraction and salt mining (from Erkens et al., 2017). Combined figure from Stouthamer et al. (2020)

below sea level, up to almost 7 m, but also river flood levels can be well above the level of the riparian land (Fig. 13).

2 Socio-economic Characteristics

2.1 Overview

The Rhine-Meuse-Scheldt Delta has a catchment area of 243,000 km^2 and an annual river water discharge of around 85 km^3, based on data from 2005. The rivers in the delta transport an estimated 1.52 million metric tons of sediment annually. About 69% of the delta consists of water surfaces, while the remaining 31% is land, as per ASTER (2011) and OSM (2019) data. The delta's total area is estimated to range between 20,000 and 25,347 km^2. Urban land cover varies from 7.9 to 17%. The population density in areas below 10 m above sea level is 519 inhabitants per square kilometer, slightly below the national average.[1] These areas contribute significantly

[1] The total population and population density in the area < 10 m a.s.l. was estimated by the authors by combining data from ASTER (2011) with the most detailed population statistics available on a municipal level in the Netherlands (CBS, 2024).

Fig. 12 The river Waal during a flood (Beeldbank Rijkswaterstaat)

to the economy, with a GDP contribution of 71%.[2] Per capita freshwater availability in the delta is 600 m^3, according to the Delta Programme of 2019.

Both in the Netherlands and in Belgium, there are major concentrations of population and economic activity in the delta area (Fig. 14), with Rotterdam and Antwerp as the principal ports. This also necessitates extensive infrastructures, to enable transport of goods and people. Rivers and estuaries are important parts of this transport system, which connects Rotterdam and Antwerp to entire European mainland (Fig. 15). The transport function puts high demands on these water systems, especially because ships are becoming ever larger with deeper draughts.

Agriculture is the predominant occupation in the RMS delta, with a substantial portion of urban land use ranging from 7.9 to 17%, depending on the measurement methods used (as per Corine Land Cover, 2012; Meyer & Nijhuis, 2014). Occupation patterns can also be assessed through population density. When combining elevation data from ASTER (2011) with population statistics from CBS (2024), it becomes evident that approximately 11 million people reside below 10 m above sea level in the RMS Delta. This population constitutes 63% of the nation's total population and results in an average population density of 511 inhabitants per square kilometer. It's important to acknowledge that the widespread settlement pattern in the RMS Delta

[2] The GDP contribution of the area < 10 m a.s.l. in the NL was estimated by the authors by combining data from digital elevation map (ASTER 2011) with economic statistics on a provincial level (CBS, 2024).

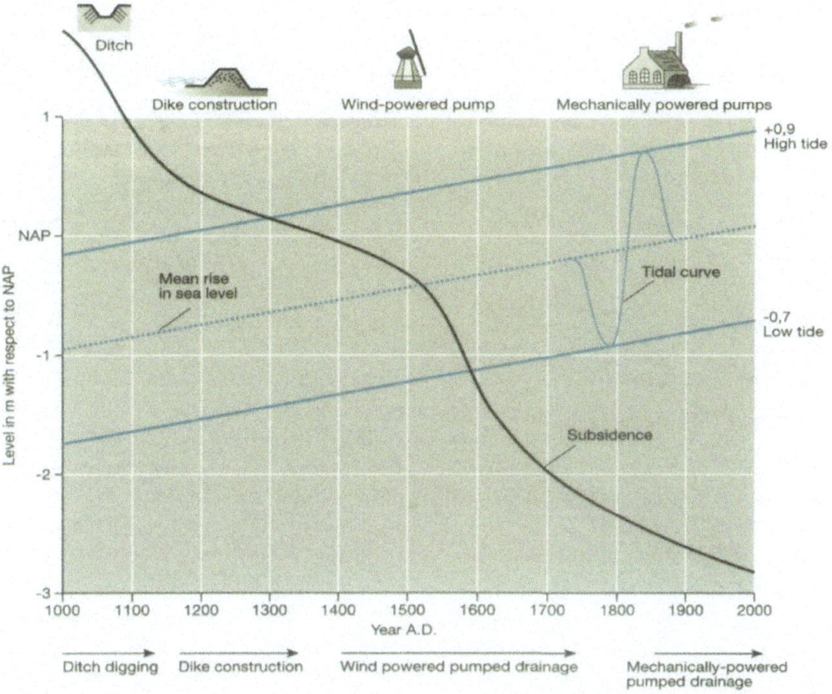

Fig. 13 Historic subsidence and seal level rise in the western part of the Netherlands (Stumpe & Tielrooij, 2000)

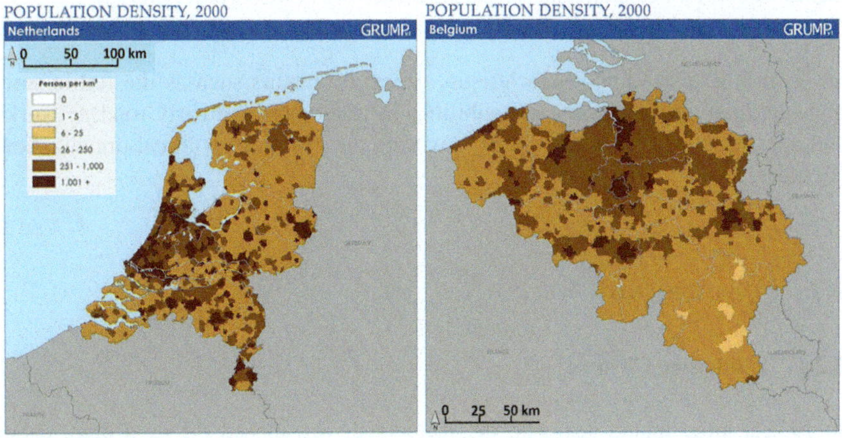

Fig. 14 Population densities in the Netherlands (SEDACMaps, 2000a) and Belgium (SEDACMaps, 2000b)

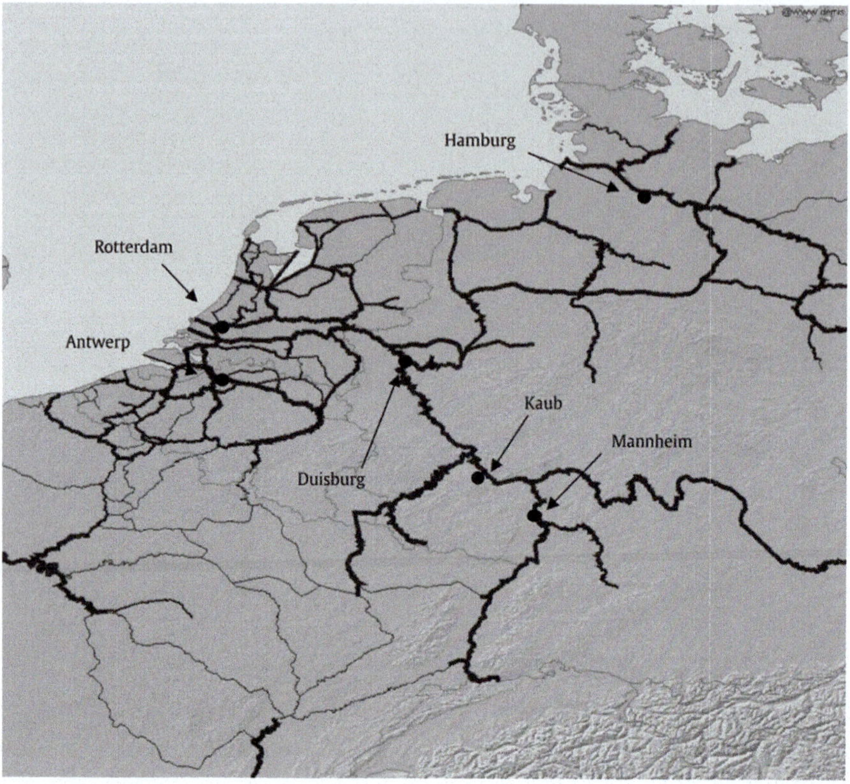

Fig. 15 Northwest-European waterborne transport network (Jonkeren et al., 2011)

has been a longstanding characteristic, driven by factors such as the fertile soil's high productivity, excellent accessibility via waterways and early road networks, and a governance model historically favoring decentralized political organizations (Boelens, 2014).

2.2 *Infrastructural Works*

2.2.1 Zuiderzee Closure

In 1932 a large inland sea arm, the Zuiderzee, was closed off by a dam, called Afsluitdijk (Fig. 16). The main purpose was land reclamation in order to provide for sufficient food, but it also gave major benefits to the country's flood protection and water management. The coastline was shortened by hundreds of kilometres and the large lake behind the dam now serves as an important freshwater reservoir. The downstream part of the river IJssel, the northernmost Rhine branch, turned from a

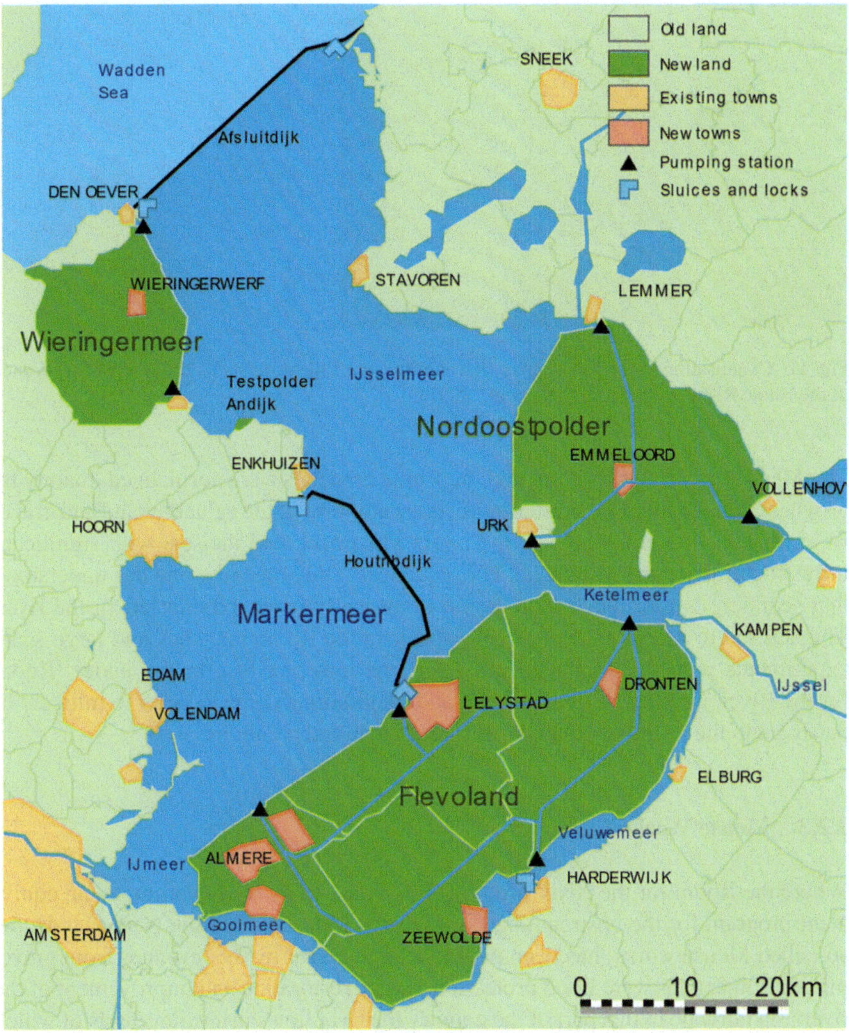

Fig. 16 Dams and land reclamations in the former Zuiderzee (Lommes, 2017)

tidal into a non-tidal river and various branches had to be shut down. At the seaside of the Afsluitdijk, the morphology of the so-called Wadden Sea is still responding to the construction of the dam (van Maren et al., 2023).

2.2.2 River Training Works

In order to improve navigability, the rivers in the delta have been subject to a variety of training works. A large part of the upstream river Meuse was canalised (Fig. 17).

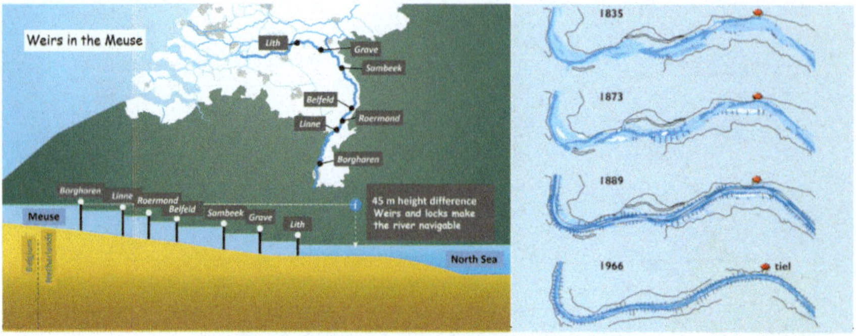

Fig. 17 Canalisation of the Maas River (left, figure from Rijkswaterstaat, 2023) and Waal branch of the Rhine River (right, from Silva et al., 2001)

The Meuse further downstream and the Rhine branches were normalized, mainly by groynes (also see Fig. 17). Furthermore, large amounts of the valuable sand and gravel were mined. As a consequence, the rivers tilted back and incised, with significant infrastructural and environmental consequences. Only recently, one has understood that sediment is a resource that should be properly managed and left in the river. This has led to a zero net abstraction policy in the Rhine branches and large-scale experiments with floodplain lowering, groyne lowering (see below under 'Room for the River'), sediment nourishment and longitudinal dams in order to mitigate or partly undo the negative effects of former human interventions.

2.2.3 Meuse Works

Where the Room for the River program was financed by the government, an equivalent program for the upstream part of the river Meuse within the Netherlands, the so-called Meuse works, has been paid to a large extent by the revenues from gravel mining. Objectives were flood protection, nature restoration and improvement of the river as a fairway. In this part of the country the river flows in a valley, parts of which have been built up and embanked. Enhancing flood safety without shifting the burden to the embanked parts further downstream was a particular challenge.

2.2.4 Delta Works

The Delta works, triggered by the 1953 flood disaster, consist of a series of major interventions in the Rhine-Meuse-Scheldt delta proper (see Fig. 18). Estuaries were closed off by dams, sluices and movable storm surge barriers. The primary objective was to enhance flood safety by shortening the coastline.

Fig. 18 Overview of the Delta works (RokerHRO, 2015)

The original plan was to close off all estuaries by dams. Only the Western Scheldt, access to the port of Antwerp, had to be left open. Public protest against the environmental impact ultimately led to the decision to leave the Eastern Scheldt open and protect it against extreme floods with a movable storm surge barrier. Clearly, the environmental impact of these engineering works was major. For example: a loss of intertidal area in the Eastern Scheldt, water quality problems in the lakes behind the dams, accumulation of polluted rivers sediment in the Hollands Diep/ Haringvliet area, erosion of the north–south oriented channels in the northern part of the delta and a loss of biodiversity in the former tidal wetland Biesbosch. This has led to a series of mitigating measures, such as allowing some tidal motion in the Haringvliet, enhancing the water circulation in the Grevelingen and nourishing sediment to intertidal flats in the Eastern Scheldt.

2.3 *Delta Management*

2.3.1 Flood and Water Management

Until Roman times, people in the lower parts of the Netherlands used to live on high grounds along the rivers, and on man-made mounds. In the Middle Ages they

Fig. 19 The Southwesterly Delta 400 years ago (left, Blaeu, 1635) and at present (right, from Bing.com)

started defending themselves against floods by dikes, often made of seaweed. Later on, they developed ever more sophisticated and efficient technology to build flood defences, remove surplus water from their land and control the groundwater level for agricultural purposes. An undesired side effect, however, was subsidence, mainly due to compaction and oxidation of peatland (also see Fig. 11).

2.3.2 Land Reclamations

The Southwesterly Delta has a long history of land reclamations and land losses (Fig. 19). Islands have been connected by land bridges and accreted further, but historic flood events led to significant losses of land, people and livestock. Figure 20 shows how the Western Scheldt has been encroached by successive land reclamations. This loss of storage area has its implications for the tidal penetration. It is one of the causes that the tidal range at Antwerp increased by about 1 m over the last century.

2.3.3 Dredging and Dumping

Since mid-twentieth century dredging activities have become important as human interference influencing the morphodynamic development of the Western Scheldt estuary. Sand mining of about 2.5 million m^3 per year between the 1950s and 2014 has caused deepening of the estuary in the order of 1 cm/year, which is much more than the present rate of sea level rise. Since 2014 sand mining is forbidden because of this serious effect. The Western Scheldt forms the entrance to Port of Antwerp. The navigation channel through the estuary has been deepened three times since the late 1970s, causing the maintenance dredging to increase to about 10 million m^3 per year. The dredged sediment for deepening and maintaining the navigation channel is deposited within the estuary to minimize sediment extraction. The Western Scheldt estuary is characterized by a two-channel system. Typically, one of the competing

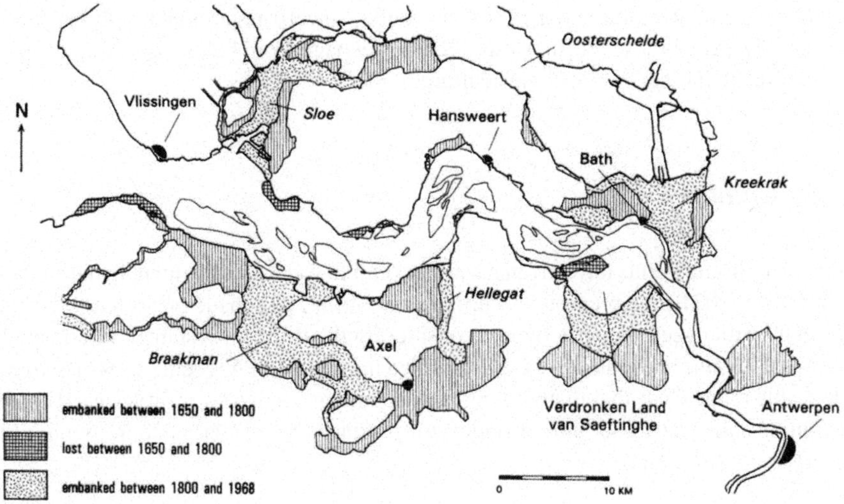

Fig. 20 Loss of storage area of the Western Scheldt mainly due to land reclamation (from Van der Spek, 1997)

channels is deepened, and dredged material is deposited in the other, impacting the multiple channel structure of the estuary (Wang & Winterwerp, 2001; Jeuken & Wang, 2010). Conserving the stability of the multi-channel system is considered as an important aspect in optimizing the strategy of dredged material disposal in the estuary.

3 Current Issues

Many of the phenomena and trends described in the foregoing constitute management challenges. Therefore, we will focus here on aspects that have not yet been mentioned.

3.1 Low Lying Land

A large part of the Netherlands lies below sea level. This in itself is a management concern, because the land needs to be protected against flooding. The flood protection approach is risk-based, where risk is defined as the probability of occurrence of a flood event multiplied by the damage done in terms of property and lives lost. Sea level rise, subsidence, climate change, increasing population density and economic activity in protected areas affect the risk level, thus necessitate regular upgrading of the flood defence system.

However, flood protection is not the only concern regarding low-lying areas, especially where these are subsiding. Saline seepage from below, for instance, may render fertile polders unfit for agriculture.

3.2 Climate Change

Recent extreme events have given a wake-up call to water management in the Netherlands. The 2021 precipitation and flooding event in 2021 made clear that flood risk cannot be managed properly by just considering the flood level statistics in the main stems of the rivers: tributaries have to be taken into account. A recent series of extreme droughts made clear that maintaining surface and groundwater reserves needs more attention and that the common policy to discharge rainwater as soon as possible needs reconsideration.

3.3 Coastal Squeeze and Loss of Intertidal Area

Typical estuarine areas like mudflats and marshes are threatened from both sides. Reclamations take away space at the land side, navigation channel dredging and (ship-) wave attack limit their extent at the water side. This so-called coastal squeeze goes at the expense of the typical estuarine habitat, hence of the high biological productivity of these systems. Initiatives such as polder abandonment, dike relocation and restoration of intertidal areas are meant to mitigate these effects (see, for instance, Fig. 21).

In the Eastern Scheldt, the loss of intertidal area is not due to coastal squeeze, but rather to the almost complete blockage of marine sediment import by the storm surge barrier. Due to the reduced tide, the tidal channels tend to become smaller and the sediment needed for this can only come from the intertidal flats. This manifests itself not only by a loss of intertidal area, but also by a loss profile of the intertidal shoals, due to which the flooding time is much shorter and birds have less time to forage. Attempts are made to compensate this by sediment nourishment on top of the intertidal flats, but as long as there is no sediment brought in from outside, this remains an internal sediment redistribution operation that does not solve the underlying problem.

3.4 Navigation Channel Maintenance

The Rhine-Meuse-Scheldt delta gives access to a Europe-wide network for waterborne transport (also see Fig. 15). This is of paramount economic importance to the Netherlands and Belgium. Natural processes such as sedimentation, but also

Fig. 21 The Perkpolder renaturalisation project two years after completion (Zuidwestelijke Delta, 2017)

the growth of waterborne transport and the ever increasing vessel size necessitate maintenance and regular upgrading of the fairways. This encompasses dredging operations, like in the Western Scheldt, but also safety measures and engineering works on bridges, locks and barriers, for instance. For some parts of the network, such as the rivers in the Netherlands, this raises the management question how far the fairway capacity can be extended without unacceptable consequences for infrastructures and river functions such as nature and riparian agriculture.

3.5 Coastal Erosion

Coastal erosion within the estuaries is not much of a problem, because most of the estuaries are embanked. Wherever necessary, erosion of the North Sea coast is compensated by regular nourishments with sand mined offshore. Clearly, accelerated sea level rise will require more and more frequent nourishments (see, for instance, Delta Commission, 2008).

3.6 Environmental Pollution

Being densely populated countries at the downstream end of river basins, the Netherlands and Belgium are facing persistent water quality problems. In the Netherlands, virtually none of the surface waters meets European quality standards. Also, the state of preservation of most aquatic ecosystems is below the standards set by the European Water Framework Directive. In the Southwesterly Delta this has led to mitigating measures, such as allowing some tidal motion into the Haringvliet/Hollands Diep, flushing the Grevelingen with sea water and making Lake Veer saline again.

Water pollution is not the only problem in the estuaries, also the bed sediment may be polluted. In many river and estuarine deposits, there is a polluted layer dating from the second half of last century. This is often covered by layer of cleaner recent deposits. Yet, there are also more recent bed pollutions. Until very recently, PFAS have been discharged into the Scheldt near Antwerp and have polluted the bed sediments of the Western Scheldt. PFAS are also found in fish and shellfish caught in these waters and people are advised not to swim there.

3.7 Biodiversity Loss

Estuaries are among the biologically most productive ecosystems of the world. Yet, their biodiversity is under threat. Not only by habitat loss due to embankments and coastal squeeze, but also by invasive species. One example is the Japanese Oyster (*Crassostrea gigas*), which covers large areas of intertidal flats in the Eastern Scheldt (Fig. 22), for instance, and has virtually outcompeted the original flat oyster (*Ostrea edulis*).

Macrofaunal studies also refer to the historical presence and decline of otters and harbor seals in the region, with pollution contributing to the latter's decline (Reijnders, 1982). The avian population in the Delta region, which had adapted to traditional farming methods, underwent significant transformations. Meadow bird populations dwindled due to the pressures of intensive land utilization, while herbivorous waterfowl thrived, primarily depending on fertilized agricultural land. This situation resulted in a noticeable clash of interests with farmers, as opportunistic species prospered in response to the increasing number of artificial habitats and the prevailing trend of afforestation (Nienhuis, 2008c).

Both fish and macro-invertebrate communities show limited species diversity, often dominated by non-native species among benthic macroinvertebrates (Nienhuis, 2008b). A significant recent development is the EU Water Framework Directive, which mandates EU member states to develop river basin management plans with the aim of achieving a good ecological status for rivers by 2015. Nevertheless, persistent eutrophication remains a significant environmental concern in the Delta's surface waters. This includes external eutrophication due to the inflow of nutrient-rich polluted water (containing nitrogen and phosphorus) and internal eutrophication driven by sulfate in systems loaded with organic material, such as floodplain and peat lakes.

Fig. 22 Japanese oysters covering a tidal flat (https://www.wur.nl/nl/show/inventarisatie-japanse-oesters.htm)

3.8 Harbour Siltation

Within the Rhine-Meuse-Scheldt delta there is a wide variety of ports, from very large like Rotterdam and Antwerp, via medium-sized ones like Sloe and Terneuzen/Ghendt, to fisheries harbours like Yerseke. As the North Sea water and the rivers and estuaries are laden with fine sediment, harbours tend to silt up with sediment from either side. The Port of Rotterdam, for instance, dredges 15–20 million m^3 of sediment from its harbour basins. Most of this sediment is clean enough to be used in nature restoration or dike building projects, for instance. Polluted sediment is stored in a separate basin (the Slufter), depending on the degree of pollution. Dredged material from the inner locks of the Port of Antwerp is polluted with TBT (a product used until 2003 in antifouling paint) and treated in a facility (Amoras) into dry matter (van Maren & van Kessel, 2015).

4 Measures to Solve the Problems

4.1 The Delta Program

In the past, many major flood protection measures, such as the closure of the Zuiderzee and the Delta Works, were taken after a major flood disaster. More recently, the Netherlands has been working proactively in finding solutions for the future problems. Since the end of the last century, various large research programmes such as Coastal Genesis I and Coastal Genesis II have been carried out to develop coastal maintenance strategies. At present the ongoing national research programme is called "Sea Level Rise".

Following the advice of the Delta Commission (2008) the Delta Fund was established for the implementation of the Delta Program, led by the Delta Commissioner. The Delta Program is meant to protect the Netherlands against high water and flooding, to ensure sufficient freshwater and to contribute to a climate-proof and water-robust organization of the country. Various solution directions (Fig. 23) and their "building blocks" (Haasnoot & Diermanse, 2022) are investigated, together with the pathways to implement them in the future.

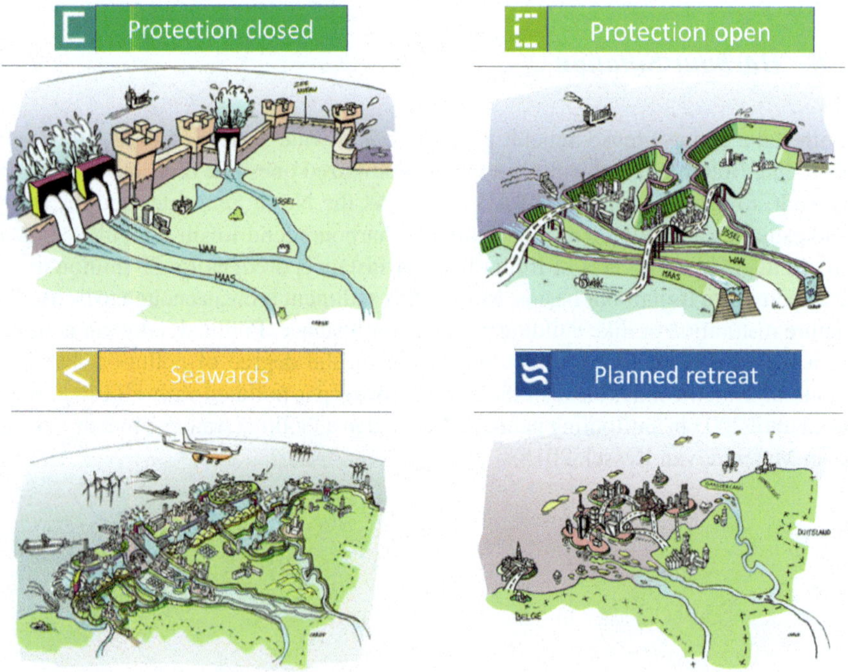

Fig. 23 The four solution directions. Adapted from Haasnoot and Diermanse (2022)

4.2 *Building with Nature*

As a low-lying delta country, the Netherlands has always been dealing with the threat of flooding. Nowadays the country is protected against flooding according to the highest safety standard in the world. With accelerated sea level rise the most important problem is how to maintain this high safety standard in the future. Another problem is to guarantee freshwater supply related to increasing salt intrusion and the changing flow regime of the rivers.

In managing the delta the Dutch have always considered safety against flooding as prerequisite, while the balance between socio-economic development and environmental protection is optimized. However, the consideration of these same three aspects has resulted into different solutions of the problems in different periods. During the construction of the Delta Works the closure of the Eastern Scheldt estuary was redesigned to a storm surge barrier, attributing more importance to the environmental protection. In the course of time, the philosophy of "building against nature" has shifted to "building with nature". As an example, the coastal protection is nowadays according to the rule "soft if possible, hard if necessary", making the soft measure sand nourishment preferable to hard structures. The long-term safety against flooding along the sandy coast is presently maintained by regular sand nourishments. The required amount of sand nourishment will increase with the increasing rate of sea level rise. As too frequent sand nourishments may cause ecological damage, experiments such as the "sand engine" are being carried out (see Fig. 24).

4.3 *Room for the River*

In the course of time, the rivers in the Netherlands have been encroached by embankments, land reclamations, building up floodplains, et cetera (see Fig. 25). This has led to a loss of fluvial habitat and higher and more rapidly propagating flood waves, hence increased flood risks. In order to mend this situation for the Rhine branches, a major river improvement program was carried out in the first two decades of this century. This Room for the River program encompassed 34 measures, such as relocating embankments, lowering floodplains, digging side channels, creating flood storage areas and building flood alleviation facilities. It has been completed in 2019.

4.4 *Sigma Plan*

In the densely populated Scheldt basin increasing flood levels are enhancing flood risks. In order to improve flood protection and in the meantime boost nature, the so-called Sigma plan is being implemented, aiming for completion in 2030. It consists

Fig. 24 Sand engine: a mega sand nourishment of more than 20 million m^3, following the principles of Building with Nature: ocean currents, wind and waves are gradually spreading the sand along the coast and into the dunes. *Source* https://dezandmotor.nl

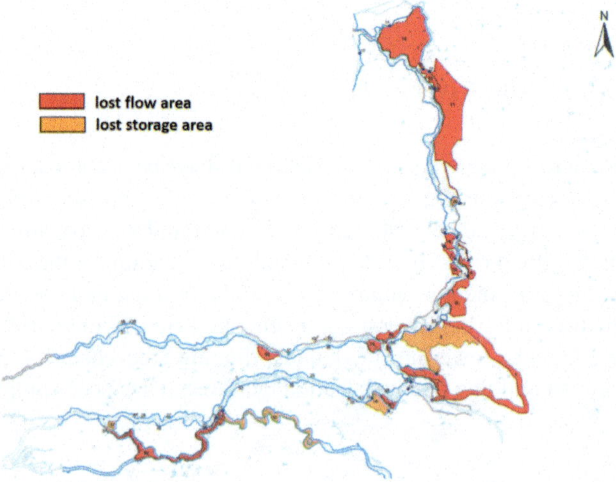

Fig. 25 Encroachment of the Rhine branches and the Meuse (Klijn & Stone, 2000)

Fig. 26 Overview of the Sigma Plan (https://www.sigmaplan.be/nl/projecten/)

of a large number of measures (see Fig. 26), such as dike reinforcements, polder abandonment and creation of storage areas.

5 Future of the Rhine-Meuse-Scheldt Delta

The Rhine-Meuse-Scheldt Delta is relatively successful in terms of ecosystem protection and biodiversity. This achievement is the result of continuous efforts in flood risk management, biodiversity protection, the promotion of nature-based solutions, and water quality control.

Efforts to conserve and restore these ecosystems will undoubtedly play a pivotal role in shaping the delta's future. This includes the preservation and rehabilitation of wetlands, the protection of critical habitats, and the implementation of sustainable land-use practices. Striking a balance between economic development and environmental conservation will be essential to ensure a sustainable future for the delta. As urbanization and industrial development in the delta region may continue to expand, these factors may impact natural habitats and water quality.

The delta will also need to adapt to rising sea levels. This adaptation may involve the promotion of ecosystem-based coastal defences, the enhancement of sandy coasts, and effective water management strategies. Ensuring a sustainable freshwater supply, robust flood control measures, and the prevention of saltwater intrusion will be vital aspects demanding ongoing attention and investment. A more resilient and sustainable future for the delta will be facilitated by advances in technology, including improved flood forecasting systems, sustainable agricultural practices, and the adoption of renewable energy solutions.

Spanning multiple countries, including the Netherlands, Belgium, Germany, and France, the Rhine-Meuse-Scheldt Delta and its catchment underscore the importance of continued international cooperation. Additionally, engaging local communities and stakeholders in decision-making processes will be crucial for addressing shared challenges and developing coordinated strategies for the delta's future. Their knowledge and input can help shape policies and projects that better align with the needs and aspirations of the people living in the region.

References

ASTER. (2011). ASTER Global Digital Elevation Map 2. Retrieved September 1, 2019, from https://asterweb.jpl.nasa.gov/gdem.asp

Blaeu, W. (1635). Zeelandia Comitatus.

Boelens, L. (2014). Delta governance: The DNA of a specific kind of urbanization. *Built Environment, 40*, 169–183.

Cox, J. R., Huismans, Y., Knaake, S. M., Leuven, J. R. F. W., Vellinga, N. E., van der Vegt, M., et al. (2021). Anthropogenic effects on the contemporary sediment budget of the lower Rhine-Meuse Delta Channel Network. *Earth's Future, 9*(7), 1–22.

Cox, J. R., Leuven, J. R. F. W., Pierik, H. J., van Egmond, M., & Kleinhans, M. G. (2022). Sediment deficit and morphological change of the Rhine-Meuse river mouth attributed to multi-millennial anthropogenic impacts. *Continental Shelf Research, 244*, 104766.

de Jong, M. P. C., & Battjes, J. A. (Eds.). (2010). Generation and prediction of seiches in Rotterdam harbor basins. In Y. C. Kim (Ed.), *Handbook of coastal and ocean engineering* (pp. 179–236). World Scientifc.

Delta-Alliance. (2024). http://www.delta-alliance.org/deltas/rhine-meuse-delta

Delta Commission. (2000). *Working together with water. A living country builds for its future* (138 pp). Secretariat Delta commission. ISBN/EAN 978-90-9023484-7.

der Spek, V. (1997). Tidal asymmetry and long-term evolution of Holocene tidal basins in The Netherlands: Simulation of palaeo-tides in the Scheldt estuary. *Marine Geology, 141*, 71–90.

Erkens, G., van der Meulen, M. J., & Middelkoop, H. (2016). Double trouble: Subsidence and CO2 respiration due to 1,000 years of Dutch coastal peatland cultivation. *Hydrogeology Journal, 24*, 551–568.

Erkens, G., Stafleu, J., & van den Akker, J. J. H. (2017). Bodemdalingvoorspellingskaarten van Nederland, versie 2017. Deltares rapport klimaateffectatlas.

Fettweis, M., Sas, M., & Monbaliu, J. (1998). Seasonal, neap-spring and tidal variation of cohesive sediment concentration in the Scheldt Estuary, Belgium. *Estuarine Coastal and Shelf Science, 47*, 21–36.

Frings, R. M., Hillebrand, G., Gehres, N., Banhold, K., Schriever, S., & Hoffmann, T. (2019). From source to mouth: Basin-scale morphodynamics of the Rhine River. *Earth-Science Reviews, 196*, 102830.

Frings, R. M., Schüttrumpf, H., & Vollmer, S. (2011). Verification of porosity predictors for fluvial sand-gravel deposits. *Water Resource Research, 47*(7), 1–15.

Haasnoot, M., & Diermanse, F. (Eds.). (2022). Analyse van bouwstenen en adaptatiepaden voor aanpassen aan zeespiegelstijging in Nederland.

Hegnauer, M., Beersma, J. J., van den Boogaard, H. F. P., Buishand, T. A., & Passchier, R. H. (2014). Generator of rainfall and discharge extremes (GRADE) for the Rhine and Meuse basins (84 pp.). Deltares.

https://help.openstreetmap.org/questions/84042/osm-historical-dump-for-2019

https://land.copernicus.eu/en/products/corine-land-cover/clc-2012

https://www.cbs.nl/en-gb
IPCC. (2014). Climate Change 2014: Synthesis Report. In Core Writing Team, R.K. Pachauri., L.A. Meyer (Eds.). *Contribution of Working Groups I, II and III to the fifth assessment report of the intergovernmental panel on climate change*. IPCC.
Jeuken, M. C. J. L., & Wang, Z. B. (2010). Impact of dredging and dumping on the stability of ebb-flood channel systems. *Coastal Engineering, 57*, 553–566.
Jonkeren, O., Jourquin, B., & Rietveld, P. (2011). Modal-split effects of climate change: The effect of low water levels on the competitive position of inland waterway transport in the river Rhine area. *Transportation Research Part A: Policy and Practice, 45*(10), 1007–1019.
Klijn, F., & Stone, K. (2000). *Past room for the Rhine branches*. Delft Hydraulics, Report R3294.67, 31 pp. Also see: https://open.rijkswaterstaat.nl/open-overheid/onderzoeksrapporten/@239208/vroegere-ruimte-rijntakkenverslag/#highlight=Klijn,%20F
KNMI. (2024). *Climate of the Netherlands*. https://www.knmi.nl/klimaat
Lommes. (2017). A map of the Zuiderzeeworks in the Netherlands. https://commons.wikimedia.org/w/index.php?curid=54839004Dedalus. CC BY-SA 3.0.
Meyer, H., & Nijhuis, S. (Eds.). (2014). *Urbanized Deltas in transition*. Techne Press.
Nienhuis, P. H. (2008a). Changing Rhine ecosystems: Pollution and rehabilitation. In P. H. Nienhuis (Ed.), *Environmental history of the Rhine-Meuse Delta* (pp. 329–353). Springer.
Nienhuis, P. H. (2008b). Changes in biodiversity: Lower organisms, vegetation and flora. In P. H. Nienhuis (Ed.), *Environmental history of the Rhine-Meuse Delta* (pp. 481–507). Springer.
Nienhuis, P. H. (2008c). Changes in biodiversity: Birds and Mammals and their use. In P. H. Nienhuis (Ed.), *Environmental history of the Rhine-Meuse Delta* (pp. 509–535). Springer.
Peelen, R. (1967). Isohalines in the delta area of the rivers Rhine, Meuse, and Scheldt: Classification of water in the delta area according to their chlorinity and the changes in these waters caused by the delta works. *Netherlands Journal of Sea Research, 3*, 575–597.
Perkpolder: twee jaar later. (2017). https://www.zwdelta.nl/nieuws/perkpolder-twee-jaar-later/
Pseudomas. (2022). CC BY-SA 3.0, https://commons.wikimedia.org/w/index.php?curid=120915437
Reijnders, J. H. (1982). On the extinction of the southern Dutch harbour seal population. *ICES, Marine Mammals Commission, 31*(1), 900357.
Rijkswaterstaat. (2017). Factsheet Southerly Delta Basin. *Rijkswaterstaat, Water, Verkeer en Leefomgeving* (26 pp.) (in Dutch).
Rijkswaterstaat. (2019). Het Verhaal van de Rijn-Maas-monding (56 pp.).
Rijkswaterstaat. (2023). De.7 Maasstuwen. https://www.rijkswaterstaat.nl/water/waterbeheer/bescherming-tegen-het-water/waterkeringen/dammen-sluizen-en-stuwen/de-7-maasstuwen
RokerHRO Map of the Delta Works. (2015). OpenStreetMap CC BY-SA 3.0. https://commons.wikimedia.org/w/index.php?curid=4446902
Schulte-Wülwer-Leidig, A., Gangi, L., Stötter, T., Braun, M., & Schmid-Breton, A. (2018). Transboundary Cooperation and sustainable development in the Rhine Basin. https://www.intechopen.com/chapters/59001
SEDACMaps. (2000a) SEDACMaps—Netherlands: Population Density, 2000, CC BY 2.0. https://commons.wikimedia.org/w/index.php?curid=85088632
SEDACMaps. (2000b). SEDACMaps—Belgium: Population Density, 2000, CC BY 2.0, https://commons.wikimedia.org/w/index.php?curid=85088904
Silva, W., Klijn, F., & Dijkman, J. P. M. (2001). Room for the Rhine branches in The Netherlands: What the research has taught us. http://resolver.tudelft.nl/uuid:12b2ad06-3469-49ea-a280-78e1dcc2fcb9
Smit, H., & Coops, H. (1991). Ecological, economic and social aspects of natural and man-made bulrush (Scirpus lacustris L.) wetlands in The Netherlands. *Landscape and Urban Planning, 20*, 33–40.
Sterl, A., van Oldenborgh, G. J., Hazeleger, W., & Dijkstra, H. (2008). Extreme maximumtemperaturen in de toekomst. *Meteorologica, 17*(3), 4–7.

Steffbauer, D. B., Riva, R. E. M., Timmermans, J. S., Kwakkel, J. H., & Bakker, M. (2022). Evidence of regional sea-level rise acceleration for the North Sea. *Environmental Research Letters, 17*, 074002.

Stouthamer, E., Erkens, G., Cohen, K., Hegger, D., Driessen, P., Weikard, H. P., et al. (2020). Dutch national scientific research program on land subsidence: Living on soft soils subsidence and society. *Proceedings of the International Association of Hydrological Sciences, 382*, 815–819.

Stumpe, J., & Tielrooij, F. (2000). Waterbeleid voor de 21ste eeuw. Eindrapport Commissie WB21 (50 pp.). http://resolver.tudelft.nl/uuid:102e013a-1357-4087-b9f3-387f877c793f

Tockner, K., Uehlinger, U., & Robinson, C. T. (2009). *Rivers of Europe*. Academic Press.

Vaas, K. F. D. (1968). Visfauna van het estuariumgebied van Rijn en Maas. *Biologisch Jaarboek, Dodonaea, 36*, 115–128.

Van de Rijt, C. W. C. J., & Coops, H. (1993). Tide creates room for vegetation in the Rhine-Meuse estuary. De Levende Natuur, 9468–72 (in Dutch with English summary).

Van Dijk, W. M., Cox, J. R., Leuven, J. R., Cleveringa, J., Taal, M., Hiatt, M. R., et al. (2021). The vulnerability of tidal flats and multi-channel estuaries to dredging and disposal. *Anthropocene Coasts, 4*(1), 36–60.

Van Hoestenberghe, T., Ferket, B., De Boeck, K., Vanlierde, E., Vanlede, J., Verwaest, T., & Mostaert, F. (2014). *Sediment load for the river Scheldt and its main tributaries (1972–2009)*. Version 5.0. WL Report 00_029. Waterbouwkundig Laboratorium/Antea Group: Antwerp.

van Maren, D. S., van Kessel, T., Cronin, K., & Sittoni, L. (2015). The impact of channel deepening and dredging on estuarine sediment concentration. *Continental Shelf Research, 95*, 1–14.

van Maren, D. S., A. Colina Alonso, A., Engels, W. Vandenbruwaene, P. L. M. de Vet, J. Vroom & Z.B. Wang. (2023). Adaptation timescales of estuarine systems to human interventions. *Frontiers in Earth Science, 11*, 1–18.

van Maren, D. S., & van Kessel, T. (2016). Long-term effects of maintenance dredging on turbidity. *Terra Et Aqua, 145*, 5–14.

Vellinga, N. E., Hoitink, A. J. F., van der Vegt, M., Zhang, W., & Hoekstra, P. (2014). Human impacts on tides overwhelm the effect of sea level rise on extreme water levels in the Rhine-Meuse delta. *Coastal Engineering, 90*, 40–50.

Vermeersen, B. L. A., Slangen, A. B. A., Gerkema, T., Baart, F., Cohen, K. M., Dangendorf, S., Duran-Matute, M., Frederikse, T., Grinsted, A., Hijma, M. P., Jevrejeva, S., Kiden, P., Kleinherenbrink, M., Meijles, E. W., Palmer, M. D., Rietbroek, R., Riva, R. E. M., Schulz, E., Slobbe, D. C., et al. (2018). Sea-level change in the Dutch Wadden Sea. *Netherlands Journal of Geosciences, 97*(3), 79–127.

Veroustraete, F. (2016). The Solomon Isles disappear in the Pacific. In *Scientific evidence for Global and Local Sea-level rise. Climate Impacts* (vol. 1, pp. 1–10). REDSTAR-CV&M Open Access Science Publishing.

Wang, Z. B., & Winterwerp, J. (2001). Impact of dredging and dumping on the stability of ebb-flood channel systems. In *Proceedings of the 2nd IAHR symposium on River, Coastal and Estuarine Morphodynamics* (pp. 515–524).

Wang, B., & Xu, Y. J. (2018). Decadal-scale riverbed deformation and sand budget of the last 500 km of the Mississippi River: Insights into natural and river engineering effects on large alluvial rivers. *Journal of Geophysical Research: Earth Surface*, 004542.

Wang, Z. B., Vandenbruwaene, W., Taal, M., & Winterwerp, H. (2019). Amplification and deformation of tidal wave in the upper Scheldt Estuary. *Ocean Dynamics, 69*(7), 829–839.

Wolff, W. J. (1973). The estuary as a habitat. An analysis of data on the soft-bottom macrofauna of the estuarine area of the rivers Rhine, Meuse, and Scheldt. Rijksmuseum van Na-tuurlijke Historie, Leiden. Zoologische Vmhandelingen, 126, 3–242.

Río de la Plata Estuary

Natalia Venturini, Ernesto Brugnoli, Ofelia Gutiérrez, Pablo Muniz, Gustavo J. Nagy, Daniel Panario, José E. Verocai, and Xianye Wang

1 Physical Setting

The Río de la Plata (RdlP) is a funnel-shaped estuarine system located between 34°00″–36°10′10′ S and 55°00″–58°10″ W on the east coast of South America (Fig. 1). It is a complex and extensive semi-enclosed inner shelf area that drains the second-largest basin of the continent to the Atlantic Ocean, comprising 3.1 million km^2 (Acha et al., 2008). The RdlP is a large (290 km in length), wide (maximum width of 220 km at its mouth) and shallow (depths ranging from 1 to 29 m) fluvial-estuarine-sea system with a very high fluvial flow and a gentle slope. It is characterized by a microtidal regime, with subtidal oscillations having a similar amplitude to astronomic tidal oscillations (Jackson et al., 2021). For example, the tidal amplitude along the Argentinean coast is about 1 m, while along the Uruguayan coast is about 0.4 m (Santoro et al., 2013).

N. Venturini (✉) · E. Brugnoli · P. Muniz · J. E. Verocai
Oceanografía y Ecología Marina, Instituto de Ecología y Ciencias Ambientales, Facultad de Ciencias, Universidad de la República, 11400 Montevideo, Uruguay
e-mail: rulo@fcien.edu.uy

O. Gutiérrez · D. Panario
Unidad de Ciencias de la Epigénesis, Instituto de Ecología y Ciencias Ambientales, Facultad de Ciencias, Universidad de la República, 11400 Montevideo, Uruguay

G. J. Nagy
Cambio Ambiental Gestión Costero Marina, Instituto de Ecología y Ciencias Ambientales, Facultad de Ciencias, Universidad de la República, 11400 Montevideo, Uruguay

X. Wang
State Key Laboratory of Estuarine and Coastal Research, East China Normal University, Shanghai 200241, China

W. Zhang and H. de Vriend (eds.), *Delta Sustainability*,
https://doi.org/10.1007/978-981-97-7259-9_15

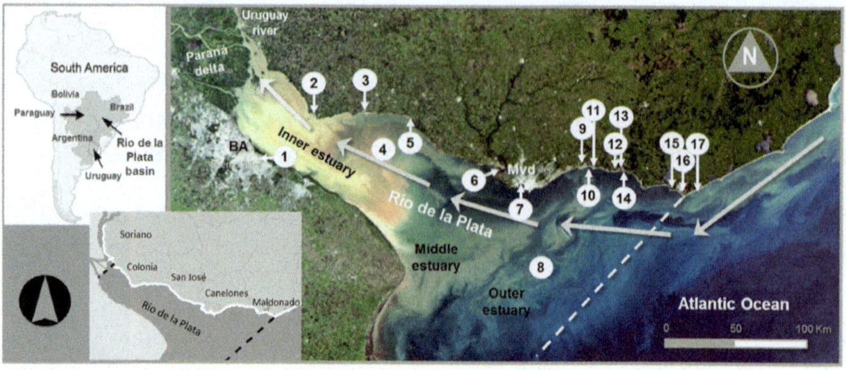

Fig. 1 The Río de la Plata estuary. The arrows show the drift transport of sediments. BA: Buenos Aires, Mvd: Montevideo, (1) Matanza Riachuelo Creek, (2) Colonia, (3) of Juan Lacaze, (4) Ortiz bank, (5) Boca del Cufré Creek mouth, (6) Santa Lucía River, (7) Pantanoso Creek, (8) Barra del Indio shoal, (9) Atlántida, (10) Arroyo (river) Solís chico, (11) Las Vegas beach, (12) Balneario Argentino beach, (13) Jaurreguiberry beach, (14) Solís grande river, (15) Punta Rasa/Punta Negra, (16) Punta del Este, (17) Punta Piedras. *Source* The Copernicus Sentinel-3 satellites on 14 February 2022 (https://www.copernicus.eu/en/media/image-day-gallery/rio-de-la-plata-estuary)

1.1 Riverine Systems

Annual mean river runoff into the RdlP amounts to ~25,000 $m^3\ s^{-1}$, with the Paraná and Uruguay rivers as the main tributaries (Acha et al., 2008). The runoff varies between less than 20,000 $m^3\ s^{-1}$ and more than 30,000 $m^3\ s^{-1}$ during dry/wet years, associated with La Niña/El Niño events, respectively (Nagy et al., 2002, 2008a). Although the contribution of the different tributaries shows seasonal, inter-annual and interdecadal variations, the Paraná River provides about three quarters of the average flow rate (Framiñan et al., 1999). Shaped at the confluence of the Paraná and Uruguay rivers is the Paraná Delta (Fig. 1), a seaward prograding delta with a rate of increase up to 75 m $year^{-1}$ and an extension of 14,000 km^2 (Sarubbi, 2007).

Sediment delivery to the RdlP is about 79.8 billion kg $year^{-1}$ (Framiñan et al., 1999). About 90% is delivered as suspended silts and clays, while the remaining 10% corresponds to very fine sands (Sarubbi, 2007). Suspended sediment concentrations (SSC) range between 60 and 300 mg L^{-1}, and high concentrations are associated with resuspension by storm events (Fossati et al., 2014). As one of the most turbid estuaries in the world, extreme suspended sediment concentrations near 400 mg L^{-1} have been reported for the RdlP (Moreira et al., 2013). Bottom sediments gradually settle along the estuary, with textures becoming progressively poorly sorted offshore. Sand is dominant at the head, silt in the intermediate estuary and clayey silt and clay at its mouth. Around the Barra del Indio shoal, clay prevails over silt and sand, water and organic matter content reach their maximum and physicochemical flocculation processes are important. In the outer region, coarse sand immediately dominates seaward of the salt wedge (Moreira et al., 2016).

Atmospheric circulation, topography, rainfall, river discharge, and prevailing wind and tide conditions are the main natural forcing factors of the RdlP dynamics (Nagy et al., 2008a; Schuerch et al., 2016). A quasi-permanent salt wedge regime develops under weak to moderate winds, which generates bottom and surface salinity fronts that play an essential role in biological processes (Mianzan et al., 2001). An estuarine turbidity maximum (ETM) develops between the transition from fresh, estuarine and marine domains, the location of which changes through the year, depending on wind direction and river discharge. Onshore winds and low river outflow prevail in summer. In contrast, offshore winds and high river outflow prevail in fall–winter periods. This variation triggers movements of the turbidity front toward the inner and outer parts of the estuary (Nagy et al., 2008b). A sharp decrease in SSC on the seaward side of the ETM relates to the increased deposition of flocculated fine-grained suspended sediments (Schuerch et al., 2016). Atmospheric circulation over the RdlP is controlled by the quasi-permanent South Atlantic high-pressure system and the episodic passage of frontal systems that propagate from the south (Nagy et al., 2008a). Rainfall regime, RdlP tributaries flow, and consequently, the RdlP discharge are strongly influenced by El Niño–Southern Oscillation (ENSO) phases (Barreiro, 2010). Under a warm ("El Niño") ENSO phase, austral spring and summer rains increase considerably, boosting the RdlP flow, while the contrary occurs under a cold ("La Niña") ENSO phase (Barreiro, 2010). ENSO impacts on RdlP flow fluctuations, salinity and turbidity fronts have been well documented (Barreiro, 2010; Brugnoli et al., 2018; Guerrero et al., 1997; Nagy et al., 2013). They also affect ecological and environmental quality related to cyanobacteria blooms (Kruk et al., 2019) and benthic organic enrichment (Brugnoli et al., 2021a).

1.2 Deltas

Due to its vast extension, bathymetry and bed geomorphology vary significantly along the RdlP. It can be classified into upper (2,700 km^2), intermediate (9,000 km^2), and outer (25,000 km^2) regions (Fig. 1). The cross-section at the intermediate zone increases from 32 km between Colonia and Buenos Aires to 100 km between Montevideo and Punta Piedras. In comparison, the limit of the outer zone between Punta del Este and Punta Rasa reaches 220 km (Jackson et al., 2021). The upper region is a shallow shelf of about 1–4 m depth and a smooth slope. Here the confluence of the Paraná and Uruguay rivers creates the principal channel, which follows the Uruguayan coast until the large and shallow Ortiz bank in the intermediate region (Figs. 1 and 2). This bank, nearly 3 m in its deepest part, is one of the most critical features and causes the channel bifurcation. The intermediate region has a depth less than 10 m, with the Intermedio channel located on the southern coast. The outer region, with depths between 10 and 20 m, displays the Oriental channel along the northern coast and the Marítimo channel on the southern coast (Fig. 2). Depths greater than 20 m are observed after Punta del Este–Punta Rasa.

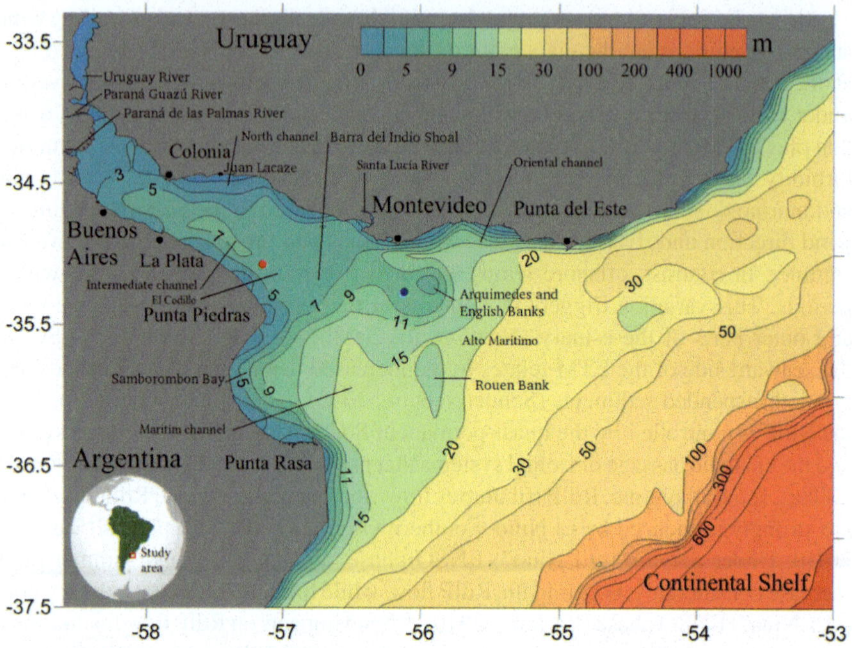

Fig. 2 Bathymetry of the Río de la Plata Estuary (in m), main geographical references and bed topography. The red dot indicates the position of Torre Oyarvide, whereas the blue dot shows the location of the Oceanographic Buoy. *Source* Cited from Moreira and Simionato (2019)

The submarine delta front, sloping towards the continental shelf, with depth greater than 100 m, is located 460 km from the head of the RdlP (Fig. 2) (Moreira et al., 2016; Moreira & Simionato, 2019; Muniz et al., 2019b).

Although the innermost stretch, called "Bajo (lower) Uruguay" (which extends between the Paraná River delta and the Uruguay River delta located in front of the city of Fray Bentos), is a freshwater system (Uruguay River), its beaches are dominated by waves and drift transport associated with said waves (Fig. 3). This sector has been strongly affected by underwater extraction of aggregates and exhibits a notable sand deficit, reflected by native tree species from the 1970s of the beach area.

The continental limit in some of these beaches presents sectors with bluffs (cliffs) in active retreat. However, except for some small urban settlements, the economic activity throughout this stretch is agricultural, with large farms and ranches (hundreds to thousands of hectares). So, except for some sites that can be considered as heritage, bank erosion is not a major threat.

This chapter describes the fundamental characteristics (geomorphology, climatic and socio-economic) of the RdlP estuary and the most relevant environmental and anthropic pressures; finally, it includes one section about solutions and paths for the future.

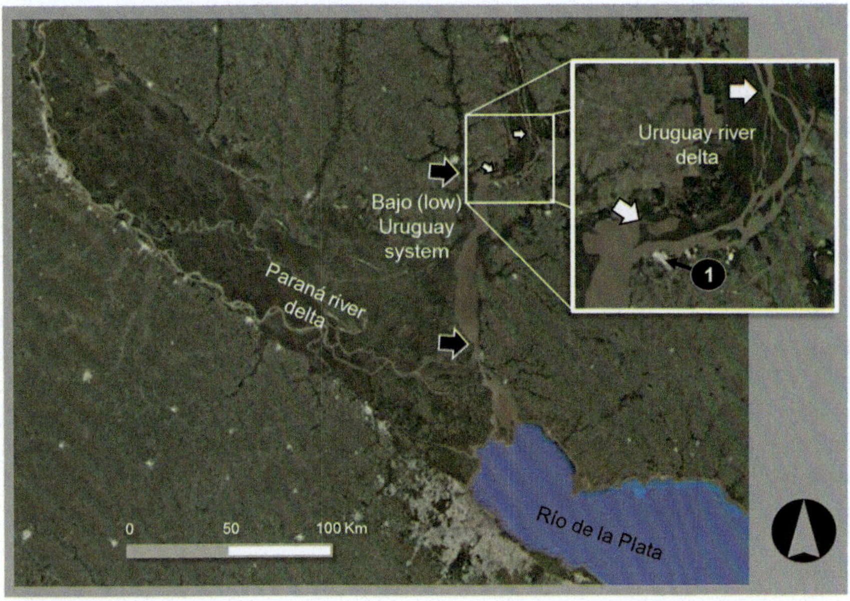

Fig. 3 Deltas of the Paraná River and the Uruguay River. The black arrow delimits the Uruguay "Bajo" system, and the white arrows delimit the delta of the Uruguay River, (1) the city of Fray Bentos. *Source* Image Landsat Copernicus on December 2020 (Google Earth)

2 Socio-economic Characteristics

2.1 Population

The RdlP is unique in terms of social and economic multiplicity. Its basin comprises southeastern Bolivia, southern and central Brazil, entire Paraguay, most of Uruguay and northern Argentina (Fig. 1), which account for more than 75% of the Gross National Product (GNP) of the five countries (Bonachea et al., 2010). Roughly 150 million people inhabit the RdlP basin, and nearly 60 cities have more than 100,000 people, including the capitals of Buenos Aires, Brasilia, Asunción, and Montevideo (Achkar et al., 2012). An agreement signed in 1973 between Uruguay and Argentina called the "Treaty of the Río de la Plata and its Maritime Front" established the maritime limits, shared use and management of bilateral waters and resources, exclusive jurisdiction zones, and a shared fishing zone in the Atlantic Ocean. Such jurisdictional issues and activities are regulated by binational administrative commissions (CARP, 2012; Marín et al., 2021).

2.2 Major Industries

La Plata-Buenos Aires (Argentina) and Montevideo (Uruguay) are the biggest harbors in the RdlP estuary, handling jointly almost 20,000 t of cargo and 550,000 passengers. The petrochemical pole located in the western coastal plain of the RdlP estuary has modified the hydrodynamics of the area through piping and landfilling actions. These changes, along with the leaking of effluent derived from industrial activity, resulted in modifications to the surface and groundwater chemistry in this coastal plain sector.

Currently, overseas ships use two channels. The backbone network begins at the Punta Indio Channel (Fig. 4), close to Montevideo harbour and ends at the Paraná de las Palmas outlet (delta). It is dredged to a depth of 10.4 m, administered and managed by Uruguay. The second, called Martín García, begins at 39 km of the RdlP and ends at 0 km of the Uruguay River. It passes close by Martín García Island and is administered by the Administrative Commission of the Río de la Plata (CARP). It is currently at a depth of 9.7 m and is dredged to 10.4 and 11.6 m in hard bottoms.

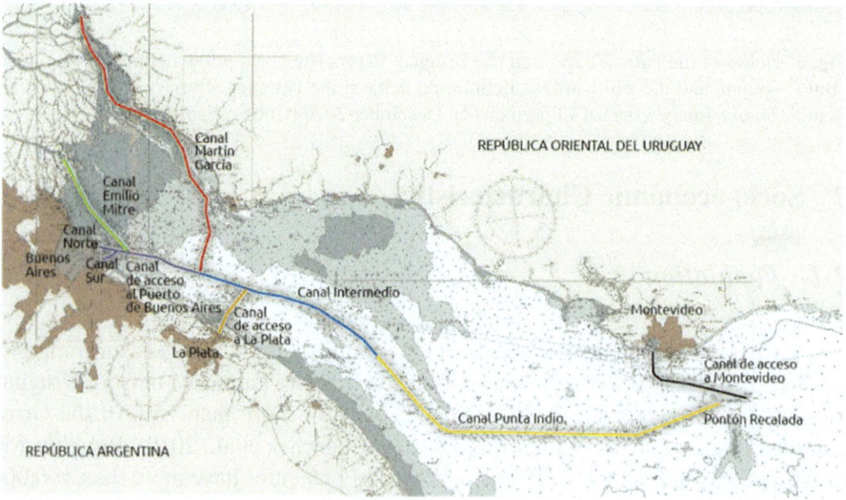

Fig. 4 Navigation channels of the Río de la Plata estuary. *Source* Cited from Portatadino (2022)

3 Current Issues

3.1 Relative Sea Level Rise

Continuous global sea level rise (SLR), likely the most severe coastal impact of anthropogenic climate change, is due to ocean warming, thermal expansion, mass loss from ice sheets and glaciers, and terrestrial water storage fluxes (IPCC, 2019; Masson-Delmotte et al., 2021). Consequently, SLR is a subject of great interest in scientific research and public policy. However, the dominant factor in changes of future local sea level extremes is relative sea-level change, that is, the level change with respect to the ground level.

SLR threatens coastal areas through a range of hazards and impacts, but mainly through the enhancement of episodic marine flooding events, due to a combination of slow continuous sea level rise, storm surges and high astronomic tides (Fiore et al., 2009). They lead to groundwater inundation, shoreline erosion, change and loss of coastal ecosystems; salinity intrusion into ground and surface waters, and impeded drainage of natural or human hydrological systems (Cooley et al., 2022; Glamore et al., 2016; Nagy et al., 2023a).

Sea level change occurs at all time scales, and at a continuous range of spatial scales, from global to regional (e.g., El Niño Southern Oscillation (ENSO), and local (e.g., wind storm surge) (Brugnoli et al., 2018; Leal Filho et al., 2022; Piecuch, 2022). Because of tidal and wind-driven effects, the sea level is constantly fluctuating. It is essential to determine the mean sea level (MSL), the average sea level at a given location over several years (Douglas, 2001). The hydro-morphological response of estuaries to rising sea levels and changes in other climatic variables (e.g., increased storminess) will vary markedly between locations. Some deep, narrow systems may undergo little change, whilst others (e.g., coastal plain estuaries with a relatively low depth, like the RdlP (Verocai et al., 2015, 2016), may be significantly altered. Several urban areas develop along the RdlP shores, with more than 15 million people. For the southern (Argentinian) coast the RdlP is the primary freshwater resource (Lecertua, 2010). Besides, it will soon be a secondary source for Uruguay.

Montevideo Harbour tide gauge station provides a long-term (since 1902) sea-level time series. This Gloss Station N°300 has been part of the Global Sea Level Observing System programme since the 1990s. Verocai et al. (2016) reported an SLR of $\geq$11 cm between 1902 and 2016, at a long term rate of 0.9 mm/year, showing significant fluctuations and a raised interannual variability partly related to ENSO events. For example, during the period 1961–2014, both the mean sea level and the acceleration rate fluctuated, showing acceleration and stabilization (Verocai et al., 2015, 2016) from 1971–1998 to 1999–2010, respectively. They were followed by a new acceleration from 2013–2016 (Gutiérrez et al., 2016; Nagy et al., 2019), which continued up to 2019 (Nagy et al., 2023b).

The Montevideo SLR time series contains the signals of global trends and ENSO-related regional influences at the same time; the latter encompass the local effects of varying river flows (Nagy et al., 2014, 2019; Piecuch, 2022) and winds

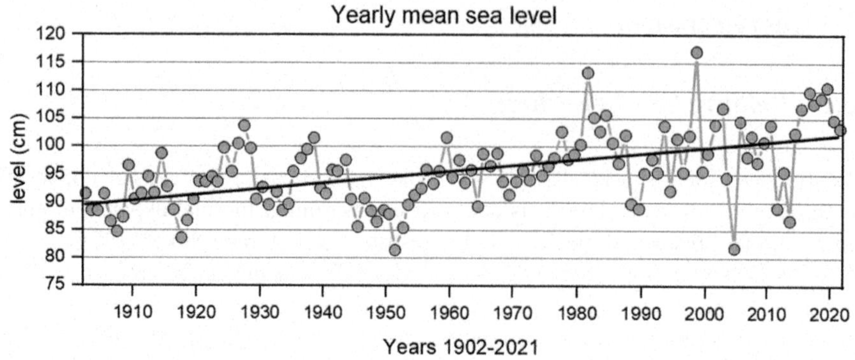

Fig. 5 Sea level time series at Montevideo harbor Gloss station from 1902 to 2021

(Gutiérrez et al., 2016; Verocai et al., 2015, 2016). The increasing rate observed since 2014 follows the global trend, with a peak triggered by "El Niño" 2015–16 (Nagy et al., 2019).

Figure 5 shows the long-term SLR at Montevideo Harbour (1902–2021). The observed trend partly follows the global SLR, with a weaker increase and a remarkable short-term variability primarily related to ENSO-related river inflow and wind changes. For instance, the recent increasing trend (Nagy et al., 2014, 2019, 2023b; Verocai et al., 2015, 2016), particularly from 2015 to 19, when the levels reached 107–111, 16–20 cm above the reference level (91 cm established in 1949) before the rise in MSL (Nagy et al., 2023b), was reversed in 2020–21 due to the decreased freshwater inflow related to the multi-year La Niña event that began in 2020–21.

The updated SLR for the period 1902–2021 is +13 cm. The new contrary fluctuation during 2020–21 interrupts the previous acceleration, keeping the increase rate below the global trend due to regional and local factors, in line with previous reports about the importance of local hydroclimatic factors (Nagy et al., 2014, 2019; Piecuch, 2022). Yet, the observed decreasing level in 2020–21 is above 102 cm, showing that, notwithstanding the negative influence of reduced river inflow, the sea level lies well above (>5–7 cm) the one observed under similar hydroclimatic negative anomalies over the past three decades (e.g., the La Niña events 1988–89, 1999–00, or 2008–09).

The RdlP is experiencing a continuous but moderate sea level rise (SLR) and, to a lesser extent, suffered an extraordinary increase in the flow of the main tributaries from the early 70s to the early 2000s, which then stabilised and even slightly decreased. Also, the frequency of extreme events, mainly wind-induced storm surges, has risen over the last three decades (Gutiérrez et al., 2016; Nagy et al., 2023b; Verocai et al., 2015), increasing climate-related concerns and vulnerability along the RdlP Uruguayan coastal zone.

As of 2015–2019, the sea level reached 107–113 cm, well above the level in 1902 (91 cm), in line with the global SLR. Moreover, there has been an increase in the occurrence of compound events (winds and river flow) leading to sea heights ≥200, 250 and 300 cm (Nagy et al., 2023b).

Therefore, the current extreme hydroclimatic anomaly due to the multi-annual La Niña event since 2020–2021, which reversed SLR in 2020–2021 due to decreased river discharges (see above), will serve as a natural laboratory to test how wind-induced exceedance levels (>200, 250 cm) behave under low discharge.

3.2 Flooding and Storm Surges

Since the 1970s, the annual mean discharges of the Paraná and the RdlP have increased significantly, as well as the flood frequency. The built environment lies in very low areas, favoring flooding by a combination of rainfall and SE storms (extratropical depressions), whose increase in recent years has been observed simultaneously with the mean sea level increase (Barros et al., 2005; Gutiérrez et al., 2018; Nagy et al., 2023b). There is evident that the construction in 1979 of a hydropower dam over the Uruguay River (Salto Grande) between the two countries has not had a significant impact downstream along the RdlP coast (Barros et al., 2005). Furthermore, a large area of the Uruguayan coast of the RdlP is below the height of 5.0 m, hence considered a priori as vulnerable to flooding during extreme events, such as storm surges combined with extraordinary positive anomalies in river discharge (Gutiérrez et al., 2016; Nagy et al., 2023b; Verocai et al., 2015).

This zone of potential impact not only covers the RdlP coast, but also extends to the banks of the Santa Lucía River, to the West of Montevideo city (Fig. 1), which provides fresh water to around 1.0 million people. This illustrates the seriousness of the potential problem if extreme events intensify.

Local short-term factors of sea level variability are essential, because temporary fluctuations in a microtidal estuary like the RdlP (<0.5 m) may locally be more critical than long-term changes. Both urban systems (coastal infrastructures, freshwater intakes, drains, sewage, logistics operations) and ecosystems (wetlands, coastal biodiversity, flora and fauna) are adapted to a certain frequency of water level changes. However, some thresholds are frequently exceeded, like $\geq$2.5 m blocking rainstorm discharge, $\geq$2.8 m covering most sandy beaches, or $\geq$3.2 m adversely impacting coastal infrastructure.

Flooding by storm waves is associated with the meteorological phenomenon called "Sudestada", or intense southeastern winds. Figure 6 shows the relationship between wind speed (in knots) and sea level at Montevideo. Given the shape of the RdlP estuary, these winds set up the sea level, (Fig. 6) the effect of which propagates upstream and leads to flooding of low-lying areas. Consequently, the economic costs (when the events are severe, e.g., 23 August 2005) are very high (Magrin et al., 2007).

For wind speeds greater than average on the coastal of Montevideo (6 m s^{-1}), e.g., 8 knots (~14 m s^{-1}) (Nagy et al., 2003, 2008a; Verocai et al., 2015), the sea level set-up often exceeds 100 cm. Therefore, winds $\geq$25 knots (55 m s^{-1}) usually exceed several of the said thresholds, the actual impact depending on their duration.

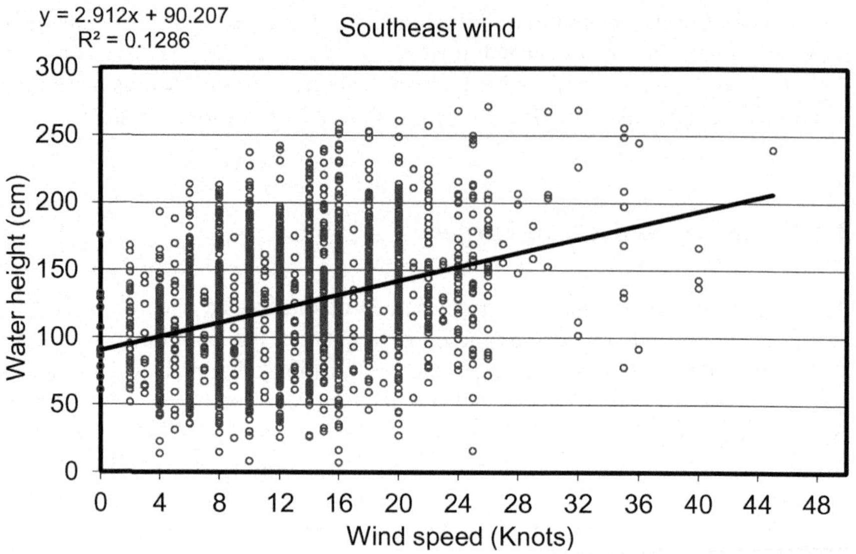

Fig. 6 Relationship wind speed versus water height at Montevideo. *Source* modified from Verocai et al. (2015)

3.3 Coastal Erosion and Sediment Transport

Although there is a retreat of the coastline in the innermost stretch of the Río de la Plata, it is not very significant due to a resulting very weak westward drift. Only west of the port of Juan Lacaze (Fig. 1), which interrupts the E-W drift, there is a risk of erosion.

The large width and the geometry of the estuary play a significant role in the sedimentation processes. Consequently, the effects of wind are diverse in the different parts of the RdlP estuary. In the outer region, winds force the resuspension of sediments through increased bottom erosion and redistribute suspended sediments by advection downstream of the Barra del Indio (Moreira & Simionato, 2019).

The coastal erosion processes caused by port facilities in different portions of the RdlP estuary were triggered before the mid-twentieth century. A lack of understanding of coastal dynamics prompted the authorities to build breakwaters. In the face of their failure, the recommended another hard measure to be implemented, such as a rockfill, which requires frequent maintenance.

To the east of the city of Colonia, there are small touristic urban settlements interspersed with rural areas. The sandy coasts are subject to erosion, due to similar causes as described for the lower Uruguay (extraction of both underwater and subaerial sand). This lead to wet beaches where herbaceous vegetation grows that must be removed in summer to enable beach tourism (Gutiérrez & Panario, 2019; Panario & Gutiérrez, 2005).

The section of the municipality of San José in the upper river estuary has been affected by an erosive process due to human intervention and natural processes. This erosion was due to the absence of hard points that can partly control the longshore drift, the extraction of aggregates in the recent past, the construction of coastal infrastructure in inconvenient places and afforestation with tall exotic species that hinder the formation of primary dunes after storms (Carro et al., 2018).

The middle estuary, which includes the metropolitan area of Montevideo, is characterized by salinity fluctuations and turbidity reduction due to flocculation of fluvial silt and clays. This region shows a sequence of pocket beaches and others of moderate drift, which has allowed determining its retreat due to sea level rise (1.70 m per cm of SLR), using a sequence of vertical aerial photos from between 1927 and 2008 (Gutiérrez et al., 2015).

Installing sand capture fences and planting native herbaceous plants has been successful since more than 20 years. Furthermore, the local governments of San José, Montevideo and Canelones (Fig. 1) have shown that the consequences of long-term Ecosystem-based Adaptation measures, even with scarce resources, have succeeded in reducing the effects of SLR, increasing wave energy and inadequate engineering. Such irreversible situations occur when the coastal interface lacks of a minimum a freedom to recover from moderate storm events. For instance, when the primary dune has been replaced by a reflective hard structure, the beach is to be lost. This leads to the installation of hard structures to defend the urban infrastructures.

In the middle and outer estuary, the most intense processes of deterioration have occurred on beaches adjacent to the mouths of drainage channels. During torrential rains in the basins, they discharge large amounts of sand, blown into them by strong SW winds, in the order of magnitude of $22m^3\ m^{-1}$ (Gutiérrez et al., 2018) that then went out to sea during torrential rains in the basins. Initial attempts to reduce the impact with breakwaters, and more recently with geo-tubes, have proven ineffective; diverting stormwater runoff, on the other hand, has proven to be a more effective option.

It is too early to assess the results of other measures, such as the recent beach refill. Such measures are only taken when sand is available at low cost from nearby projects, such as the construction of an underground parking or an excavation for road building.

The outer estuary, characterized by saline water, high wave energy and depth-controlled wave propagation, extends from the limit of the middle estuary to Punta Negra (Fig. 1) (Gutiérrez & Panario, 2019), where the Atlantic Ocean begins.

The coast of the outer estuary is covered with an almost uninterrupted sequence of resorts, some of which are on relatively stable beaches. Other parts of the coast, however, have retreated so far that they hit a sedimentary paleo-bluff which causes so much wave reflection that the beach tends to disappear (Fig. 1).

Part of the outer estuary, between Balneario Argentino and Jaureguiberry, is subject to neo-tectonic subsidence. The rest undergoes moderate to low erosion, without high risk in the short term, or requires mitigation interventions such as the reconstruction of dunes, the elimination of invasive species such as *Carpobrotus edulis*, *Lampranthus multiradiatus* and *Acacia longifolia* and eventually, the diversion of rainwater. At some points, the sediment drift in this part of the estuary was 70,000 m^3 $year^{-1}$ (MTOP/PNUD/UNESCO, 1979).

3.4 Sedimentation/Erosion at Harbors and Navigation Channels

In the section between Boca del Cufré Creek mouth and Juan Lacaze (Fig. 1), the inner estuary coast suffered intense erosion due to the interruption of the eastward by a breakwater to build a marina. The interruption of 35,000–70,000 m^3 $year^{-1}$ (IMFIA, 1998) led to downdrift erosion, spreading eastward to Juan Lacaze. A claim of the academic community and mobilized residents to shorten the breakwater was settled in court, in favor of the claimants. But only after 30 years of controversy between the two municipalities bordering the breakwater (San José to the East and Colonia to the West) (Fig. 1), where the former accumulated the sand stopped by the breakwater and the latter lost it due to the decrease in littoral drift.

The removal of part of the breakwater after the transfer of a significant volume of sand (bypass) in the direction of the drift has, at least temporarily, reduced the impact of the structure. The longshore drift, however, has not or at best partly been restored. Monitoring is needed to properly evaluate long-term impacts.

Due to sedimentation of the material carried by the Paraná and Uruguay Rivers, their navigation channels are dredged to allow access to larger vessels. For example, in Uruguay, the National Port Administration (ANP) dredges the access channel to Montevideo harbour (Fig. 6) to accommodate the increasing draft of container ships and bulk carriers calling on Montevideo (1.66 m and 1.01 m, respectively), particularly since 2007.

Dredging and material extraction negatively impact the physical, chemical and biological environment due to the resuspension of contaminated sediments, increased turbidity, removal and coverage of living organisms, water contamination and substrate, and changes in the circulation patterns and bottom morphology (Ecoplata, 1996; Marcomini & López, 2014). Vulnerability of this coastal ecosystem increases during low river levels due to an intensification of dredging activities. Despite the binational and multi-sectoral administration and management of the RdlP, the requirement of a multidisciplinary and joint approach to this issue, including a systematic treatment of dredged materials before deposition, is broadly recognized (Defeo et al., 2009; Ecoplata, 1996).

3.5 Water Resources/Hazards and Environmental Pollution

The environmental status of the RdlP is the result of actions and development policies undertaken since the establishment of its bordering countries, but also is affected by activities carried out in the drainage basin. The Río de la Plata basin is heavily industrialized and densely populated, comprising the most inhabited cities of the continent (Buenos Aires and São Paulo) and showing increasing agricultural activity in recent years. Consequently, the Paraná and Uruguay rivers and their tributaries discharge xenobiotics, nutrients, heavy metals, persistent organic pollutants (POPs), hydrocarbons, and other organic compounds (García-Alonso et al., 2019). Also, many industrial centres and essential economic activities in Argentina and Uruguay, including the two principal harbors of Buenos Aires and Montevideo, are located along the RdlP. As a result, estuarine pollution induces rapid, increasing and lasting environmental modifications. In the present review, special attention is given to estuaries in South America (SA), where legislation, policies and actions to guarantee environmental quality remain ineffective.

Eutrophication processes in the RdlP resulting from nutrient inputs linked to human activities in the del Plata basin are regulated by hydrological, oceanographic, meteorological factors and climatic events, such as RdlP flow, salinity, turbidity, water temperature, wind intensity and direction, ENSO strength and phase (Brugnoli et al., 2018; Kruk et al., 2019; Muniz et al., 2019b; Nagy et al., 2002). Today, anthropogenic eutrophication is one of the most important environmental impacts on the RdlP ecosystem's health and services. Point sources (domestic sewage and industrial discharge) and diffuse sources (agriculture activities) throughout the whole del Plata basin and the estuary itself strongly increased the concentration of nutrients, affecting the whole quality of the system, but in particular, the recreational use of shore areas (Muniz et al., 2019a; Nagy et al., 2002). Massive cyanobacteria blooms promoted by nutrient enrichment associated with hydrological, oceanographic and meteorological factors have been reported in the RdlP, which have generated a loss of water quality, cultural ecosystem services and tourism revenues (Aubriot et al., 2020; Kruk et al., 2019). Additionally, spatial and temporal variations in trophic conditions of both water column (based on IET and TRIX, Brugnoli et al., 2019, 2021b) and bed sediments (García-Rodríguez et al., 2014) have been reported in the Uruguayan RdlP coast linked to different phases of ENSO events.

The Argentinian metropolitan coastal zone receives discharges of untreated wastewater (e.g., main Buenos Aires receives approximately 2 million $m^3\ d^{-1}$) also containing POPs, hydrocarbons, trace metals and organic matter (Colombo et al., 2005a, 2005b, 2006; Tatone et al., 2009). One of the historically contaminated watersheds emblematic of the metropolitan area of Buenos Aires that drains into the RdlP is the Matanza-Riachuelo Creek (Fig. 1). This basin, having an area of 2200 km^2 and 3.5 million residents, is severely contaminated by heavy metals (Cr, Hg, Cd, Cu, Pb, Zn, Ni, Al, As and Co), hydrocarbons and pathogens, the concentrations of which exceed the limits proposed by the Hazardous Waste Law24051 (Decree 831/93) (ACUMAR, 2011). Furthermore, in other areas of the Argentinean metropolitan coastal zone, heavy metal levels are higher than the limits suggested for the protection of biota established by the Secretary of Water Resources of Argentina (FREPLATA, 2014).

The north coast of the RdlP belongs to Uruguay. Montevideo (the most polluted area of the Uruguayan coastal zone) and the metropolitan coastal region are the most altered ecosystems. They concentrate most of the country's economic activities (accounting for 59% of the GNP of Uruguay, including 71% of its industry and 66% of services). Montevideo's most serious problem is the urban expansion, to the detriment of the natural environment. The current understanding of anthropogenic impacts and local pollution effects on the biota at the Uruguayan margin of the RdlP still needs to be improved (Muniz & Venturini, 2019). The first studies regarding heavy metal and hydrocarbon pollution in the Montevideo coastal zone were performed in the early nineties. They found sediments severely contaminated by chromium derived from leather factories near the mouth of the Pantanoso Creek (Fig. 1). In the late nineties, many studies were carried out to evaluate the pollution degree and its effects on ecosystem health and biota along the Montevideo coastal zone (Brugnoli et al., 2019, 2021a, 2021b; Corrales et al., 2016; Danulat et al., 2002; Hutton et al., 2015; Muniz & Venturini, 2011, 2015, 2019; Muniz et al., 2002, 2004, 2005a, 2011, 2015, 2019a, 2019b; Pedelacq et al., 2022; Venturini et al., 2012, 2015; Viana et al., 2005). Together, these studies have improved our understanding of natural and anthropogenic disturbances and the structure and interactions among this system's abiotic and biotic components. This enables creating a scientific-technical basis for correctly implementing an environmental monitoring plan for the Montevideo coastal zone. Marine traffic is another important source of pressure at the ecosystem level, produced by harbor operations and by ships, impacting directly the water body or, indirectly, air and terrestrial areas (García-Alonso et al., 2019). Currently, the most common environmental concerns of shipping refer to chemical discharges (oil, wastewater, antifouling paint, ballast water, and marine litter) and physical pollution (underwater noise, ports, fairways and channels, dredging, and shipwrecks). A recent review of these impacts and their ecosystem alterations in land reclamations can be found in García-Alonso et al. (2019). Plastic debris is another critical environmental problem in the RdlP; urban wastes and landfills are the primary sources and need improved management. In the Montevideo metropolitan area, plastics encompass the second most important type of waste (14%), arising from land- and sea-based activities (Lozoya et al., 2015, 2016). National policies are well suited for minimizing the impacts of plastic pollution in Uruguay, but management for reducing waste and sewage will not be enough. Plastic is cheap and durable and is part of our life; new technologies and infrastructures are required to mitigate this problem jointly with citizen education and behavioral changes.

The artisanal fishery operates up to seven nautical miles offshore. The ships have a gross registered tonnage (GRT) of less than 10 t and generally use gill nets or longline methods. The products are primarily supplied to the domestic markets (Defeo et al., 2011). The main landing sites for artisanal fisheries in Uruguay are located in the middle and outer regions of the estuary, with some places on the Atlantic coast. The most important species in this fishery are the Chondrichthyes *Mustelus schmitti* and the Squatina spp. in the marine region and the bony fishes *Micropogonias furnieri*,

Cynoscion guatucupa, Macrodon atricauda and *Prochilodus lineatus* in the estuarine region (Defeo et al., 2009). The white mouth croaker (*M. furnieri*) always has the greatest annual catch, with a registered capture of 4,160 t in 2015. Among the chondrichthyes, the shark *Mustelus schmitti* is the most captured species, with 176 tons landed in 2015. However, as García-Alonso et al. (2019) highlighted, catches have declined, reaching in Uruguay their lowest value in the last 35 years for the most important species.

3.6 Biodiversity Loss

The RdlP benthic ecosystem is likely to be affected by bottom fishing trawls where invertebrates are damaged and suffer an increase in mortality during the fishing manoeuvre (García-Alonso et al., 2019). In addition, fishing trawls remove and partially resuspend contaminants previously deposited on bottom sediments, which can represent a mechanism of transfer of contaminants to other communities besides the benthic (plankton and nekton). Therefore, the possible effects of the resuspension and transport of sediments and the effects of human activity (fishing trawl, introduction of exotic species and pollution) must be evaluated in more detail to achieve an ecosystem understanding enabling adequate management and restoration actions of the RdlP estuary.

Habitat loss is the primary concern identified for all South American ecosystems. It may be caused by damming of rivers, deforestation, water pollution, mining, poor agricultural practice or inadequate management (Barletta et al., 2016). Actions should be focused on habitat conservation and social rescue of the less well-off populations of indigenous and non-indigenous people. Energy and freshwater demands must also be rescaled to control habitat loss.

The recent establishment of invasive alien species (IAS) (aquatic or terrestrial), which have a high capacity to modify the invaded environment, has been identified as the second cause of global biodiversity loss, and the Uruguayan coastal zone is not an exception to this widespread environmental problem (Brazeiro et al., 2021; Brugnoli et al., 2006, 2007; Masciadri et al., 2010; Muniz et al., 2019b, 2021). Coastal departments of Uruguay (Colonia, San José, Canelones, Montevideo, Maldonado and Rocha) contain the majority of the IAS in the country, with 222 terrestrial and 17 aquatic species recorded (Brugnoli et al., 2009; Masciadri et al., 2010). Aiming the terrestrial IAS, herbaceous vascular plants are the most frequently occurring, followed by trees and shrubs, while among the aquatic IAS fish and molluscs are the most relevant animal groups (Masciadri et al., 2010). Vascular IAS plants are one of the main threats to coastal biodiversity (Defeo et al., 2008; Panario & Gutierrez, 2006). Figure 7 shows the most relevant terrestrial and aquatic IAS. They substantially affect the dunes area, promoting dune fixation, which leads to beach erosion, and colonisation, which leads to homogenized landscapes and a loss of the native community's structure. This ultimately leads to a decrease in habitats available for native flora and fauna (Masciadri et al., 2010). The exotic aquatic species reported in

Fig. 7 Most relevant terrestrial and aquatic invasive alien species in the Río de la Plata estuary. Photo by the Authors

coastal ecosystems of Uruguay are found mainly in the RdlP (inter or subtidal zone and secondary estuaries). They comprise invertebrates (4 molluscs, one polychaeta, and four crustaceans) and one vertebrate (*Cyprinus carpio*). Among the invertebrates, *Ligia exotica* (crustacean), *Ficopomatus enigmaticus* (polychaete), *Limnoperna fortunei* and *Rapana venosa* (molluscs) occur most frequently (Brugnoli et al., 2021b). Since the invasion of the RdlP estuary by these molluscs, a substantial amount of research has been generated about their ecology, distribution, biodiversity effects and economic impacts (Brugnoli et al., 2006, 2014, 2021b; Carranza et al., 2007, 2010; Giberto et al., 2011; Lanfranconi et al., 2009, 2013; Muniz et al., 2005b, 2021).

4 Solutions and Paths for the Future of RdlP

The main threats to the RdlP basin (including the Paraguay, Paraná, and Uruguay rivers) include 27 applied dams, dredging, overfishing, climate warming, and pollution. In addition, the region is shared by multiple countries, with high investment and economic growth indexes, which generates a series of environmental issues. The proposed solutions are to (1) strengthen coordination between countries; (2) improve agricultural cultivation methods to conserve soil and water; (3) adopt non-engineering measures to reduce flood risks, such as flood forecasting and insurance services in flood-prone areas; (4) draft strict development and protection plans for the Pantanal

Wetland and implement them effectively; (5) strengthen the management of water pollution prevention and control; 6) protect the ecological environment of the watershed; (7) develop corresponding laws, regulations and international conventions; (8) establish institutions for legislative supervision and law enforcement.

4.1 Construction of Capture Fences and Other Solutions

The construction of capture fences has been effective in recovering the prism of beaches (Carro et al., 2018; Nagy & Gutiérrez, 2018; Panario et al., 2008). The diversion of rainwater towards drainage ways or rocky points has been another widely successful intervention in reducing backwash by infiltration-exfiltration.

The Uruguayan coast of the Río de la Plata has several climate reasons for concern whose importance depends on climatic, economic, physical, biological, socio-economic and cultural factors dependent on land use and its associated infrastructure (Nagy et al., 2023b).

The risks associated with global change are strongly dependent on physical, biotic or social aspects depending on the region of the estuary, local governments and the awareness and social organization reached by local inhabitants, as well as population density, the presence of infrastructure at risk, and associated with it, socio-economic activities in the coastal zone (Gutiérrez & Panario, 2019, 2020).

4.2 Conceptual Framework

The Pressure-State-Impact-Response conceptual framework was developed in recent decades and is used in various environmental approaches. It allows the identification of relations between the pressures due to human activities or changing environmental conditions. These influence the state of the system, which has certain impacts on inhabitants, users and stakeholders. The response encompasses human interventions (laws, management, environmental actions) that enable preventing, compensating, mitigating or adapting changes in the state of the system (Elliot, 2014; Smith et al., 2016). In Table 1, anthropic and environmental pressures identified by different authors in the del Plata basin and RdlP are synthesized, including possible responses and the intervention scale of the response.

Table 1 Anthropic and environmental Pressures in the del Plata basin and RdlP estuary with relevant local references. Response and Spatial and political scales of intervention. References: del Plata basin[1], RdlP estuary[2], Scales: subnational[3], national[4], regional[5], international[6]. The order of different anthropic pressures is not prioritized. The order of anthropic and environmental pressures was according to the order presented during the chapter

Pressures	Response	Intervention scale
Climate change[(1,2)] (Barreiro, 2010; Nagy et al., 2023b)	– Mitigation, adaptation	3, 4, 5 and 6
Coastal urbanisation[(2)] (Gutiérrez & Panario, 2019, 2020)	Land use planning, exclusion zones Territorial Planning Law and National Coastal Directive (Laws N°s 18308 and 19772)	3 and 4
Coastal erosion[(2)] (Panario & Gutiérrez, 2006)	– Nature-based solutions (barriers, barreras, captor fences)	3 and 4
Water pollution (chemical, physical)[(1,2)] (Muniz et al., 2019a; Pedelacq et al., 2022; Venturini et al., 2015)	– Reduce pesticides use and hydrocarbons emissions – Increase treatment of industrial effluents and sewage – Long-term monitoring	3, 4 and 5
Eutrophication[(1, 2)] (Aubriot et al., 2020; Brugnoli et al., 2019; Kruk et al., 2019; Muniz et al., 2019b; Nagy et al., 2023b; Venturini et al., 2012)	– Reduce nutrients concentration (effluents and diffuse sources) – Water quality continuum monitoring – Evaluate using trophic and biotic indexes (water column and sediments) – Environmental water quality laws, regulations and controls	3, 4 and 5
Dams and hydro-power generation[(1)] (Aubriot et al., 2020; Kruk et al., 2019)	– Sediment management – Renewable energies	3 and 4
Maritime traffic[(2)] (Marin et al., 2021)	– Marine spatial planification	5 and 6
Overfishing[(2)] (Defeo et al., 2009; Horta & Defeo, 2012)	– Citizen science – Participatory monitoring	4, 5 and 6
Habitat loss (agriculture, forestation and deforestation)[(1)] (Brazeiro et al., 2003)	– Conservation, remediation, reforestation, restoration	3, 4 and 5
Invasive alien species (aquatic and terrestrial)[(1,2)] (Brazeiro et al., 2021; Brugnoli et al., 2009)	– Regional, national and subnational strategies for prevention and control of invasive alien species	3, 4, 5 and 6

References

Acha, M. E., Mianzan, H., Guerrero, R., Carreto, J., Giberto, D., Montoya, N., & Carignan, M. (2008). An overview of physical and ecological processes in the Rio de la Plata Estuary. *Continental Shelf Research, 28*(13), 1579–1588.

Achkar, M., Domínguez A., & Pesce, F. (2012). Cuenca del Río Santa Lucía-Uruguay. Aportes para la discusión ciudadana (25 pp.). Programa Uruguay Sustentable, Redes-Amigos de la Tierra.

ACUMAR. (2011). Cuenca Matanza Riachuelo. Medición del estado del agua superficial y subterránea análisis e interpretación de los resultados. Informe Trimestral Calidad de Agua. Buenos Aires: Autoridad de Cuenca Matanza Riachuelo (ACUMAR), Dirección General Técnica. Coordinación de Calidad Ambiental. https://www.acumar.gob.ar/wp-content/uploads/2016/12/CDCA_ACUMAR_INFO_AGUA_JULIO_2011.pdf

Aubriot, L., Zabaleta, B., Bordet, F., Sienra, D., Risso, J., Achkar, M., & Somma, A. (2020). Assessing the origin of a massive cyanobacterial bloom in the Río de la Plata (2019), Towards an early warning system. *Water Research, 181*, 115944.

Barletta, M., Blaber, S. J. M., & Craig, J. F. (2016). Fish and aquatic habitat conservation in South America. *Journal of Fish Biology, 89*(1), 1–3.

Barreiro, M. (2010). Influence of ENSO and the South Atlantic Ocean on climate predictability over Southeastern South America. *Climate Dynamics, 35*(7–8), 1493–1508.

Barros, V., Menéndez, A., & Nagy, G. J. (2005). El cambio climático en el Río de la Plata. Project "Assessments of impacts and adaptation to climate change (AIACC)". Centro de Investigaciones del Mar y la Atmósfera - CIMA/CONICET, Buenos Aires, Argentina. START-TWAS-UNEP. https://www.cima.fcen.uba.ar/~lcr/libros/Cambio_Climatico-Texto.pdf

Bonachea, J., Bruschi, V. M., Hurtado, M. A., Forte, L. M., da Silva, M., Etcheverry, R., et al. (2010). Natural and human forcing in recent geomorphic change; case studies in the Rio de la Plata basin. *Science of the Total Environment, 408*(13), 2674–2695.

Brazeiro, A., Achkar, M., Mianzan, H., Gómez-Erache, M., & Fernández, V. (2003). Áreas prioritarias para la conservación y manejo de integridad biológica del Rio de la Plata y su Frente Marítimo. FREPLATA PNUD/GEF/RLA99/G31.

Brazeiro, A., Bresciano, D., & Brugnoli, E. (2021). Panorama general de las invasiones biológicas en Uruguay. In A. Brazeiro, D. Bresciano, E. Brugnoli, & M. Iturburu (Eds.), Especies exóticas invasoras de Uruguay: distribución, impactos socioambientales y estrategias de gestión (pp.13–38). Ministerio de Ambiente, CEEI, UdelaR.

Brugnoli, E., Clemente, J., Riestra, G., Boccardi, L., & Borthagaray, A. (2006). Especies acuáticas exóticas en Uruguay: situación, problemática y gestión. In R. Menafra, L. Rodríguez, F. Scarabino, & D. Conde (Eds.), Bases para la conservación y manejo de la costa uruguaya. Vida Silvestre Uruguay (pp. 351–362).

Brugnoli, E., Muniz, P., Venturini, N., & Burone, L. (2007). Environmental perturbation and coastal benthic biodiversity in Uruguay. In I. C. Willis (Ed.), *Progress in environmental research* (pp. 75–126). Nova Publishers.

Brugnoli, E., Masciadri, S., & Muniz, P. (2009). Base de Datos de Especies Exóticas e Invasoras en Uruguay, un instrumento para la Gestión Ambiental y Costera. Programa EcoPlata. http://www.inbuy.fcien.edu.uy/fichas_de_especies/

Brugnoli, E., Giberto, D., Lanfranconi, A., Schiariti, A., Aguilera, F., Bremec, C. S., Barrero, G., & Muniz, P. (2014). El gasterópodo invasor Rapana venosa (VALENCIENNES 1846) y sus posibles efectos en el ecosistema costero estuarial del Río de la Plata. In: Nuevas Miradas a la Problemática de los ambientes costeros. Sur de Brasil, Uruguay y Argentina. Compilador y Revisor, Gosso, C. DIRAC, Facultad de Ciencias, Montevideo (pp. 211–228).

Brugnoli, E., Verocai, J., Muniz, P., & García-Rodríguez, F. (2018). Weather, hydrological and oceanographic conditions of the northern coast of the Rio de la Plata estuary during ENSO 2009–2010. In W. Froneman (Ed.), *Estuary* (pp. 19–38). InTech.

Brugnoli, E., Muniz, P., Venturini, N., Brena, B., Rodríguez, A., & García-Rodríguez, F. (2019). Assessing multimetric trophic state variability during an ENSO event in a large regional estuary (Rio de la Plata, South America). *Regional Studies in Marine Science, 28*, 100565.

Brugnoli, E., Muniz, P., Venturini, N., & García-Rodríguez, F. (2021a). Benthic community responses to organic enrichment during an ENSO event (2009–2010), in the north coast of Rio de la Plata estuary. *Journal of Marine Systems, 222*, 103597.

Brugnoli, E., Arocena, R., Cabrera-Lamanna, L., & Muniz, P. (2021b). Management and monitoring of eutrophication: trophic state indexes on the Rio de la Plata northern coast. In W. Filho, A. Marisa Azul, L. Brandli, A. Lange Salvia, & T. Wall (Eds.), *Life below water, encyclopedia of the UN sustainable development goals* (pp. 1–13). Springer Nature.

CARP. (2012). Comisión Administradora del Río de la Plata. Informe anual. https://www.gub.uy/ministerio-ganaderia-agriculturapesca/comunicacion/publicaciones/caru-carp-informe-anual-2012-0

Carranza, A., Scarabino, F., & Ortega, L. (2007). Distribution of large benthic gastropods in the Uruguayan continental shelf and Río de la Plata estuary. *Journal of Coastal Research, 24*, 161–168.

Carranza, A., de Mello, C., Ligrone, A., González, S., Píriz, P., & Scarabino, F. (2010). Observations on the invading gastropod Rapana venosa in Punta del Este, Maldonado Bay, Uruguay. *Biological Invasions, 12*(5), 995–998.

Carro, I., Seijo, L., Nagy, G. J., Lagos, X., & Gutiérrez, O. (2018). Building capacity on ecosystem-based adaptation strategy to cope with extreme events and sea-level rise on the Uruguayan coast. *International Journal of Climate Change Strategies and Management, 10*(4), 504–522.

Colombo, J. C., Cappelletti, N., Barreda, A., Migoya, M. C., & Skorupka, C. (2005a). Vertical fluxes and accumulation of PCBs in coastal sediments of the Río de la Plata estuary, Argentina. *Chemosphere, 61*(9), 1345–1357.

Colombo, J. C., Capelletti, N., Lasci, J., Migoya, M. C., Speranza, E., & Skorupka, C. N. (2005b). Sources, vertical fluxes and accumulation of aliphatic hydrocarbons in coastal sediments of the Rio de la Plata Estuary, Argentina. *Environmental Science and Technology, 39*(21), 8227–8234.

Colombo, J. C., Capelletti, N., Lasci, J., Migoya, M. C., Speranza, E., & Skorupka, C. N. (2006). Sources, vertical fluxes and equivalent toxicity of aromatic hydrocarbons in coastal sediments of the Río de la Plata estuary, Argentina. *Environmental Science and Technology, 40*(3), 734–740.

Cooley, S., Schoeman, D., Bopp, L., Boyd, P., Donner, S., & Ghebrehiwet, D. Y., Ito, S.-I., Kiessling, W., Martinetto, P., Ojea, E., Racault, M.-F., Rost, B., & Skern-Mauritzen, M. (2022). Ocean and Coastal Ecosystems and their Services. In H.-O. Pörtner, D. C. Roberts, M. Tignor, E. S. Poloczanska, K. Mintenbeck, A. Alegría, M. Craig, S. Langsdorf, S. Löschke, V. Möller, A. Okem, & B. Rama (Eds.). (2022). Climate change 2022: impacts, adaptation and vulnerability. *Climate change 2022: impacts, adaptation and vulnerability. Contribution of working group II to the sixth assessment report of the intergovernmental panel on climate change* (pp. 379–550). Cambridge University Press.

Corrales, D., Acuña, A., Salhi, M., Saona, G., & Brugnoli, E. (2016). Copper, zinc, mercury and arsenic content in Micropogonias furnieri and Mugil platanus of the Montevideo coastal zone, Río de la Plata. *Brazilian Journal of Oceanography, 64*, 57–65.

Danulat, E., Muniz, P., García-Alonso, J., & Yannicelli, B. (2002). First assessment of the highly contaminated Harbour of Montevideo (Uruguay). *Marine Pollution Bulletin, 44*(6), 554–565.

Defeo, O., Vincent, P., Acuña, A., Amestoy, F., de Álava, A., Castiñeira, E., et al. (2008). Zona Costera. In M. Gómez & D. Martino (Coord.), *GeoUruguay* (pp. 118–170). Mosca.

Defeo, O., Horta, S., Carranza, A., Lercari, D., de Álava, A., Gómez, J., et al. (2009). Hacia un Manejo Ecosistémico de Pesquerías. Áreas Marinas Protegidas en Uruguay. Facultad de Ciencias-DINARA.

Defeo, O., Puig, P., Horta, S., & de Álava, A. (2011). Coastal fisheries of Uruguay. In S. Salas, R. Chuenpagdee, A. Charles, & J. C. Seijo (Eds.), *Coastal Fisheries of Latin America and the Caribbean. FAO Fisheries and Aquaculture.* Technical Paper No. 544 (pp. 357–384). Food and Agriculture Organization of the United Nations (FAO).

Douglas, B. C. (2001). Sea level change in the era of the recording tide gauge. In B. C. Douglas, M. S. Kearney, & S. P. Leatherman (Eds.), *Sea level Rise: History and consequences* (pp. 37–64). Academic Press.

Ecoplata. (1996). Hacia el desarrollo sostenible de la zona costera del Río de la Plata. Libro de Resumenes Montevideo, Conferencia Internacional ECOPLATA '96. Uruguay. http://hdl.handle.net/10625/15751

Elliot, M. (2014). Integrated marine science and management: Wading through the morass. *Marine Pollution Bulletin, 86*(1–2), 1–4.

Fiore, M. M., D'Onofrio, E. E., Pousa, J. L., Schnack, E. J., & Bertola, G. R. (2009). Storm surges and coastal impacts at Mar del Plata, Argentina. *Continental Shelf Research, 29*(14), 1643–1649.

Fossati, M., Cayocca, F., & Piedra-Cueva, I. (2014). Fine sediment dynamics in the Río de la Plata. *Advanced Geosciences, 39*, 75–80.

Framiñan, M. B., Etala, M. P., Acha, E. M., Guerrero, R. A., Lasta, C. A., & Brown, O. B. (1999). Physical characteristics and processes of the Río de la Plata estuary. In G. M. Perillo, M. C. Piccolo, & M. Pino-Quivira (Eds.), *Estuaries of South America: Their geomorphology and dynamics* (pp. 161–194). Springer.

FREPLATA. (2014). Estrategia para la implementación del Programa Binacional de Monitoreo del Río de la Plata y su Frente Marítimo. Grupo Binacional de Monitoreo. Documento Técnico. FREPLATA.

García-Alonso, J., Lercari, D., & Defeo, O. (2019). Río de la Plata: A Neotropical Estuarine system. In E. Wolanski, J. W. Day, M. Elliott, & R. Ramachandran (Eds.), *Coasts and Estuaries: The future* (pp. 45–56). Elsevier.

García-Rodríguez, F., Brugnoli, E., Muniz, P., Venturini, N., Burone, L., Hutton, M., Rodríguez, M., Pita, A. L., Kandratavicius, N., Pérez, L., & Verocai, J. (2014). Warm-phase ENSO events modulate the continental freshwater input and the trophic state of sediments in a large South American estuary. *Marine and Freshwater Research, 65*, 1–11.

Giberto, D. A., Schiariti, A., & Bremec, C. S. (2011). Diet and daily consumption rates of Rapana venosa (Valenciennes, 1846) (Gastropoda: Muricidae) from the Río de la Plata (Argentina-Uruguay). *Journal of Shellfish Research, 30*(2), 349–358.

Glamore, W., Rayner, D., Miller, B., Rahman, P., & Dieber, M. (2016). Estuaries in a changing climate. In *25th NSW Coastal Conference 2016*. NSW. https://www.coastalconference.com/2016/papers2016/Will%20Glamore.pdf

Guerrero, R. A., Acha, E., Framiñan, M. B., & Lasta, C. A. (1997). Physical oceanography of the Río de la Plata estuary, Argentina. *Continental Shelf Research, 17*(7), 727–742.

Gutiérrez, O., Panario, D., Nagy, G. J., Piñeiro, G., & Montes, C. (2015). Long-term morphological evolution of urban pocket beaches in Montevideo (Uruguay): Impacts of coastal interventions and links to climate forcing. *Journal of Integrated Coastal Zone Management, 15*(4), 467–484.

Gutiérrez, O., & Panario, D. (2019). Caracterización y dinámica de la costa uruguaya, una revisión. In P. Muniz, D. Conde, N. Venturini, & E. Brugnoli (Eds.), *Ciencias Marino-Costeras en el Umbral del Siglo XXI, Desafíos en Latinoamérica y El Caribe*, (pp. 61–91). Editorial AGT S.A.

Gutiérrez, O., Panario, D., & Nagy, G. J. (2018). Relationships between the sand cycle and the behaviour of small river mouths: A neglected process. *Journal of Sedimentary Environments, 3*(4), 307–325.

Gutiérrez, O., & Panario, D. (2019). Caracterización y dinámica de la costa uruguaya, una revisión. In P. Muniz, D. Conde, N. Venturini, & E. Brugnoli (Eds.), Ciencias Marino-Costeras en el Umbral del Siglo XXI, Desafíos en Latinoamérica y El Caribe (pp. 61–91). Editorial AGT S.A.

Gutiérrez, O., & Panario, D. (2020). Zona costera, buscando respuestas a un futuro incierto. In M. Dazzini Langdon & H. Navarrete (Eds.), *Bosques Azules: Humedales en Riesgo* (pp. 52–67). Una Mirada Latinoamericana. Publicaciones PUCE.

Hutton, M., Venturini, N., García-Rodríguez, F., Brugnoli, E., & Muniz, P. (2015). Assessing the ecological quality status of a temperate urban estuary by means of benthic biotic indices. *Marine Pollution Bulletin, 91*(2), 441–453.

Horta, S., & Defeo, O. (2012). The spatial dynamics of the whitemouth croaker artisanal fishery in Uruguay and interdependencies with the industrial fleet. *Fisheries Research, 125*, 121–128.

IMFIA. (1998). Estudio de las obras de navegación en el Arroyo Cufré y sus impactos sobre la zona costera. Informe final. Facultad de Ingeniería, UdelaR.

IPCC. (2019). Summary for policymakers. In H.-O. Pörtner, D. C. Roberts, V. Masson-Delmotte, P. Zhai, M. Tignor, E. Poloczanska, K. Mintenbeck, A. Alegría, M. Nicolai, A. Okem, J. Petzold, B. Rama, & N. M. Weyer (Eds.), *IPCC special report on the ocean and cryosphere in a changing climate* (pp. 3–35). Cambridge University Press.

IPCC. (2021). Summary for policymakers. In V., Masson-Delmotte, P. Zhai, A. Pirani, S. L. Connors, C. Péan, S. Berger, N. Caud, Y. Chen, L. Goldfarb, M. I. Gomis, M. Huang, K. Leitzell, E. Lonnoy, J. B. R. Matthews, T. K. Maycock, T. Waterfield, O. Yelekçi, R. Yu, & B. Zhou (Eds.) *Climate change 2021: The physical science basis. contribution of working group I to the sixth assessment report of the intergovernmental panel on climate change*. Cambridge University Press.

Jackson, M., Sienra, G., Santoro, P., & Fossati, M. (2021). Temporal and spatial variability scales of salinity at a large microtidal estuary. *Journal of Marine Science and Engineering, 9*(8), 860.

Kruk, C., Martínez, A., Martínez de la Escalera, G., Trinchin, R., Manta, G., Segura, A., et al. (2019). Floración excepcional de cianobacterias tóxicas en la costa de Uruguay, verano 2019. INNOTEC (pp. 36–68).

Lanfranconi, A., Hutton, M., Brugnoli, E., & Muniz, P. (2009). New record of the alien mollusc Rapana venosa (Valenciennes 1846) in the Uruguayan coastal zone of Río de la Plata. *Pan-American Journal of Aquatic Sciences, 4*(2), 216–221.

Lanfranconi, A., Brugnoli, E., & Muniz, P. (2013). Preliminary estimates of consumption rates of Rapana venosa (Gastropoda, Muricidae); a new threat to mollusk biodiversity in the Río de la Plata. *Aquatic Invasion, 8*, 437–442.

Lecertua, E. A. (2010). *Análisis de riesgo de duración de inundaciones en las áreas costeras del Río de la Plata considerando el Cambio Climático, Facultad de Ingeniería*. Universidad de Buenos Aires, Buenos Aires.

Leal Filho, W., Nagy, G. J., Martinho, F., Saroar, M., Gómez-Erache, M., Primo, A. L., Pardal, M. L., & Li, C. L. (2022). Influences of climate change and variability on estuarine ecosystems: An impact study in selected European, South American and Asian countries. *International Journal of Environmental Research and Public Health, 19*(1), 585.

Lozoya, J. P., Carranza, A., Lenzi, J., Machin, E., Teixeira de Mello, F., González, S., et al. (2015). Management and research on plastic debris in Uruguayan Aquatic Systems: Update and perspectives. *Journal of Integrated Coastal Zone Management, 153*, 377–393.

Lozoya, J. P., Texeira de Mello, F., Carrizo, D., Weinstein, F., Olivera, Y., Cedrés, F., Suarez, M. P. A., & Fossati, M. (2016). Plastics and microplastics on recreational beaches in Punta del Este (Uruguay): Unseen critical residents? *Environmental Pollution, 218*, 931–941.

Magrin, G., Gay García, C., Cruz Choque, D., Giménez, J. C., Moreno, A. R., Nagy, G. J., et al. (2007). Latin America. In M. L. Parry, O. F. Canziani, J. P. Palutikof, P. J. van der Linden, & C. E. Hanson (Eds.), *Climate change 2007: Impacts, adaptation and vulnerability. Contribution of working group II to the fourth assessment report of the intergovernmental panel on climate change* (pp. 581–615). Cambridge University Press.

Marcomini, S. C., & López, R. A. (2014). Impacto ambiental en el estuario del río de la Plata y delta del Paraná. In C. Goso (Ed.), *Nuevas miradas a la problemática de los ambientes costeros, sur de Brasil, Argentina y Uruguay* (pp. 303–320). Montevideo DIRAC, Facultad de Ciencias, UdelaR.

Marín, Y. H., Defeo, O., & Horta, S. (2021). So far and so close: Opportunities for marine spatial planning in the Southwest Atlantic Ocean. *Ocean and Coastal Management, 211*, 105737.

Masciadri, S., Brugnoli, E., & Muniz, P. (2010). InBUy database of Invasive and Alien Species (IAS) in Uruguay: A useful tool to confront this threat to biodiversity. *Biota Neotrópica, 10*(4), 205–213.

Masson-Delmotte, V., Zhai, P., Pirani, S., Connors, C., Péan, S., Berger, N., et al. (2021). IPCC, 2021: Summary for policymakers. In *Climate change 2021: The physical science basis. contribution of working group I to the sixth assessment report of the intergovernmental panel on climate change*. Cambridge University Press.

Mianzan, H., Lasta, C., Acha, E., Guerrero, R., Macchi, G., & Bremec, C. (2001). The Río de la Plata Estuary, Argentina-Uruguay. In U. Seeliger & B. Kjerfve (Eds.), *Coastal Marine ecosystems of Latin America, ecological studies* (pp. 185–204). Springer.

Moreira, D., Simionato, C. G., Gohin, F., Cayocca, F., & Luz Clara Tejedor, M. (2013). Suspended matters mean distribution and seasonal cycle in the Río de la Plata estuary and the adjacent shelf from MODIS and in situ observations. *Continental Shelf Research, 68*, 51–66.

Moreira, D., Simionato, C. G., Dragani, W., Cayocca, F., & Tejedor, M. L. C. (2016). Characterisation of bottom sediments in the Río de la Plata Estuary. *Journal of Coastal Research, 32*(6), 1473–1494.

Moreira, D., & Simionato, C. G. (2019). Modeling the suspended sediment transport in a very wide, shallow, and microtidal estuary, the Río de La Plata, Argentina. *Journal of Advances in Modeling Earth Systems, 11*(10), 3284–3304.

MTOP/PNUD/UNESCO. (1979). Conservación y mejora de playas - URU. 73.007. Unesco.

Muniz, P., Venturini, N., & Martínez, A. (2002). Physico-chemical characteristics and pollutants of the benthic environment in the Montevideo coastal zone, Uruguay. *Marine Pollution Bulletin, 44*(9), 962–968.

Muniz, P., Danulat, E., Yannicelli, B., García-Alonso, J., Medina, G., & Bícego, M. C. (2004). Assessment of contamination by heavy metals and petroleum hydrocarbons in sediments of Montevideo Harbour (Uruguay). *Environment International, 29*(8), 1019–1028.

Muniz, P., Venturini, N., Pires-Vanin, A. M. S., Tommasi, L. R., & Borja, A. (2005a). Testing the applicability of a Marine Biotic Index (AMBI) to assessing the ecological quality of soft-bottom benthic communities, in the South America Atlantic region. *Marine Pollution Bulletin, 50*(6), 624–637.

Muniz, P., Clemente, J., & Brugnoli, E. (2005b). Benthic invasive pests in Uruguay: A new problem or an old one recently perceived? *Marine Pollution Bulletin, 50*(9), 1014–1018.

Muniz, P., & Venturini, N. (2011). Uruguay: Environmental conditions in the coast of Montevideo: Historical aspects, present status and perspectives of habitat degradation and uses. In J. O. Nriagu (Ed.), *Encyclopedia of environmental health* (pp. 590–601). Elsevier Science.

Muniz, P., Venturini, N., Hutton, M., Kandratavicius, N., Pita, A., Brugnoli, E., et al. (2011). Ecosystem health of Montevideo coastal zone: A multi approach using some different benthic indicators to improve a ten-year-ago assessment. *Journal of Sea Research, 65*(1), 38–50.

Muniz, P., & Venturini, N. (2015). Macrobenthic communities in a temperate urban estuary of high dominance and low diversity: Montevideo Bay (Uruguay). *Oceánides, 30*, 9–20.

Muniz, P., Venturini, N., Martins, C. C., Munschi, A., Garcia-Rodriguez, F., Brugnoli, E., Lindroth, A. L., Bícego, M. C., & García-Alonso, J. (2015). Integrated assessment of contaminants and monitoring of an urbanized temperate harbor (Montevideo, Uruguay): A twelve-year comparison. *Brazilian Journal of Oceanography, 63*, 311–330.

Muniz, P., & Venturini, N. (2019). Environmental conditions in the estuarine coast of Montevideo (Uruguay): Historical aspects and present status: An update. In J. Nriagu (Ed.), *Encyclopedia of Environmental Health* (pp. 408–418). Elsevier.

Muniz, P., Marrero, A., Brugnoli, B., Kandratavicius, N., Rodríguez, M., Bueno, C., Venturini, N., & Figueira, R. C. L. (2019a). Heavy metals and as in surface sediments of the north coast of the Río de la Plata estuary: Spatial variations in pollution status and adverse biological risk. *Regional Studies in Marine Sciences, 28*, 100625.

Muniz, P., Venturini, N., Brugnoli, E., Gutiérrez, J. M., & Acuña, A. (2019b). Río de la Plata: Uruguay. In C. Sheppard (Ed.), *World seas: An environmental evaluation, volume I: Europe, the Americas and West Africa* (pp. 703–724). Academic Press.

Muniz, P., Góngora, N., Sánchez, M., Lago, V., Antuña, D., Correa, P., Chiesa, E., & Brugnoli, E. (2021). Rapana venosa (Muricidae, Rapaninae): un invasor "exitoso" en el estuario del Río de la Plata. In A. Brazeiro, D. Bresciano, E. Brugnoli, & M. Iturburu (Eds.), *Especies exóticas invasoras de Uruguay: distribución, impactos socioambientales y estrategias de gestión* (pp. 149–161). MVOTMA, CEEI, UdelaR.

Nagy, G. J., Gómez-Erache, M., López, C. H., & Perdomo, A. C. (2002). Distribution patterns of nutrients and symptoms of eutrophication in the Río de la Plata River estuary system. *Hydrobiología, 475–476*, 125–139.

Nagy, G. J., Pshennikov, V., & Robatto, P. (2003). Monthly variability of salinity in the frontal zone of the Rio de la Plata off Montevideo in response to consecutive ENSO fluctuations and the flow of the Uruguay River (1998–2000). Contribution Nº 1. In D. Vizziano, P. Puig, C. Mesones, & G. J. Nagy (Eds.), *The Rio de la Plata: Research to Manage the Environment, Fish Resources and the Fishery in the Saline Front*. EcoPlata Programme.

Nagy, G. J., Severov, D. N., Pshennikov, V. A., De los Santos, M., Lagomarsino, J. J., Sans, K., et al. (2008a). Rio de la Plata estuarine system: Relationship between river flow and frontal variability. *Advances in Space Research, 41*(11), 1876–1881.

Nagy, G. J., Bidegain, M., Norbis, W., Ponce, A., Pshennikov, V., & Severov, D. N. (2008b). Fishing strategies for managing climate variability and change in the estuarine front of the Río de la Plata. In J. Adejuwon, V. Barros, I. Burton, J. Kulkarni, R. Lasco, & N. Leary (Eds.), *Climate change and adaptation* (pp. 353–370). Earthscan.

Nagy, G. J., Seijo, L., Verocai, J. E., Brugnoli, E., & Bidegain, M. (2013). Enfoque, conocimiento y medidas para enfrentar las amenazas del clima presente en la zona frontal del Río de la Plata, Uruguay. *Costas, Revista Iberoamericana De Manejo Costero Integrado, 2*(2), 69–87.

Nagy, G. J., Muñoz, N., Verocai, J., Bidegain, M., & Seijo, L. (2014). Adjusting to current climate threats and building alternative future scenarios for the Río de la Plata coast and estuarine front, Uruguay. *International Journal of Integration Coastal Management, 14*(4), 553–568.

Nagy, G. J., & Gutiérrez, O. (2018). Scenario planning toward climate adaptation: The Uruguayan coast. In W. Leal Filho, & L. Esteves de Freitas (Eds.), *Climate change adaptation in Latin America managing vulnerability, fostering resilience* (pp. 457–476), Springer.

Nagy, G. J., Gutiérrez, O., Brugnoli, E., Verocai, J. E., Gómez-Erache, M., Villamizar, A., et al. (2019). Climate vulnerability, impacts and adaptation in Central and South America coastal areas. *Regional Studies in Marine Science, 29*, 100683.

Nagy, G. J., Krishnapillai, M., Saroar, M., & Olivares-Aguilera, I. C. (2023a). Editorial: Climate risks, resilience and adaptation in coastal systems. *Frontiers in Climate, 4*, 1090577.

Nagy, G. J., Verocai, J. E., Capurro, L., Gómez-Erache, M., Gutiérrez, O., Panario, D., et al. (2023b). Climate risks and reasons for concern along the Uruguayan coast of the Rio de la Plata Estuary. In A. J. Manning (Ed.), *Estuary research—Recent advances*. InTechOpen.

Panario, D., & Gutiérrez, O. (2005). La vegetación en la evolución de playas arenosas. El caso de la costa uruguaya. *Ecosistemas, 14*, 150–161.

Panario, D., & Gutierrez, O. (2006). Dinámica y fuentes de sedimentos de las playas uruguayas. In R. Menafra, L. Rodríguez-Gallego, F. Scarabino, & D. Conde (Eds.), Bases para la conservación y manejo de la costa uruguaya (pp. 21–34). Vida Silvestre Uruguay.

Panario, D., Gutiérrez, O., & Piñeiro, G. (2008). Propuesta técnica para la conservación y recuperación de la playa del Buceo y playas certificadas de Montevideo. Informe final. Montevideo: Convenio UNCIEP, Facultad de Ciencias y Departamento de Desarrollo Ambiental, Intendencia Municipal de Montevideo. Setiembre and annexes (76 pp.).

Pedelacq, M. E., Garaffo, G., Llanos, E., Venturini, N., & Muniz, P. (2022). Pollution has negative effects on macrozoobenthic traits diversity in a large subtropical estuary. *Marine Pollution Bulletin, 184*, 114101.

Piecuch, C. G. (2022). River effects on sea-level rise in the Río de la Plata during the past century. *Egusphere, 7*(1), 57–75.

Portatadino, A. V. (2022). Magdalena canal and the strategic projection of Tierra del Fuego, Antarctica and South Atlantic Islands. *Malvinas En Cuestión, 1*, e008.

Santoro, P., Fossati, M., & Piedra-Cueva, I. (2013). Study of the meteorological tide in the Río de la Plata. *Continental Shelf Research, 60*(1), 51–63.

Sarubbi, A. (2007). Análisis del avance del Frente del Delta del río Paraná. Degree thesis Universidad de Buenos Aires, Facultad de Ingeniería, Argentina (136 pp.).

Schuerch, M., Scholten, J., Carretero, S., García-Rodríguez, F., Kumbier, K., Baechtiger, M., et al. (2016). The effect of long-term and decadal climate and hydrology variations on estuarine marsh dynamics: An identifying case study from the Río de la Plata. *Geomorphology, 269*, 122–132.

Smith, C. J., Papadopoulou, K.-N., Barnard, S., Mazik, K., Elliot, M., Patricio, J., et al. (2016). Managing the marine environment, conceptual models and assessment considerations for the European Marine Strategy framework directive. *Frontiers in Marine Science, 3*, 144.

Tatone, L., Bilos, C., Skorupka, C. N., & Colombo, J. C. (2009). Vertical fluxes and accumulation of trace metals in superficial sediments of the Río de la Plata Estuary, Argentina. *Bulletin of Environmental Contamination and Toxicology, 83*, 913–920.

Venturini, N., Pita, A. L., Brugnoli, E., García-Rodríguez, F., Burone, L., Kandratavicius, N., et al. (2012). Benthic trophic status of sediments in a metropolitan area (Rio de la Plata estuary): Linkages with natural and human pressures. *Estuarine, Coastal and Shelf Science, 112*, 139–152.

Venturini, N., Bícego, M. C., Taniguchi, S., Sasaki, S. T., García-Rodríguez, F., Brugnoli, E., & Muniz, P. (2015). A multi-molecular marker assessment of organic pollution in shore sediments from the Río de la Plata Estuary, SW Atlantic. *Marine Pollution Bulletin, 91*(2), 461–475.

Verocai, J., Gómez-Erache, M., Nagy, G. J., & Bidegain, M. (2015). Addressing climate extremes in coastal management: The case of the Uruguayan coast of the Rio de la Plata system. *International Journal of Integration Coastal Zone Management, 15*(1), 91–107.

Verocai, J., Nagy, G. J., & Bidegain, M. (2016). Sea-level trends along freshwater and seawater mixing in the Uruguayan Rio de la Plata estuary and Atlantic Ocean coast. *International Journal of Marine Science, 6*(7), 1–18.

Viana, F., Huertas, R., & Danulat, E. (2005). Heavy metal levels in fish from coastal waters of Uruguay. *Archives of Environmental Contamination and Toxicology, 48*, 530–537.

Yellow River Delta

Houjie Wang, Xiao Wu, Naishuang Bi, and Peng Li

1 Physical Setting

The total global sediment flux from rivers to the global oceans has been estimated as ~20×10^9 metric tons per year (Gt/yr), of which ~31.5% is contributed by large rivers in Asia, particularly by those originating from the Himalaya–Tibetan Plateau (Milliman & Meade, 1983; Milliman & Syvitski, 1992; Syvitski et al., 2005). Among the large rivers created by the tectonic uplift of Himalayas-Tibet ~55 million years ago and the corresponding climate shift (e.g., birth of the Asian Monsoon) (Clift et al., 2008; Tapponnier et al., 2001), the Yellow River annually provides ~50 km^3/yr of freshwater and 1.08 Gt/yr of sediment to the sea (Table 1), which is of great significance to both regional social community and physical–ecological environments in the Western Pacific Ocean.

1.1 Geographic and Climatic Settings

The Yellow River originates in the eastern Qinghai–Tibet Plateau. By the time the river pours into the Bohai Sea, a shallow epicontinental sea northwest of the Shandong Peninsula, the river has spanned 5,464 km, making it the sixth-longest river system in the world (Fig. 1). It drains a wide basin that covers more than 750,000 km^2 with a variety of geological and climatic features.

The Yellow River can be divided into three sections according to its geographical settings: (1) The upper reaches extend over a length of 3471 km from the river source to Toudaoguai and drain an area of 385,996 km^2; (2) The middle reaches stretch 1206 km from Toudaoguai to Huayuankou and drain an area of 343,751 km^2.

H. Wang (✉) · X. Wu · N. Bi · P. Li
College of Marine Geosciences, Ocean University of China, Qingdao 266237, China
e-mail: hjwang@ouc.edu.cn

W. Zhang and H. de Vriend (eds.), *Delta Sustainability*,
https://doi.org/10.1007/978-981-97-7259-9_16

Table 1 Information of the Yellow River

Items	Information
Basin area (10^6 km^2)	0.77
River length (km)	5464
Water discharge (km^3/yr)	49
Runoff (mm/yr)	64
Sediment Load (10^6 t/yr)	1080
Sediment yield (t/km^2/yr)	1403
Annual precipitation (mm)	300
Average elevation (m)	1547
Maximum elevation (m)	4200
Delta area (km^2)	36272
Climatic zone	Mountains-arid-temperate
Average population density (person/km^2)	162

A considerable number of tributaries join the mainstream in the middle reaches; And (3) the lower reaches extend over a length of 786 km from Huayuankou through the flat North China alluvial plains to the river mouth and drain an area of 22,726 km^2 (Fig. 1). The river descends from an altitude of approximately 4500 m at its source with a variable gradient to 1000 m at Toudaoguai, and then with a relatively continuous gradient to 110 m at Huayuankou. Changes from steep to gentler gradients near Lanzhou and Huayuankou correspond to the geographic boundaries between the Qinghai–Tibet Plateau (upstream Lanzhou), the Loess Plateau, and the flat alluvial plains of the lower reaches (Fig. 1).

The climate condition within the Yellow River basin varies spatially with distinct seasonality (Wang et al., 2007a, 2007b; YRCC, 1989). Annual mean temperature in the upper basin is approximately 1–4 °C and then increases in the middle reaches to 12–14 °C. In addition, temperature varies seasonally with the high in July and the low in January (Chen et al., 2005). The upper and middle reaches are mostly located in the arid and semi-arid regions, corresponding to low annual precipitation of ~100–400 mm/yr (Fig. 2). Controlled by different climate settings, the annual precipitation generally increases towards the southeast, up to a magnitude of 600–900 mm/yr (Fig. 2). The basin-wide precipitation exhibits a clear seasonal variation, as ~73% of the annual precipitation happens in summer (June to October) owing to the combined impacts from Southeast Asian summer monsoon and Southwest Indian monsoon (YRCC, 1989; Zhao et al., 2014). Besides the water supply from ice melting in the highly elevated upper region, the river water mostly comes from the precipitation in the drainage basin, which indicates that the river water discharge is sensitive to the change of precipitation (Lu, 2004; Wang et al., 2007a, 2007b). Since heavy rainfall happens frequently in summertime in the middle reaches, the erodible loess deposits are easily eroded and transported to the main stem, resulting in the high sediment yield (Wang et al., 2006a, 2006b).

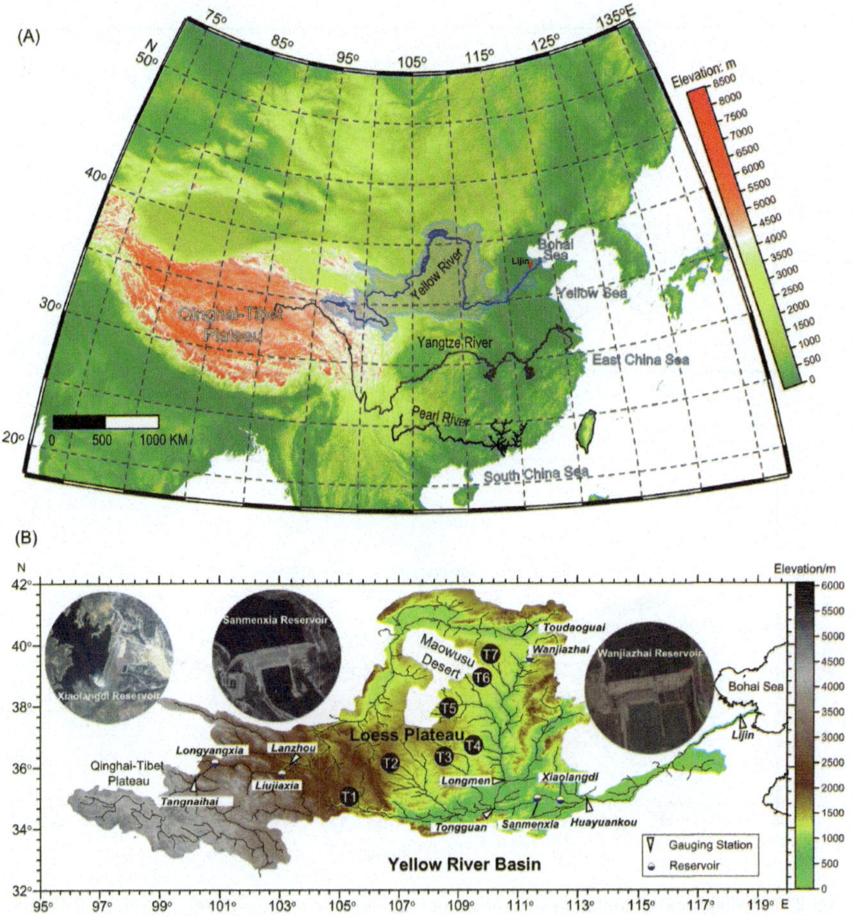

Fig. 1 Location of the Yellow River in China (**a**), and map of the Yellow River Basin with the locations of major reservoirs and gauging stations (**b**). The pictures of Wanjiazhai, Sanmenxia, and Xiaolangdi that jointly operated to implement the water-sediment regulation scheme were acquired from Google Earth. Major tributaries in the loess region are: T1–Weihe River, T2–Jinghe River, T3–Beiluohe River, T4–Yanhe River, T5–Wudinghe River, T6–Kuyehe River and T7–Huangfuchuan River

1.2 River Water and Sediment

The Yellow River has been well known for its high sediment load and low water discharge. The river's name (Huanghe in Chinese) comes from the abundant yellowish silt the water picks up and transports eastward as it winds through the dusty, friable landscapes of the Loess Plateau. The average annual water discharge from the Yellow River to the sea between 1950 and 2005, as recorded at station Lijin, the last hydrographic gauging station ~100 km upstream the river mouth (Fig. 1), was

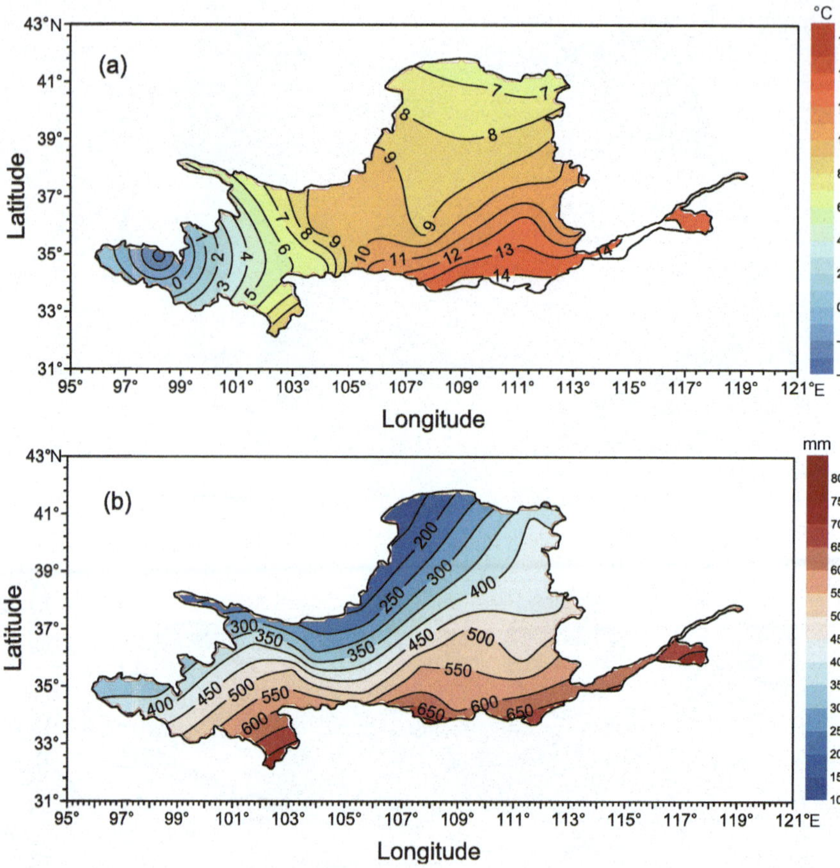

Fig. 2 Distributions of annual mean temperature (**a**) and precipitation (**b**) in the Yellow River basin

32.0 km^3/yr. This represents only 5% of that of the Yangtze (Yang et al., 2006), or 0.5% of that of the Amazon (Milliman & Meade, 1983). Most of the Yellow River's water is sourced from the upper reaches, as the average annual water discharge (1950–2005) at Toudaoguai, near the boundary between the upper and middle reaches, is 22.0 km^3/yr, and accounts for 56.4% of the annual discharge at Huayuankou. The annual discharge at station Lanzhou is somewhat higher than that at Toudaoguai because water is used for agricultural irrigation along the river between Lanzhou and Toudaoguai (Chen et al., 2005). Downstream station Toudaoguai, the water discharge increases as several tributaries of various sizes join the mainstream in the middle reaches. The highest annual river discharge (39.0 km^3/yr) occurs at Huayuankou, downstream of which the river is characterized by an elevated riverbed ~10 m above the surrounding area due to cumulative sedimentation in the past. At station Lijin, the annual water discharge has decreased to 32.0 km^3/yr, which is 82.0% of the discharge at Huayuankou. This decrease in discharge between Huayuankou and Lijin is mainly caused by water abstraction for agricultural irrigation and domestic use in the lower reaches (Wang et al., 2006a, 2006b). Because of climate change (e.g., decrease in precipitation related to ENSO events) and increasing water consumption facilitated by dams and reservoirs, discharge from the Yellow River to the sea has decreased continuously over the past half century (Wang et al., 2006a, 2006b, 2011a, 2011b). This illustrates a complicated transformation of the river system as results of both natural and human-induced influences.

The spatial distribution of the Yellow River's sediment load differs from that of the water discharge. Nearly 90% of the sediment load comes from the middle reaches, where the dominant geological feature is the wind-deposited loess, which is acknowledged as the region of highest erodibility in the world (Huang & Zhang, 2004; Ren & Shi, 1986). In contrast, the upper reaches upstream Toudaoguai provide only 10% of the total sediment load, whereas an increase of sediment by more than 1.0 Gt/yr at Huayuankou indicates the dominant contribution of the major tributaries in the middle reaches. As the river flows over the flat alluvial plain, a considerable amount of sediment settles down and forms a unique 'suspended river'. This has led to catastrophic flooding disasters in the surrounding area over a long period in Chinese history (Chen et al., 2005). Finally, the river discharges 800 mt/yr (1950–2005) of sediment load to the sea, as recorded at station Lijin.

The climatic regime of the Loess Plateau during the middle Holocene was characterized by a warm and wet period (the so-called Holocene Climatic Optimum, He et al., 2006), with a temperature approximately 3 °C higher than at present. The humid and warm weather nourished a lush vegetation, with over 40% of the Loess Plateau covered by a mixed deciduous-coniferous forest from ~7000 to 3000 cal yr BP. Correspondingly, palaeosol was predominant in the sediment eroded from the Loess Plateau, with a percentage as high as 80–88% (Wang et al., 2014). Although precipitation was plentiful during this period, the soil erosion was still very weak due to the well-vegetated land surface (He et al., 2006; Zheng, 2006). Therefore, the riverine sediment discharge to the sea was less than 0.05 Gt/yr from 7000 to 3000 cal yr BP, a period in which the Yangshao and Longshao cultures developed and the Xia, Shang, and Western Zhou dynasties were established (Zhao et al., 2013; Wu et al., 2020b).

After 3000 cal yr BP, the East Asian Summer Monsoon became weakened, and the continental interior experienced increasing chill and aridity (He et al., 2006). In addition, this period witnessed a population increase and the development of the Feudal Society. The Qin and Han dynasties unified much of North China, and their social, economic, and political centers were centralized in the middle Yellow River basin (Loewe, 2006). Under the combined effects of natural and anthropogenic forces, the middle Yellow River basin experienced a decrease in forest coverage, and the Loess Plateau became easily erodible due to the destruction of the natural vegetation and palaeosol (Ren & Zhu, 1994). Consequently, the sediment load in the Lowe Yellow River increased to 0.23 Gt/yr at 1400 cal yr BP.

During the period from 1400 cal yr BP to AD 1855, there was a cooling and drying trend coinciding with several large cold-warm alterations in the loess region. Both the increasing population and destruction of land cover in the middle Yellow River increased the sediment yield, which resulted in enhanced soil erosion in the loess region and the increasing accumulation rate in the flood plain (Shi et al., 2002, 2010). During 1855–1949, vegetation on the Loess Plateau was continuously impacted by frequent warfare, and extensive agricultural activities led to the cultivation of remaining wilderness areas in the middle region. Thus, the sediment discharge from the Yellow River to the sea averaged to be a relatively high value (1.31 Gt/yr) with fluctuations of ±0.61 Gt/yr. With the explosive growth of population and advances in mechanized agricultural practices, human activities began to affect the environment at a scale comparable with powerful natural forces from 1400 cal yr BP to the late 1950s.

The past seven decades witnessed stepwise decreases in both water and sediment discharge from the Yellow River to the sea as dominated by dam regulation and soil conservation practices in the river basin. Prior to the operation of Liujiaxia Reservoir (Fig. 1b) in 1968, the average water and sediment discharge to the sea were 50 km^3/yr and 1.2 Gt/yr, respectively (Fig. 3); nevertheless, high interannual variability was due to less regulation of dams except for the temporary effect of the Sanmenxia Reservoir. The joint operations of Liujiaxia Reservoir (1968) in the upper reaches (Fig. 1b) and the soil-conservation practices in the middle reaches since the mid-1970s subsequently caused a stepwise decrease in water and sediment discharge. As a result, the water and sediment discharge at station Lijin decreased by 35% and 30%, respectively, during the period of 1969–1985. Furthermore, the combined impacts from the Liujiaxia Reservoir, Longyangxia Reservoir (1985), and the soil-conservation practices continuously reduced the water and sediment to ~15 km^3/yr and 0.4 Gt/yr, respectively, so to only some 30% of those in the 1950s (Fig. 3). The stepwise decreases in both water and sediment discharge were similar until the year 2000. From 2000 to 2020, the Xiaolangdi Reservoir that controls the water and sediment discharge into the Lower Yellow River had different effects on these discharges, as appears from the evident departure of sediment from the water discharge, particularly after the initiative of Water-Sediment Regulation in 2002. The recent water discharge to the sea is 20.7 km^3/yr, comparable to that of 1986–1999, whereas the sediment load decreased to 0.15 Gt/yr, approximately 12% of that in the 1950s.

1.3 The Yellow River Delta

The Yellow River's delta and estuary is an unusually dynamic, fast-changing place. As river water flows into the sea, large amounts of sediment drop to the shallow ocean floor, building up new land if there is sufficient material to overcome the erosive forces of waves and tides. The river, historically with sediment loads exceeding 1.6 Gt/yr, has earned a reputation for being one of the most turbid rivers in the world. As a result of rapid sedimentation on the floodplain, the river's lower course has continuously shifted over the past 6000 years and formed 10 delta lobes on the coasts of the Bohai and Yellow seas. In 1855, a particularly abrupt avulsion swept the channel hundreds of miles to the north, to its current location near the Shandong Peninsula as the

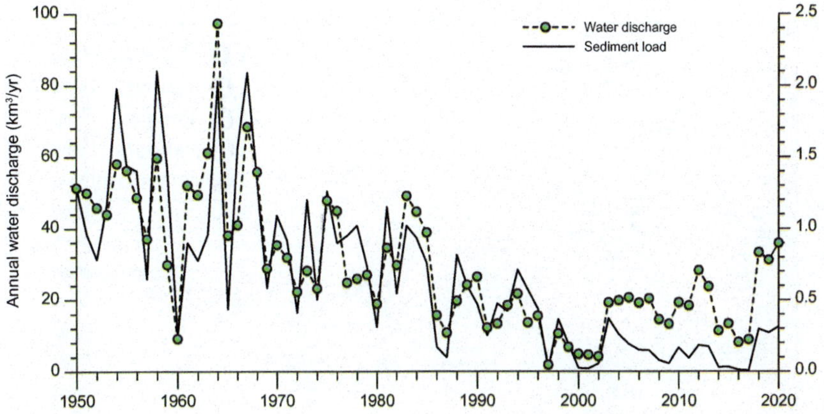

Fig. 3 Stepwise decreases of the Yellow River's water discharge and sediment load to the sea (1950–2020)

river course migrated from the Yellow Sea coast northward to the present Bohai Sea coast (Wang & Aubrey, 1987). This event triggered such extensive flooding and food shortages that it fueled civil unrest and rebellion. The modern Yellow River Delta of more than 5000 km^2, with the world's most rapid land-building rate of 22.5 km^2/yr (Pang & Si, 1980), began to form in 1855 (Fig. 4). The mouth of the river has shifted or avulsed at least 55 times since 1855, often triggering destructive floods. The last major shift of the deltaic course from Diaokou to Qingshuigou was in 1976, starting the formation of the active delta lobe. As a result, the abandoned delta lobes (formed during 1947–1964) were subject to strong erosion. Meanwhile, the active delta lobe prograded seaward rapidly at a mean rate of 0.83 km/yr. However, this fast seaward progradation completely reversed in the late 1990s due to the decline of river sediment discharge to the sea, which is critical for delta construction and coastal protection (Cui & Li, 2011; Kong et al., 2015; Zhang et al., 2018; Wu et al., 2023).

In recent decades, engineering efforts have made to tame the river, as the last two major course migrations were engineered primarily to control flooding and protect coastal development. In 1976, authorities shifted the river's course away from the Diaokou lobe in the northern part of the delta toward a newer channel to the east, which potentially favoured the exploitation of coastal gas and oil reserves. After the 1976 diversion, new land was rapidly built to the east in the Qingshuigou delta lobe, whereas the ocean dynamics eroded the unstable lands around the old river channel outlet in the north.

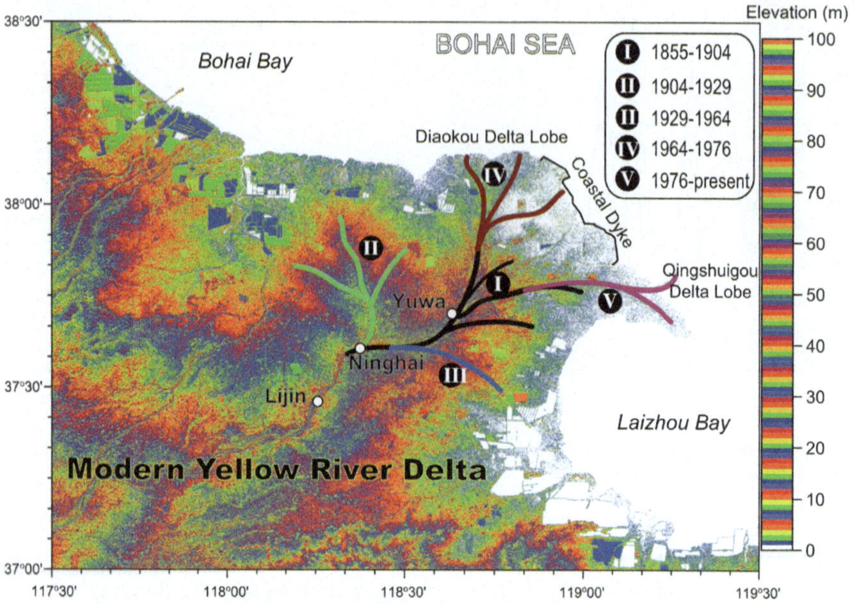

Fig. 4 Map of the modern Yellow River Delta

1.4 Hydrodynamics and Suspended Sediment Transport

The tidal regime in the Bohai Sea is dominated by mixed semi-diurnal tide with two counter-clockwise semi-diurnal amphidromic systems (M_2 and S_2) formed (Sun & Xi, 1988). The surface current velocity in the Bohai Sea has evident spatial variability due to the constraining of coastline. The mean tidal currents vary from 20 cm s^{-1} at the mouth of Laizhou Bay to 80 cm s^{-1} in the northern Bohai Strait and eastern part of Liaodong Bay (Huang et al., 1999). The tidal residual currents are generally weak with velocity less than 5 cm s^{-1} except for those in the Bohai Strait where a pair of cyclonic and anticyclonic headland eddies are significant. The residual currents in the Bohai Sea are also impacted by winds and density structures that are significantly dominated by seasonal monsoon activities and thermal/saline stratifications (Jiang et al., 2000).

Wind waves are a dominant feature in the Bohai Sea that play an important role in sediment transport. The significant wave height varies from 0.3 to 0.7 m in the nearshore of the Bohai Sea to ~1.0 m in the Bohai Strait and the central basin (Jiang et al., 2000). Due to the dominance of monsoon activities over wind waves, northerly waves prevail in the winter seasons and southerly waves dominate in the summer seasons but with relatively low significant wave heights. Waves are critical to the sediment resuspension and transport in the shallow region. There is strong seasonal variability of suspended sediment distribution along the coast of Yellow River Delta, which illustrated that the suspended sediment concentration in the shallow area in

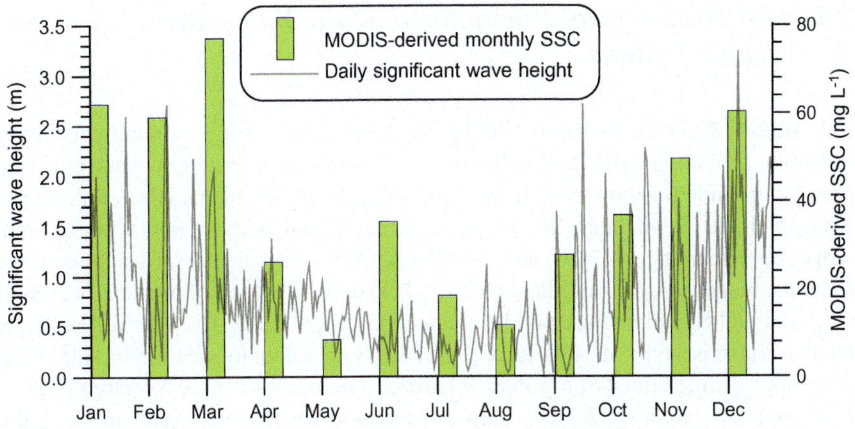

Fig. 5 Spatially averaged daily significant wave height versus monthly surface suspended sediment concentration (SSC) derived from MODIS data in the southern Bohai Sea in 2010 (Wang et al., 2014)

winter season seems to be one order in magnitude higher than that in the summer season, potentially associated with the wave actions induced by winds in winter season (Fig. 5).

The suspended sediment was mostly concentrated in the southern Bohai Sea around the Yellow River Delta in the summer season as impacted by the river plume but confined within a very limited area near the river mouth, and in winter–spring season with much higher concentration as a result of strong sediment resuspension induced by energetic wave actions in the shallow water (Fig. 5). The spatial pattern of suspended sediment in the Bohai Sea was dominated by the river input and resuspension that depends on the type of local surface sediment and the strength of hydrodynamics. Large amount of sediment delivered by the Yellow River in summer season and strong resuspension of fine-grained sediment in winter–spring season make the southern Bohai Sea to be a highly turbid coastal environment at most time of a year. Monsoon and its associated hydrodynamics such as wave actions, coastal currents and mixing-stratification of water column predominantly induce strong seasonal variability of sediment distribution in the Bohai Sea. As a result, the subaqueous Yellow River delta acts as a sink in summer season that traps most of the sediment delivered from the river within the delta front, whereas it becomes a primary source for sediment dispersal in the Bohai Sea in winter–spring seasons through active coastal resuspension in shallow water by highly energetic wave actions in accord with monsoon activities.

1.5 Eco-Hydrological Environment of the Yellow River Deltaic Wetland

The Yellow River Delta has the largest and youngest coastal wetland ecosystem in China, providing habitat to millions of wild birds and serving as spawning and nursery grounds for numerous freshwater and marine organisms (Cui et al., 2009; Kong et al., 2015; Ma et al., 2014). The deltaic wetland, with various types of landscape, covers an area of 4167 km^2, including 3131 km^2 of natural wetlands (e.g., swamps, marshes, tidal mudflats, rivers, lakes) and 1036 km^2 of artificial wetlands (e.g., rice fields, aquaculture ponds, channels) (Qi & Luo, 2007). The Yellow River Delta wetland is a typical estuarine wetland ecosystem with four dominant salt marsh plant species, including *Phragmites australis*, *Tamarix chinensis*, *Spartina alterniflora*, and *Suaeda salsa* (Fig. 6). *Suaeda salsa* and *Tamarix chinensis* are the halophytic plants in heavy saline soil and tidal flats. As the soil salinity decreases, the vegetation type changes to *Phragmites australis*. A frequent succession among vegetation communities exists in the newly created wetlands (Yue et al., 2003), due to the rapidly changing environmental conditions such as delta progradation, salinity and human disturbance.

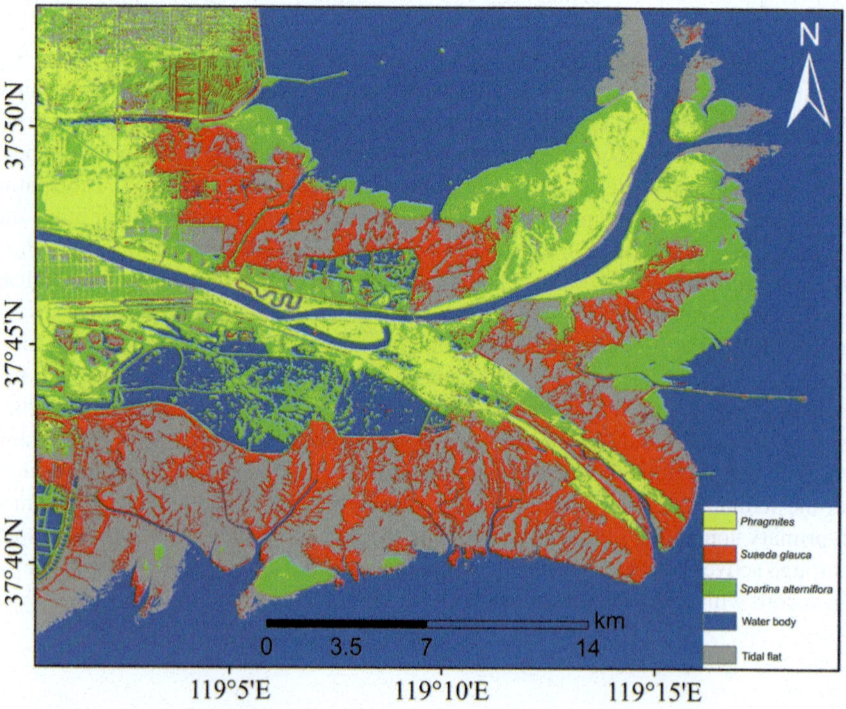

Fig. 6 Coastal salt marsh classification map on the Yellow River Delta wetland in 2020 from Sentinel-1 and Sentinel-2 Images

Highly valuable ecosystem services including nutrient cycling, carbon storage, as well as tourism and recreation are provided by the deltaic wetlands (Yu et al., 2011, 2013). They support an abundant biodiversity and are home to 220 seed plant species such as endangered wild soybean, more than 800 wild animal species, and 283 bird species, many of which are listed as endangered species (Cui et al., 2009). They provide an important habitat and transfer area for millions of migrating birds, such as the red-crowned crane, hooded crane, Siberian crane, oriental stork, black stork, and golden eagle (Wang et al., 2014; Xu et al., 2015).

Economically, the Yellow River Delta has the second largest oil field in China and the energy production and related industrial activities have had an enormous impact on the surface and subsurface environment in the region (Kuenzer et al., 2014). Agricultural production has occupied a large fraction of the newly formed delta lands and used a substantial amount of riverine and underground freshwater resources (Han et al., 2014; Xu et al., 2015).

2 Socio-economic Characteristics

2.1 *Population and Infrastructure*

The present delta supports the livelihood of 5.2 million people in the northeast of Shandong Province. The major cities include Dongying and Binzhou. Dongying is situated in the center of the Yellow River Delta. Established in 1983, Dongying exercises jurisdiction over three districts—Dongying, Hekou and Kenli District, and two counties—Guangrao and Lijin. It covers an area of 8,243 km^2, with a population of 2.15 million.

Dongying is the central city on the Yellow River Delta, a major node city along the Bohai Rim region, and is of importance to the coordinated and integrated development of Beijing-Tianjin-Hebei region. The city's infrastructure is experiencing constant improvement. A three-dimensional transport network has been established, consisting of Dongqing Highway, Weiwu Highway, Huangda Railway, Delongyan Railway, Dongying Port, Guangli Port and Dongying Airport.

2.2 *Major Industries*

The rich reserves of geothermal energy and bittern produce a good development prospect for the Yellow River Delta. The Yellow River Delta is endowed with rich oil and gas resources and is the birthplace and main production area of Shengli Oilfield, the second largest petroleum industry base in China. At present, the oilfield's verified reserve of petroleum amounts to 4.8 billion tons and the reserve of natural gas to 230 billion m^3. The exploitation of Shengli Oil Field was the starting point of an intense industrialization and urbanization of the delta. The delta is also rich in valuable minerals, rock salt, and halogens. The favourable geographic location and the resource deposits make the delta region to one of the most important industrial centres in North-East Asia.

The Yellow River Delta has relatively abundant land resources with an average per capita land area of 0.48 ha, which is 5.33 times that in the Changjiang (Yangtze) River Delta. Aquatic production is another important economic value for Dongying City with an average annual per capita aquatic production of 46.3 kg, which is 2.59 times the national level.

The Yellow River Delta adheres to the development orientation of high efficiency and ecology, and keeps optimizing the regional economic structure. The leading industries include petroleum, petro-chemicals, salt chemicals, petroleum equipment manufacturing, textile and clothing, rubber tires, paper-making and timber processing, and food processing. Focused efforts were made to promote the development of emerging industries, including new energy, new materials, information technology, new medicine, pharmaceutical and biology industry. In the year 2017, Dongying realized a total GDP of 380.2 billion yuan RMB, registered public budget revenue of 23.3 billion yuan RMB and import–export value of 13.12 billion yuan RMB.

On the other hand, the Yellow River Delta represents one of the youngest wetlands on earth and it is the habitat for numerous species. The area is a particularly important resting place for millions of migrating birds of which many are under protection. The Yellow River Delta with abundant tourism resources including landscapes and seascapes has become an attractive place to domestic and international visitors.

3 Current Issues

Due to the abundant sediment supply and rapid sedimentation on the lowlying floodplain over thousands of years, the Yellow River's lower course has continuously shifted since the Holocene, and eventually formed 10 superlobes bounded by coastal cheniers (shelly ridges) on the coasts of the Bohai and Yellow seas. Among these nine ones are located around the western Bohai Sea and one is on the western Yellow Sea. A chronology of cheniers on the delta plain around the NW Bohai Sea with radiocarbon ages and culture site ages illustrated that chenier formation and delta progradation are linked, and controlled by the course shifts of the lower reaches of the Yellow River. For the last 3000 years, more than 1500 floods and 26 course shifts are recorded in Chinese literature (Cheng, 1991). In 1128, the most remarkable man-made breaching of dikes was done for military reasons as the river was diverted southward to the Yellow Sea. Except for the presently active delta lobe, other deltaic coasts of the Yellow River have been subjected to coastal erosion and subsequent chenier development. The past shoreline configuration, estimated from the cheniers, was probably a smooth and linear pattern. However, delta lobe 9 along the Yellow Sea, which formed between 1128 and 1855, has a cuspate topography.

3.1 *Frequent Course Shifts of the Modern Yellow River Delta*

The modern Yellow River Delta, formed after 1855 with a subaerial area of >5000 km^2, has experienced >50 channel shifts due to rapid channel siltation and resultant channel instability, with major shifts occurring approximately every 10 years (Fig. 3) (Wu et al., 2015a, 2015b).

The most recent major channel shift happened in 1976 as the river channel on the delta plain was artificially migrated southward from the Diaokou course to the Qingshuigou course (Fig. 3), which consequently formed an abandoned delta lobe in the northern part and an active delta lobe in the southeastern part. In August 1996, a minor channel shift was artificially implemented as the ending channel shifted northward to form the present delta lobe. This was done not only to prolong the duration of the Qingshuigou course but also to facilitate oil-extraction of Xintan and Kendong Oil Field from sea to land (Chu et al., 2006).

In contrast to the continuous delta progradation during the initial 100 years (1855–1950s), the Yellow River Delta has experienced a regime shift from accretive to erosive over the past 60 years, primarily due to the stepwise decrease of the sediment supply from the river (Figs. 3 and 7). The Diaokou and Shenxiangou promontories, were generally eroded during the period 1976–2000. During this period, their total area was reduced by -141.3 km^2 with a mean net erosion rate of -6.03 km^2/a. During the period 1976–1981, their total area was reduced by -97.12 km^2 with a net erosion rate of -20.9 km^2 /a in less than five years. As a result, the coastline retreated landward rapidly. Previous investigations showed that submarine landslides offshore of the Diaokou Promontory occurred frequently due to the instability of delta front slope. Since 1984, erosion rates have been decreased and the coastline gradually smoothened. Under the protection of artificial structures built from 1986 to 1987 along most sections of the Shenxiangou Promontory, the coastal retreat nearly stopped at the foot of the artificial structures after 1987 (Chu et al., 2006). The delta lobes that were abandoned after 1976 and 1996 presently suffer significant coastal erosion particularly in the upper slope shallower than 10–15 m, whereas the presently active delta lobe since 1996 has prograded seaward due to the accumulation of river sediment.

At present, coastal erosion is a dominant feature of the Yellow River Delta, except for the local deposition around the presently active river mouth that receives most of the river sediment delivered to the sea during the WSRS. The WSRS-induced coarsening of suspended sediment was expected to play a critical role in maintaining the present river mouth, as the coarser sediments are expected to deposit closer at the present river mouth, which consequently contributes to the delta accumulation in the area shallower than 10 m. The erosion–accretion patterns of the active delta lobe in the subaerial and subaqueous portions have changed significantly since 2002. The active delta lobe and the Q8 lobe (formed after 1996) shifted from erosive to accretive in 2002 (Fig. 6). The slope of the subaqueous delta off the river mouth also increased markedly after 2002, corresponding to the changes in the erosion–accretion patterns of the land area. The sediment particle size increased due to the

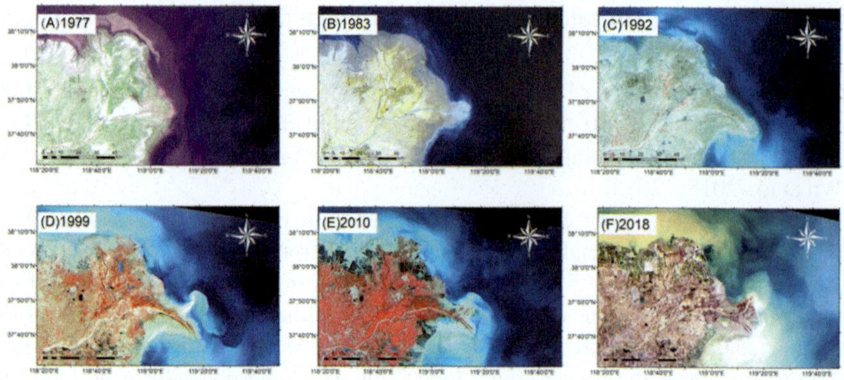

Fig. 7 The changing Yellow River Delta from 1977 to 2018 based on Landsat images

scour in the lower Yellow River, which resulted in more sediment deposition in the nearshore area and caused most of the changes in the erosion–accretion patterns of the active delta lobe. Efforts to flush sediment from reservoirs and to scour sediment from the downstream riverbed led to a spurt of accelerated land formation in the delta between 2002 and 2014. However, the volume of sediment reaching the delta began dwindling in 2014 as coarser sediments coated and "armored" the lower river channel in key areas, preventing additional scouring (Wang et al., 2017). Since 2014, the delta has once again begun to lose more land each year than it gains. Under the impact of recent intense human activity, approximately 50 million tons of sediment, less than one-third of that before the WSRS, is needed each year to maintain the equilibrium of the active Yellow River delta lobe in the short term (Bi et al., 2021).

3.2 Delta Destruction and Wetlands Degradation

Relative sea-level rise in deltas is collectively determined by the aggradation rate associated with the river sediment delivery, the eustatic sea-level rate determined by the volume of the global ocean over time, and natural and accelerated compaction or subsidence of the delta, as well as impacted by the tectonic movement over geological time (Syvitski et al., 2009). Given the continuous decreases in water and sediment discharged by the Yellow River and the enhanced coastal erosion in the delta region, the relative sea-level rise of the Yellow River Delta is estimated to be 8–23 mm/yr with considerable uncertainty. Among the factors that impact the relative sea-level rise, the decreasing aggradation rate, the eustatic sea-level rise and the accelerated subsidence are the most influential. Compared with the fast aggradation in the early twentieth century (~50 mm/yr), the aggradation of the Yellow River Delta has presently stopped or reversed to erosion. The eustatic sea level is rising at a rate of 4 mm/yr as identified from the long-term tidal gauging records at several stations around the delta, approximately to the global average rate. However, the land subsidence induced by human activities has been accelerated with funnels subsiding at more than 50 mm/yr, mainly near the coastal brine mining clusters (Wang et al., 2022). Delta erosion owing to the reduced sediment supply and continued delta subsidence has increased the flood risk along the delta coast (Syvitski et al. 2009). A large portion of the Yellow River Delta (~3,400 km^2, Syvitski et al. 2009) newly formed during the past 100 years is lower than 2 m above mean sea level, due to significant compaction. This makes the Yellow River delta much more vulnerable to episodic events such as storm surges and long-term rise of the sea level, especially if these are exacerbated by a rapidly changing climate at global scale.

Although the Bohai Sea is less susceptible to typhoons, cold air outbreaks and extratropical cyclones usually cause significant storm surges in the southern Bohai Sea, particularly in the winter half year (October–March). There is a statistically significant correlation between storm surge intensity and the Arctic Oscillation and the Siberian High.

Like many of the large deltas around the world (Blum & Roberts, 2009; Syvitski et al., 2009), the Yellow River Delta is now facing increasing risks of degradation due to anthropogenic and natural forces (Bi et al., 2014a, 2014b, Cui et al., 2009, Gao et al., 2014, Kong et al., 2015, Wang et al., 2007a, 2007b). The fundamental changes in land-building and land loss in the delta plain are rooted in the supply of water and sediment from the Yellow River. Superimposed on the natural processes, population growth, oil and gas extraction, and agricultural development have put high demands on the land and water resources, and in combination they modified the delta's natural sedimentologic, hydrologic, and ecologic systems. Furthermore, global climate change and sea level rise has modified ecohydrological processes within the delta and the effect is expected to be intensified in the future (Wang et al., 2006a, 2006b).

In addition, elevated concentrations of heavy metals, polycyclic aromatic hydrocarbons (PAHs), petroleum hydrocarbons, and pesticides have been observed in soil and plant samples on the deltaic wetland, potentially threatening the wetland ecosystem (Li et al., 2014; Yuan et al., 2014; Zhu et al., 2015). The dam-orientated WSRS has become a critical time window for the delivery of heavy metals (e.g., Mn, Pb, Zn, V, Cr, and Ni) to the sea, as more than 57% of the annual quantity of particulate heavy metals was transported to the sea during the WSRS (Bi et al., 2014a, 2014b), These pollutants were transported and accumulated along with the fine-grained sediment, resulting in heavily contaminated coastal wetlands. Recent investigations reported the WSRS-induced accumulation of arsenic (As) and cadmium (Cd) in the wetland soils around the river mouth, indicating a degraded quality of deltaic wetlands and potential threats to the ecosystem (Bai et al., 2012).

The ecological evolution of the wetlands and its interactions with changing climate and hydrology is complex and intractable upon the long-term monitoring data of soils, plants, and animals. Since the channel shift in 1976, the extension of wetland is well correlated with the cumulative river freshwater discharge and sediment load, particularly after 2002 when the WSRS delivered a large amount of coarse-grained sediment

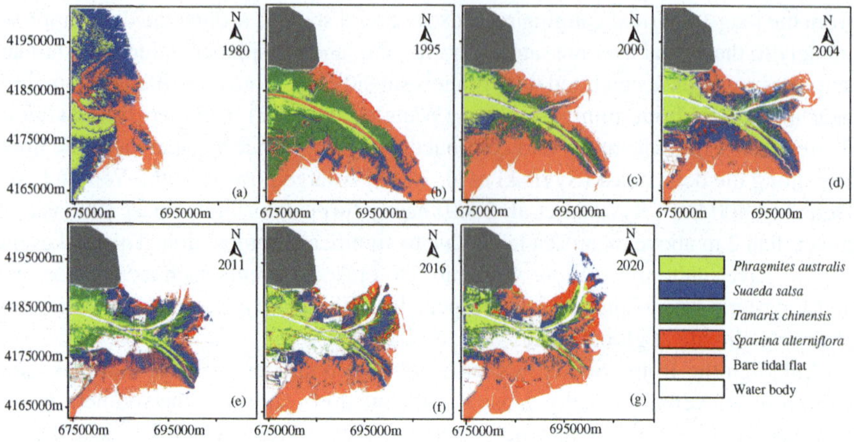

Fig. 8 Spatial distribution of vegetation on the Yellow River deltaic wetland at different stages (after Li et al., 2022)

to the river mouth (Li et al., 2022). *Phragmites australis* is generally distributed along the river channel as favored more by freshwater, whereas *Suaeda salsa* is more salt-tolerant and is distributed further away from the river channel (Fig. 8). Since 2010 *Spartina alterniflora*, as an invasive species, has explosively spread and occupied the tidal flat on the Yellow River Delta, which suggests the potential risk of the wetland ecosystem.

3.3 Loss of Biodiversity

The wetland ecosystem is seriously stressed due to depleting water resources, degradation of soil quality, water pollution, and agricultural and industrial activities on the Yellow River Delta. The reed marsh, salt meadow, and tidal wetland decreased by 17%, 37%, and 38%, respectively, from 1986 to 2001 due to the decreased runoff and sediment load and the conversion to aquacultural ponds and agricultural fields (Li et al., 2009). Construction of oil wells and roads has decreased landscape patch connectivity (Yue et al., 2003).

The Yellow River delta is home to the local Shengli Oilfield (Fig. 9). With sufficient supply of sediment in 1970s, the coastline at Gudong propagated seaward more than 5 km from 1976 to 1986. However, since 1987, the coastline has been subject to erosion due to the reduced sediment supply. An artificial coastal dike was first built to slow down the coastal erosion in 1987. Three groins were successively built along Gudong Dike in the 2000s. Despite the coastal defenses the overall Gudong area continued erode. During 1996–2016, more than 94% of Gudong littoral zone suffered from erosion, with an average erosion rate of 0.1 m/yr (Ji et al., 2018).

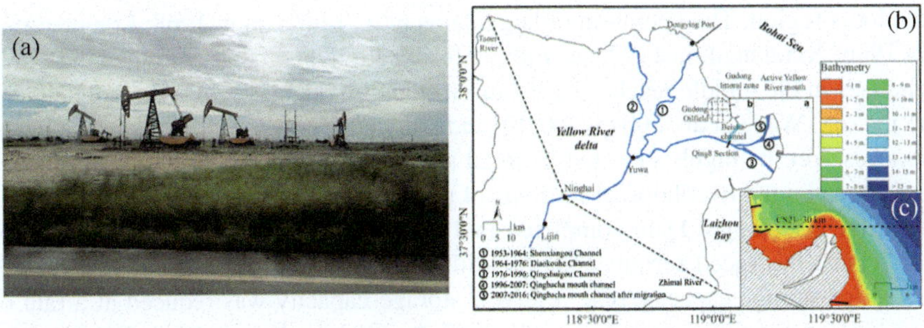

Fig. 9 **a** Oil pumps in Gudong Oil Field, **b** Modern Yellow River delta, **c** Bathymetry of active YRSD (after Ji et al., 2018)

As a result, the areas of red-crowned crane habitat decreased by 5935 ha from 1992 to 2008 (Wang et al., 2014). It is reported that there are severe decreases in bird populations, a sharp decrease in indigenous plant communities accompanied by an increasing population of invasive species. In addition, coastal construction and aquaculture activities may have destroyed the habitat of many wild species of aquatic organisms.

The rapid environmental change also significantly impacts the biodiversity of the wetland ecosystem. Recent investigations documented evident changes in species diversity and spatial-temporal dynamics of macrobenthic community on tidal flats of the Yellow River Delta (Li et al., 2019). Compared to historical records, the dominant species on the tidal flats have been replaced by small-sized polychaete, bivalve and crustacean species, instead of the large-sized crustacean and mollusc species. These changes were possibly ascribed to the decreases in freshwater discharge and sediment load and the resultant increasing salinity in estuarine environment as well as the human activities in the coastal sea (e.g., overfishing and increasing exploitation). For example, the Chinese Shrimp (*Penaeus chinensis*), one of the traditional economic products in the Bohai Sea (Jin, 2001), particularly around the Yellow River Mouth, has almost disappeared since the late 1990s. This is primarily because the salinity at the river mouth where its spawning and nursery grounds are located, has increased along with the decrease in spring freshwater discharge from the Yellow River to the sea (Wang et al., 2017).

4 Management Measures

4.1 *Efforts of Dam-Orientated Water-Sediment Regulation*

Since the 1950s, China has made great efforts to manage the Yellow River, including the dam constructions in the upper and middle reaches, and the effective water-soil conservation program in the middle reaches in order to control the sediment yield in the loess region. The operation of large dams (e.g., Liujiaxia in 1968; Longyangxia in 1986; Xiaolangdi in 1999, see Fig. 1b for locations) together with the land cover changes in the middle reaches have caused stepwise decreases in sediment delivery to the sea (Wang et al., 2007a, 2007b), and reduced the risk of flooding in the lower reaches. Yet the highly elevated riverbed and shallow channel in the lower reaches still threaten the livelihood of millions of people living in the fluvial plain. In addition, the Xiaolangdi Dam completed in 1999 with a reservoir storage capacity of 12.7 km^3, has faced serious siltation within its reservoir and a significant loss of effective storage capacity. The reservoir storage capacity was reduced at a rate of 0.3 km^3/yr during the period of 1999–2007. In 2002 the Yellow River Conservancy Committee (YRCC), as the official administrative department for the Yellow River, initiated the water-sediment regulation scheme (WSRS) as facilitated by the reservoirs of Xiaolangdi, Sanmenxia and Wanjiazhai (see Fig. 1b). The objectives of the WSRS include scouring the lower riverbed to mitigate the siltation and expelling the sediment that previously accumulated within the reservoir to maintain the storage capacity by the dam-regulated artificial flood water (Wang et al., 2005).

The WSRS was designed to have two phases with different purposes. During the first one (so-called water-regulation period) clear water is utilized to scour the sediment accumulated within the Xiaolangdi Reservoir. The water discharge gradually decreases to ~2000–3000 m^3/ s. Within a few days, however, the SSC increases rapidly to ~60–100 kg/m^3. As a result, sediment is exported from the reservoir as an artificial hyperpycnal flow with high sediment concentration and fluid density. The sediment-regulation usually lasts for 4–6 days, but is responsible for several to tens of million tonnes of sediment export from the Xiaolangdi Reservoir. As the WSRS is completed, the water discharge is brought back to less than 500m^3/s, which represents a normal condition. During the second phase of regulation the sediment exported from the Xiaolangdi Reservoir is fine silt with a median grain size of ~0.006–0.010 mm, which renders the sediments delivered to the sea differ largely from those during the first phase, in terms of both sediment texture and sediment composition.

The WSRS has been implemented annually since 2002, usually lasting for ~20 days from mid-June to early July, and greatly altering the natural seasonal rhythm of the hydrological cycle, as ~30 and ~50% of annual water and sediment is discharged within a short period (Wu et al., 2020a). Such an impulsive discharge of water and sediment has impacts on the coastal ocean, such as changes of the coastal dynamics, the quality of the wetlands and the nutrient delivery—all closely associated with the health of the coastal ecosystem.

The pulse floods created by the WSRS have essentially scoured the riverbed downstream the Xiaolangdi reservoir, and induced an evident increase in grain size of the suspended sediment transport. This has significant impacts on the morphology of the river mouth since a larger portion of sediment rapidly deposits at the active river mouth and contribute to river mouth accretion. Bi et al. (2014a) confirmed that the present river mouth accreted at a rate of ~4 km^2/yr since 2002 in contrast with 0.01 km^2/yr for the period of 1996–2002, even though the average annual sediment load to the sea since 2002 was 0.17 Gt/yr, almost the same as during the period of 1996–2002 (~0.16 Gt/yr). Therefore, the coarser sediment derived during WSRS played an important role in shaping the river mouth morphology (Wu et al., 2017). In addition, WSRS has important influences on the estuarine biogeochemical cycles since ~35% and 56% of the annual dissolved organic carbon (DOC) and particulate organic carbon (POC) are discharged into the sea during WSRS, respectively (Zhang et al., 2013). Nutrient fluxes from the Yellow River to the Bohai Sea during the WSRS period increased to 8–10 times of those during the non-WSRS period, which exacerbates the nutrient imbalance and has significant impacts on the coastal ecosystem (Liu et al., 2012). Besides the rapid exports of carbon and nutrients, heavy metals such as lead (Pb) and zinc (Zn), and pollutants such as polycyclic aromatic hydrocarbon (PAH) were also flushed to the sea during the WSRS, originating either from the crop lands along the lower river reaches, or directly from the reservoir (Bi et al., 2014b; Dong et al., 2015). Consequently, the soil in the coastal wetlands around the river mouth is highly polluted after the WSRS (Bai et al., 2012). Thus the quality of the coastal wetlands degrades, even though the WSRS increases the wetland area.

4.2 Wetland Restoration Projects

Wetland restoration projects have been initiated since the start of new millennium, in order to prevent the wetlands from further deterioration and to protect endangered birds (Cui et al., 2009, Wang et al., 2014, Yang et al., 2011). During WSRS-operations in July or August, a total volume of 1.0×10^8 m^3 of freshwater from the Yellow River has been distributed over the wetlands through a system of reservoirs and channels, to ensure sediment supply. Dikes have been constructed along the coast to prevent seawater intrusion. The freshwater replenishment has significantly improved the structure and function of the wetland ecosystem as indicated by rising water table, improving soil quality, increasing plant cover, and return of migrating and breeding birds by 2007 (Cui et al., 2009). However, recent surveys suggested that operations at the current level may not be sufficient to fully restore the plant and bird communities. A systematic review and assessment of the management practices of the Yellow River Delta wetlands is urgently needed to find an effective remedy for the ecosystem health in the future (YRCC, 2013).

5 Future Challenges

Although the evolution of the Yellow River Delta has been studied extensively, the present literature review shows that large knowledge gaps still exist on the dynamics of ecohydrology of Yellow River Delta wetland. Especially the impacts of large-scale anthropogenic activities including oil and gas production, agricultural development, as well as sediment and water regulation are insufficiently known. From the ecohydrological perspective, we recommend to carry out following studies: (1) Interaction between river flow, sea water, and groundwater analyzed through in situ monitoring and modeling; (2) Biological and ecological response of wetlands to different water and sediment regimes studied by well-designed plot experiments; (3) Ecosystem health assessment based on comprehensive field monitoring of water, soils, plants, birds, and aquatic organisms; (4) Future projection of the change of the Yellow River Delta wetland system due to climate changes and anthropogenic forces. However, it should be emphasized that the current issues in the Yellow River Delta wetlands are intimately connected with the regional socio-economic development. Competition for limited land and water resources by industrialization and urbanization has become the dominant stressor of the wetland ecosystem. Therefore, a cross-disciplinary discussion and policy development is essential to restore the ecological integrity and to improve the livelihood of people in Yellow River Delta.

References

Bai, J., Xiao, R., Zhang, K., & Gao, H. (2012). Arsenic and heavy metal pollution in wetland soils from tidal freshwater and salt marshes before and after the flow-sediment regulation regime in the yellow river delta, China. *Journal of Hydrology, 450-451*, 244–253.

Bi, N., Wang, H., & Yang, Z. (2014a). Recent changes in the erosion–accretion patterns of the active Huanghe (Yellow River) delta lobe caused by human activities. *Continental Shelf Research, 90*, 70–78.

Bi, N., Yang, Z., Wang, H., Xu, C., & Guo, Z. (2014b). Impact of artificial water and sediment discharge regulation in the Huanghe (Yellow River) on the transport of particulate heavy metals to the sea. *CATENA, 121*, 232–240.

Bi, N., Wang, H., Wu, X., Saito, Y., Xu, C., & Yang, Z. (2021). Phase change in evolution of the modern Huanghe (Yellow River) Delta: Process, pattern, and mechanisms. *Marine Geology, 437*, 106516.

Blum, M. D., & Roberts H. H. (2009). Drowning of the mississippi delta due to insufficient sediment supply and global sea-level rise. *Nature Geoscience, 2*(7), 488–491.

Chen, J., Wang, F., Meybeck, M., He, D., Xia, X., & Zhang, L. (2005). Spatial and temporal analysis of water chemistry records (1958–2000) in the Huanghe (Yellow River) basin. *Global Biogeochemical Cycles, 19*(3), GB3016, doi:10.1029/2004GB002325.

Cheng, G. (Ed.). (1991). Modern sedimentation and its model of the Huanghe Delta. Geological Publishing House. (in Chinese).

Chu, Z. X., Sun, X. G., Zhai, S. K., & Xu K. H. (2006). Changing pattern of accretion/erosion of the modern Yellow River (Huanghe) subaerial delta, China: Based on remote sensing images. *Marine Geology, 227*(1), 13–30.

Clift, P. D., Hodges, K. V., Heslop, D., Hannigan, R., Van Long, H., & Calves G. (2008). Correlation of Himalayan exhumation rates and Asian monsoon intensity. *Nature Geoscience, 1*(12), 875–880.

Cui, B., Yang, Q., Yang, Z., & Zhang K. (2009). Evaluating the ecological performance of wetland restoration in the Yellow River Delta, China. *Ecological Engineering, 35*(7), 1090–1103.

Cui, B.-L., & Li, X.-Y. (2011). Coastline change of the Yellow River estuary and its response to the sediment and runoff (1976–2005). Geomorphology, 127(1), 32–40.

Dong, J., Xia, X., Wang, M., Lai, Y., Zhao, P., Dong, H., Zhao, Y., & Wen, J. (2015). Effect of water–sediment regulation of the Xiaolangdi Reservoir on the concentrations, bioavailability, and fluxes of PAHs in the middle and lower reaches of the Yellow River. *Journal of Hydrology, 527*, 101–112.

Gao, M., Liu, S. Zhao, G., Yuan, H., Wei, C., Wu, Y., & Tang, J. (2014). Vulnerability of eco-hydrological environment in the Yellow River delta Wetland. *Journal of Coastal Research, 30*(2), 344–350.

Han, G., Xing, Q., Yu, J. Luo, Y., Li, D., Yang, L., Wang, G., Mao, P., Xie, B., & Mikle, N. (2014). Agricultural reclamation effects on ecosystem CO^2 exchange of a coastal wetland in the Yellow River Delta. *Agriculture, Ecosystems and Environment, 196*, 187–198.

He, X., Zhou, J., Zhang, X., & Tang K. (2006). Soil erosion response to climatic change and human activity during the Quaternary on the Loess Plateau, China. *Regional Environmental Change, 6*(1-2), 62–70.

Huang, D., Su, J., & Backhaus, J. O. (1999). Modelling the seasonal thermal stratification and baroclinic circulation in the Bohai Sea. *Continental Shelf Research, 19*(11), 1485–1505.

Huang, M., & Zhang L. (2004). Hydrological responses to conservation practices in a catchment of the Loess Plateau, China. *Hydrological Processes, 18*(10), 1885–1898.

Ji, H., Chen, S., Pan, S., Xu, C., Jiang, C., & Fan, Y. (2018). Morphological variability of the active Yellow River mouth under the new regime of riverine delivery. *Journal of Hydrology, 564*, 329–341.

Jiang, W., Pohlmann, T., Sündermann, J., & Feng, S. (2000). A modelling study of SPM transport in the Bohai Sea. *Journal of Marine Systems, 24*(3), 175–200.

Kong, D., Miao, C., Borthwick, A.G.L., Duan, Q., Liu, H., Sun, Q., Ye, A., Di, Z., &Gong, W. (2015). Evolution of the Yellow River Delta and its relationship with runoff and sediment load from 1983 to 2011. *Journal of Hydrology, 520*, 157–167.

Kuenzer, C., Ottinger, M., Liu, G., Sun, B., Baumhauer R., & Dech, S. (2014). Earth observation-based coastal zone monitoring of the Yellow River delta: Dynamics in China's second largest oil producing region over four decades. *Applied Geography, 55*, 92–107.

Jin, X. (2001). The dynamics of major fishery resources in the Bohai Sea. *Journal of Fishery Sciences of China, 7*(4), 22–26. (in Chinese with English abstract).

Li, S.-N., Wang, G.-X., Deng, W., Hu, Y.-M., & Hu W.-W. (2009). Influence of hydrology process on wetland landscape pattern: A case study in the Yellow River Delta. *Ecological Engineering, 35*(12), 1719–1726.

Li, Y., Zhang, H., Chen, X., Tu, C., Luo, Y., & Christie, P. (2014). Distribution of heavy metals in soils of the Yellow River Delta: Concentrations in different soil horizons and source identification. *Journal of Soils and Sediments, 14*(6), 1158–1168.

Li, H., Chi, Z., Li, J., Wu, H., & Yan, B. (2019). Bacterial community structure and function in soils from tidal freshwater wetlands in a Chinese delta: Potential impacts of salinity and nutrient. *Science of The Total Environment, 696*, 134029.

Li, Y., Li, P., Wu, X., Bi, N., & Wang, H. (2022). Tempo-spatial variation of wetlands at the Yellow River Mouth and its control factors. *Marine Geology and Quaternary Geology, 42*(1), 68–80.

Liu, S. M., Li, L. W., Zhang, G. L., Liu, Z., Yu, Z., & Ren, J. L. (2012). Impacts of human activities on nutrient transports in the Huanghe (Yellow River) estuary. *Journal of Hydrology, 430–431*, 103-110.

Loewe, M. (2006). The government of the Qin and Han Empires: 221 BCE–220 CE. Hackett Publishing.

Lu, X. X. (2004). Vulnerability of water discharge of large Chinese rivers to environmental changes: an overview. *Regional Environmental Change, 4*(4), 182–191.

Ma, X., Yan, J, & Fan, F. (2014). Morphology of submarine barchans and sediment transport in barchans fields off the Dongfang coast in Beibu Gulf. *Geomorphology, 213*, 213–224.

Milliman, J. D., & Meade, R. H. (1983). World-wide delivery of river sediment to the oceans. *The Journal of Geology, 91*(1), 1–21.

Milliman, J. D., & Syvitski, J. P. M. (1992). Geomorphic/tectonic control of sediment discharge to the ocean: The importance of small mountainous rivers. *The Journal of Geology, 100*(5), 525–544.

Pang, J. Z., & Si, S. H. (1980). Fluvial process of the Huanghe River estuary, Part II: The hydrological features and distribution of sedimentation at the river mouth. *Oceanologia et Limnologia Sinica, 11*(4), 295–305. (in Chinese with English abstract).

Qi, S., & Luo, F. (2007). Environmental Degradation in the Yellow River Delta, Shandong Province, China. *Ambio, 36*, 610–611.

Ren, M., & Shi, Y. (1986). Sediment discharge of the Yellow River (China) and its effect on the sedimentation of the Bohai and the Yellow Sea. *Continental shelf research, 6*, 785–810.

Ren, M., & Zhu, X. (1994). Anthropogenic influences on changes in the sediment load of the Yellow River, China, during the Holocene. *The Holocene, 4*(3), 314–320.

Shi, C., Dian, Z., & You L. (2002). Changes in sediment yield of the Yellow River basin of China during the Holocene. *Geomorphology, 46*(3-4), 267–283.

Shi, C., Zhang, L., Xu J., & Guo L. (2010). Sediment load and storage in the lower yellow river during the late holocene. *Geografiska Annaler: Series A, Physical Geography, 92*(3), 297–309.

Syvitski, J. P., Vorosmarty, C. J., Kettner A. J., & Green P. (2005). Impact of humans on the flux of terrestrial sediment to the global coastal ocean. *Science, 308*(5720), 376–380.

Syvitski, J. P. M., Kettner, A. J., Overeem, I., Hutton, E. W. H., Hannon, M. T., Brakenridge, G. R., et al. (2009). Sinking deltas due to human activities. *Nature Geoscience, 2*(10), 681–686.

Sun, X., & Xi, P. (1988). The results of the 3D nonlinear tidal boundary problem and its analysis. *Journal of Shandong College of Oceanography, 18*, 51–52. (in Chinese with English abstract).

Tapponnier, P., Zhiqin, X., Roger, F., Meyer, B., Arnaud, N., Wittlinger, G., & Jingsui, Y. (2001). Oblique stepwise rise and growth of the Tibet Plateau. *Science, 294*(5547), 1671–1677.

Wang, H., Yang, Z., Bi, N., & Li, H. (2005). Rapid shifts of the river plume pathway off the Huanghe (Yellow) River mouth in response to water-sediment regulation scheme in 2005. *Chinese Science Bulletin, 50*(24), 2878–2884.

Wang, H., Yang, Z., Li, G., & Jiang, W. (2006a). Wave Climate Modeling on the Abandoned Huanghe (Yellow River) Delta Lobe and Related Deltaic Erosion. *Journal of Coastal Research, 224*, 906–918.

Wang, H., Yang, Z., Saito, Y., Liu, J. P., & Sun, X. (2006b). Interannual and seasonal variation of the Huanghe (Yellow River) water discharge over the past 50 years: Connections to impacts from ENSO events and dams. *Global and Planetary Change, 50*(3–4), 212–225.

Wang, H., Yang, Z., Li, Y., Guo, Z., Sun, X., & Wang, Y. (2007a). Dispersal pattern of suspended sediment in the shear frontal zone off the Huanghe (Yellow River) mouth. *Continental Shelf Research, 27*(6), 854–871.

Wang, H., Yang, Z., Saito, Y., Liu, J. P., Sun, X., & Wang, Y. (2007b). Stepwise decreases of the Huanghe (Yellow River) sediment load (1950–2005): Impacts of climate change and human activities. *Global and Planetary Change, 57*(3–4), 331–354.

Wang, H., Saito, Y., Zhang, Y., Bi, N., Sun, X., & Yang, Z. (2011a). Recent changes of sediment flux to the western Pacific Ocean from major rivers in East and Southeast Asia. *Earth-Science Reviews, 108*(1–2), 80–100.

Wang, Y., Wang, H., Bi, N., & Yang, Z. (2011b). Numerical modeling of hyperpycnal flows in an idealized river mouth. *Estuarine, Coastal and Shelf Science, 93*(3), 228–238.

Wang, H., Wang, A., Bi, N., Zeng, X., & Xiao, H. (2014). Seasonal distribution of suspended sediment in the Bohai Sea, China. *Continental Shelf Research, 90*, 17–32.

Wang, H., Wu, X., Bi, N., Li, S., Yuan, P., Wang, A., Syvitski, J. P. M., Saito, Y., Yang, Z., Liu, S., & Nittrouer, J. (2017). Impacts of the dam-orientated water-sediment regulation scheme on the lower reaches and delta of the Yellow River (Huanghe): A review. *Global and Planetary Change, 157*, 93–113.

Wang, Y., & D. G. Aubrey (1987). The characteristics of the China coastline. Continental Shelf Research, 7, 329–349.

Wang, G., Li, P., Li, Z., Liang, C., & Wang, H. (2022). Coastal subsidence detection and characterization caused by brine mining over the Yellow River Delta using time series InSAR and PCA. *International Journal of Applied Earth Observation and Geoinformation, 114*, 103077.

Wu, X., Bi, N., Kanai, Y., Saito, Y., Zhang, Y., Yang, Z., Fan, D., & Wang, H. (2015a). Sedimentary records off the modern Huanghe (Yellow River) delta and their response to deltaic river channel shifts over the last 200 years. *Journal of Asian Earth Sciences, 108*, 68–80.

Wu, X., Bi, N., Yuan, P., Li, S., & Wang, H. (2015b). Sediment dispersal and accumulation off the present Huanghe (Yellow River) delta as impacted by the Water-Sediment Regulation Scheme. *Continental Shelf Research, 111*, 126–138.

Wu, X., Bi, N., Xu, J., Nittrouer, J. A., Yang, Z., Saito, Y., & Wang, H. (2017). Stepwise morphological evolution of the active Yellow River (Huanghe) delta lobe (1976–2013): Dominant roles of riverine discharge and sediment grain size. *Geomorphology, 292*, 115–127.

Wu, X., Wang, H. J., Bi, N. S., Saito, Y., Xu, J. P., Zhang, Y., Lu, T., Cong, S. & Yang, Z. S. (2020b). Climate and human battle for dominance over the Yellow River's sediment discharge: From the Mid-Holocene to the Anthropocene. *Marine Geology, 425*, 106188.

Wu, X., Bi, N. S., Syvitski, J., Saito, Y., Xu, J. P., Nittrouer, J., Bianchi, T., Yang, Z. S. & Wang, H. J. (2020a). Can reservoir regulation along the yellow river be a sustainable way to save a sinking delta? *Earth's Future, 8*, e2020EF001587.

Wu, X., Yue, Y., Borthwick, A., Slater, L., Syvitski, J., Bi, N. S., Yang, Z. S., & Wang, H. J. (2023). Mega-reservoir regulation: A comparative study on downstream responses of the Yangtze and Yellow rivers. *Earth-Science Reviews, 245*, 104567.

Xu, B., Bianchi, T. S., Allison, M. A., Dimova, N. T., Wang, H., Zhang, L., Diao, S., Jiang, X., Zhen, Y., Yao, P., Chen, H., Yao, Q., Dong, W., Sui, J., & Yu, Z. (2015). Using multi-radiotracer techniques to better understand sedimentary dynamics of reworked muds in the Changjiang River estuary and inner shelf of East China Sea. *Marine Geology, 370*, 76–86.

Yang, Z., Wang, H., Saito, Y., Milliman, J. D., Xu, K., Qiao, S., & Shi, G. (2006). Dam impacts on the Changjiang (Yangtze) River sediment discharge to the sea: The past 55 years and after the Three Gorges Dam. *Water Resources Research, 42*(4), W04407.

Yang, Z., Ji, Y., Bi, N., Lei, K., & Wang, H. (2011). Sediment transport off the Huanghe (Yellow River) delta and in the adjacent Bohai Sea in winter and seasonal comparison. *Estuarine, Coastal and Shelf Science, 93*(3), 173–181.

Yellow River Conservancy Commission (YRCC). (1989). Yellow River Sediment bulletin. Yellow River Conservancy Press.

Yellow River Conservancy Commission (YRCC). (2013). Comprehensive plan of the Yellow River watershed (2012–2030). Yellow River Conservancy Press.

Yu, Y., Wang, H., Shi, X., Ran, X., Cui, T., Qiao, S., & Liu, Y. (2013). New discharge regime of the Huanghe (Yellow River): Causes and implications. *Continental Shelf Research, 69*, 62–72.

Yu, J., Fu, Y., Li, Y., Han, G., Wang, Y., Zhou, D., Sun, W., Gao, Y., & Meixner, F. X. (2011). Effects of water discharge and sediment load on evolution of modern Yellow River Delta, China, over the period from 1976 to 2009. *Biogeosciences, 8*(9), 2427–2435.

Yuan, H., Li, T., Ding, X., Zhao, G., & Ye S. (2014). Distribution, sources and potential toxicological significance of polycyclic aromatic hydrocarbons (PAHs) in surface soils of the Yellow River Delta, China. *Marine Pollution Bulletin, 83*(1), 258–264.

Yue, T. X., Liu, J. Y., Jørgensen, S. E., & Ye, Q. H. (2003). Landscape change detection of the newly created wetland in Yellow River Delta. *Ecological Modelling, 164*(1), 21–31.

Zhang, L. J., Wang, L., Cai, W. J., Liu, D. M., & Yu Z. G. (2013). Impact of human activities on organic carbon transport in the Yellow River. *Biogeosciences, 10*(4), 2513–2524.

Zhang, X., Lu, Z., Jiang, S., Chi, W., Zhu, L., Wang, H., Lv, K., Wang, B., & Yang, Z. (2018). The progradation and retrogradation of two newborn Huanghe (Yellow River) Delta lobes and its influencing factors. *Marine Geology, 400*, 38–48.

Zhao, G., Mu, X., Wen, Z., Wang, F., & Gao P. (2013). Soil erosion, conservation, and eco-environmet changes in the loess plateau of China. *Land Degradation & Development, 24*(5), 499–510.

Zhao, G., Tian, P., Mu, X., Jiao, J., Wang, F., & Gao, P. (2014). Quantifying the impact of climate variability and human activities on streamflow in the middle reaches of the Yellow River basin, China. *Journal of Hydrology, 519*, 387–398.

Zheng, F.-L. (2006). Effect of vegetation changes on soil erosion on the loess plateau. *Pedosphere, 16*(4), 420–427.

Zeng, X., He, R., Xue, Z., Wang, H., Wang, Y., Yao, Z., Guan, W., & Warrillow, J. (2015). River-derived sediment suspension and transport in the Bohai, Yellow, and East China Seas: A preliminary modeling study. *Continental Shelf Research, 111*, 112–125.

Zhu, L., Wang, Y., Jiang, L., Lai, L., Ding, J., Liu, N., Li, J., Xiao, N., Zheng, Y., & Rimmington, G. M. (2015). Effects of residual hydrocarbons on the reed community after 10 years of oil extraction and the effectiveness of different biological indicators for the long-term risk assessments. *Ecological Indicators, 48*, 235–243.

Summary

Weiguo Zhang and Huib de Vriend

This chapter provides an overview of the prevailing sustainability concerns associated with 16 river deltas across the globe (Table 1). These deltas, heterogeneous in geographical and economic conditions, are all confronted with hazard and pollution issues, posing threats to both human quality of life and delta sustainability. Global climate change, sea level rise, and human activities are compounded drivers to this challenging situation, which calls for collaborative effort among the deltas and even at a global scale, to cope with the shared uncertain future.

1 Status of the Deltas

1.1 Natural Environment

The natural environment of a delta depends on its climatic zone, which influences its temperature, precipitation, livability, and environmental carrying capacity. Out of the 16 deltas, the majority (10) are located in tropical and subtropical area, with 4 in temperate areas and 2 in arctic and subarctic zones. Most subtropical deltas lie in monsoon regions, where precipitation varies greatly by season, increasing the risk of droughts and floods.

The water supply to a delta is not only influenced by its climatic zone but also by the upstream water sources. The river discharge of the 16 deltas varies greatly

W. Zhang (✉)
State Key Laboratory of Estuarine and Coastal Research, East China Normal University, Shanghai 200241, China
e-mail: wgzhang@sklec.ecnu.edu.cn

H. de Vriend
Faculty of Civil Engineering and Geosciences, Delft University of Technology, Delft 2628 CN, The Netherlands

W. Zhang and H. de Vriend (eds.), *Delta Sustainability*,
https://doi.org/10.1007/978-981-97-7259-9_17

Table 1 Natural and social-economic characteristics of 16 selected deltas

	Basin area (× 10^3 km^2)[a]	Transboundary river	Climate[a]	River water discharge (km^3/yr)[a]	Sediment discharge-Historical (Mt/yr)[b]	Sediment discharge-Current (Mt/y)[b]	Population (millions)[c]	Population density (people/km^2)[c]	National GDP per capita (current US$)[d]	Deltaic GDP per capita (current US$)[e]
Amazon	6300	√	Tr-W-S	6300	570	667	4.21	39	6924	3906
Changjiang	1800		Te-H-S	900	512	133	42.45	2141	10409	24726
Danube	820	√	Te-H-S	210	67	16	0.09	20	13048	10177
Fly	76		Tr-W-S	180	81	119	0.01	2	2446	2363
Ganges-Bramaputra-Meghna	1730	√	STr-W-S	1270	1044	522	139.36	1589	2198	1977
Indus	980	√	STr-A-S	90	401	13	1.24	107	1322	605
Irrawaddy	430	√	Tr-W-S	430	600	364	13.78	419	1478	1485
Lena	2500		Ar-SA-S	520	20	18.4	0.0009	0.04	10194	5894
Mackenzie	1800		SAr-SA-S	310	103	69	0.001	0.09	43350	51974
Mekong	800	√	Tr-H-S	550	97	40	17.94	463	3586	2403
Mississippi	3300		Te-SA-Sp	490	398	190	1.97	54	63529	50618
Nile	2900	√	STr-A-W	80	155	3	61.24	2317	3572	3842
Pearl	490		STr-H-Sp	260	80	24	37.09	3709	10409	19366
Rhine-Meuse-Scheldt	243	√	Te-H-C	85	62	17	2.93	606	52163	52721

(continued)

Table 1 (continued)

	Basin area (× 10^3 km^2)[a]	Transboundary river	Climate[a]	River water discharge (km^3/yr)[a]	Sediment discharge-Historical (Mt/yr)[b]	Sediment discharge-Current (Mt/y)[b]	Population (millions)[c]	Population density (people/km^2)[c]	National GDP per capita (current US$)[d]	Deltaic GDP per capita (current US$)[e]
Río de la Plata	2600	√	STr-SA-C	530	89	89	1.26	90	8496	5436
Yellow	750		Te-A-S/Au	43	1311	126	2.50	265	10409	12487

[a] Data from Milliman and Farnsworth (2011). The Climate classification is modified from the Köppen-Geiger system, of which the symbol is represented by three groups of letters: first group refers to mean basin temperature: Tropical (Tr), Subtropical (STr), Temperate (Te), Subarctic (SAr), and Arctic (A); the second group refers to mean annual runoff: Arid (<100 mm/yr: A), Wet (>750 mm/yr: W), and Semiarid (SA); and the third group refers to season of maximum discharge: Summer (S), Autumn (Au) and Winter (W), Continuous (C), and Desert (D) (Milliman & Farnsworth, 2011)

[b] Historical and current represent the sediment discharge alterations between the mid-20th and early 21st centuries (Syvitski et al., 2022)

[c] Extracted 2020 population data from Gridded Population of the World, Version 4 (GPWv4) dataset (CIESIN, 2018)

[d] Retrieved the 2020 GDP per capita data from World Bank DataBank (2023)

[e] Calculated from the GDP and population data of the subnational region where the delta is located, which are obtained from the national Statistic Yearbook

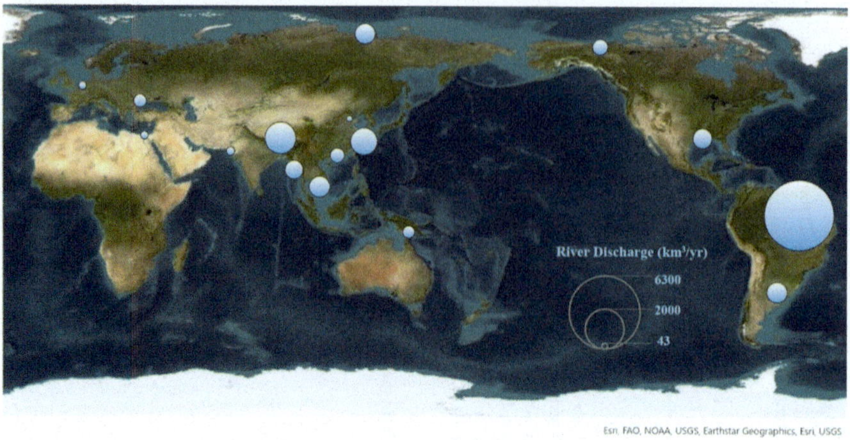

Fig. 1 River water discharges in the selected deltas. *Data source* From Milliman and Farnsworth (2011), Nienhuis et al. (2020)

(Fig. 1), reflecting differences in basin area and climatic characteristics. Among them, the Amazon River has the highest water discharge, while the Yellow River has the lowest, with a difference of 150 times.

Supply of sediment from upstream is essential for delta formation and resilience to coastal erosion and sea level rise. Fluvial sediment load, which depends on basin characteristics and human activities, differs greatly among deltas (Fig. 2). In recent years, more than half of the deltas have seen their river sediment discharge drop by over 50%. This decline is especially severe in Asia and Africa.

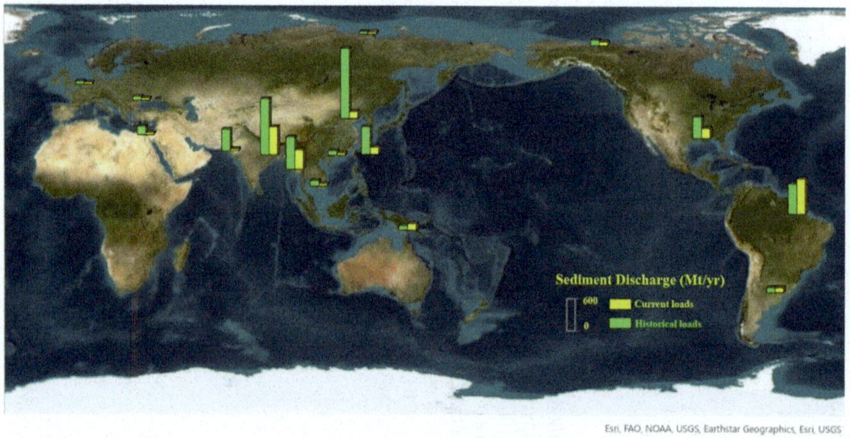

Fig. 2 River sediment discharge and alterations from the mid-20th to early 21st centuries in the selected deltas (Syvitski et al., 2022)

1.2 Socio-economic Status

Population density serves as an indicator of the livability and level of development in a delta. According to the GPW v4 population grid data (CIESIN, Fig. 3), subtropical deltas have high population densities, while tropical and arctic deltas have low population densities. The Pearl River Delta has the highest population density, with 3700 people/km^2, whereas the Lena River Delta and Mackenzie Delta have the lowest population densities, with less than 1 people people/km^2.

GDP per capita reflects the level of economic development and the ability to cope with climate change and disasters in a region. The 16 deltas studied show a wide gap in GDP per capita: those in Europe and North America are much richer than those in Asia, Africa, and South America. The Indus River Delta is the poorest of all (Fig. 4).

Water is a fundamental resource for agricultural and industrial production, as well as for sustaining life. It is also one of the key objectives of the United Nations Sustainable Development Goals (SDG6). The level of water stress, expressed by freshwater withdrawal as a proportion of available freshwater resources, is primarily affected by population density. Among the underdeveloped deltas, the Nile River suffers from the most severe water scarcity (Fig. 5).

2 Environmental Issue

This report examines the environmental challenges that deltas encounter, such as floods, saltwater intrusion, sediment reduction, coastal erosion, land subsidence, and pollution (Table 2). These challenges result from both natural and human-induced

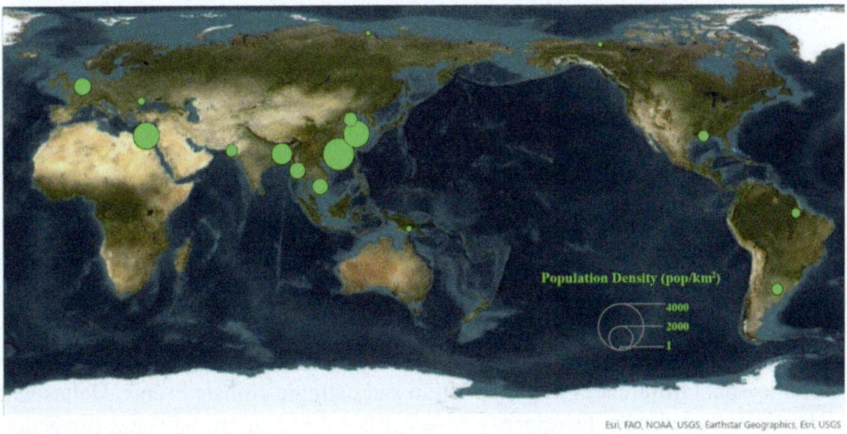

Fig. 3 Population density distribution among the selected deltas in 2020. Original data is from GWP v4 gridded population dataset at 30 arc-seconds spatial resolution (CIESIN, 2018)

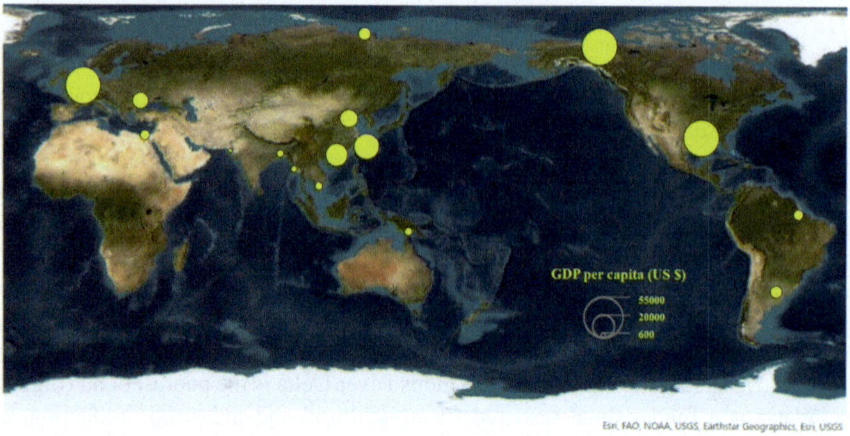

Fig. 4 Difference in GDP per capita among the selected deltas. Original data are derived from the national statistical bureaus of the countries where the deltas are located

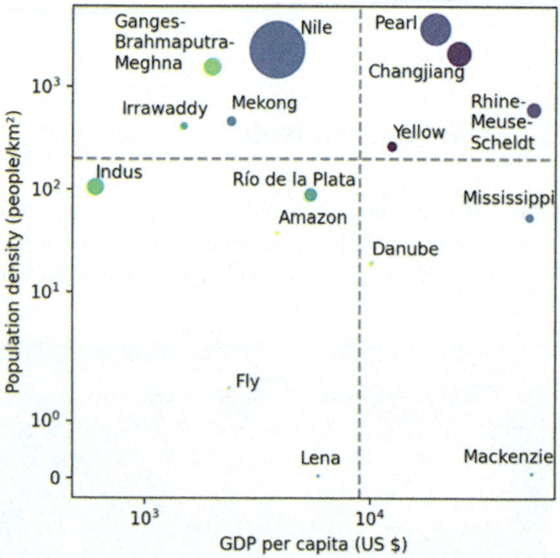

Fig. 5 The relationship between GDP per capita and population density in the selected deltas. The size of each circle indicates the water stress level based on the data from the World Bank DataBank (2023)

factors at global (e.g., climate change), regional (e.g., dams in the catchment), and local scales.

Deltas located in monsoonal regions experience more frequent floods and droughts due to seasonal differences in precipitation and extreme climate events. Deltas with high population density face increased water demand and groundwater extraction, which, coupled with drought events, exacerbate human-induced land subsidence and saltwater intrusion. Continuous upstream damming has been found to have a significant impact on the sediment flux entering the sea, which may potentially accelerate

Table 2 Current environmental issues faced by the selected deltas

	Flooding	Storm surge	Saline water intrusion	Reduced Sediment input	Coastal erosion	Land subsidence	Pollution
Amazon		√					
Changjiang	√	√	√	√	√	√	√
Danube		√	√	√	√		
Fly		√					
Ganges-Bramaputra-Meghna	√	√	√	√	√	√	√
Indus	√	√	√	√			√
Irrawaddy	√	√	√	√			
Lena							
Mackenzie				√		√	
Mekong	√	√	√	√	√	√	√
Mississippi	√	√	√	√	√	√	√
Nile			√	√	√	√	√
Pearl	√	√	√	√		√	√
Rhine-Meuse-Scheldt	√	√	√	√	√	√	
Río de la Plata			√			√	√
Yellow			√	√	√	√	

coastal erosion and land subsidence. Given the scenarios of sea level rise and future climate change, the lack of sediment inputs and a higher frequency of storm surges, the risk of flooding in deltas will further increase.

Pollution problems are prevalent in deltas with high population density. Inadequate investment in pollution control exacerbates these issues, particularly in underdeveloped areas. As shown in Table 2, these problems often co-exists in the deltas, which demonstrates the interconnectedness between hydrogeomorphic processes, ecological processes, and economic development.

3 Future Pathways

3.1 Monitoring and Modeling of Coupled Human-Natural System

Effective monitoring and in-depth research are crucial for addressing sustainable development challenges in deltas. The existing observation data and research exhibit inconsistencies in data quality, scale and resolution. Therefore, a unified and standardized observing method should be established to provide a comprehensive understanding of the delta and cope with future uncertainties, and to create a dedicated database for each delta. Furthermore, it is imperative to improve the coupled human-natural models that can represent the complex nexus between human and natural systems and enable accurate projection of delta evolution under diverse policy and climate change scenarios. These endeavors will inform decision makers and the general public of the key tradeoffs involved in enhancing regional resilience and provide them with scenarios for better planning for climate change.

3.2 Local Adaption

The selected deltas exhibit varying levels of vulnerability and adaptability to the impending impacts of climate change. These result from their unique geographical and socio-economic characteristics, as well as cultural practices. There is no single way to cope with problems in different deltas. Developed deltas are usually densely populated, highly urbanized, and well-equipped with key infrastructure. It is essential to enhance the early warning and forecasting system, and to enhance the self-adaptation capacities of the urban communities, such as ensuring sufficient water, energy, and daily supplies, to cope with complex emergency situations in the uncertain future. Deltas that are predominantly rural often face inadequate infrastructure, sparse population distribution, low connectivity among communities, and significant cultural diversity. National and regional governments need to collaborate and capitalize on local wisdom and make the most of indigenous resources, to promote community engagement in disaster prevention and mitigation planning.

3.3 Global Collaboration

Deltas play a crucial role in connecting terrestrial and marine systems. More than half of the rivers examined in this report are transboundary, as shown in Table 1. Therefore, the environmental issues of deltas are not only determined by the local conditions, but also influenced by the environmental and socioeconomic changes in the entire watershed. It is imperative to foster international collaboration among the countries that share the river basin, including the establishment of joint management institutions.

The process of globalization and trades further intensify the interconnectedness and destiny among all deltas. A crisis occurring in one delta can propagate and have far-reaching consequences on others. Hence, it is imperative to promptly initiate international cooperation to facilitate the exchange of governance experiences. Initiatives, forums, and cooperating institutions should be set up to encourage multi-directional communication among all partners and stakeholders to seek solutions for the shared future of deltas.

3.4 Financial Support

The experience of developed countries has demonstrated that achieving sustainable development in delta regions necessitates consistent and substantial financial investment. For instance, the Netherlands has established a dedicated fund for the sustainable development of its low-lying areas, with an annual allocation exceeding 1.25 billion euros. Tessler et al. (2015) have indicated that without continuous and increased investment in deltas, the risk of flooding would escalate. However, the significant disparity in national GDP per capita (Table 1) reveals the lack of financial capacity to address future challenges in underdeveloped delta regions, where inhabitants face even precarious living conditions such as water, food, and sanitation insecurity. These risks would exacerbate by social vulnerability and high exposure to natural disasters. The recent United Nations Climate Change Conference (COP27) has reached an agreement to establish an International Loss and Damage Fund, representing a substantial stride towards addressing the broader issues of inequality between the global north and south. Financial support, including the establishment of specialized funds to support delta sustainability and the implementation of insurance planning to transfer risks, remains crucial in overcoming the future challenges associated with sustainable development in delta regions.

References

Center for International Earth Science Information Network (CIESIN). (2018). Gridded Population of the World, Version 4 (GPWv4): Population Count, Revision 11. Palisades, New York: NASA Socioeconomic Data and Applications Center (SEDAC).

Milliman, J. D., & Farnsworth, K. L. (2011). River discharge to the coastal Ocean. Cambridge University Press.

Nienhuis, J. H., Ashton, A. D., Edmonds, D. A., Hoitink, A. J. F., Kettner, A. J., Rowland, J. C., & Törnqvist, T. E. (2020). Global-scale human impact on delta morphology has led to net land area gain. *Nature, 577*(7791), 514–518.

Syvitski, J., Ángel, J. R., Saito, Y., Overeem, I., Vörösmarty, C. J., Wang, H., & Olago, D. (2022). Earth's sediment cycle during the Anthropocene. *Nature Reviews Earth & Environment, 3*(3), 179–196.

Tessler, Z. D., Vörösmarty, C. J., Grossberg, M., Gladkova, I., Aizenman, H., Syvitski, J. P. M., & Foufoula-Georgiou, E. (2015). Profiling risk and sustainability in coastal deltas of the world. *Science, 349*(6248), 638–643.

World Bank. (2023). National GDP per capita (current US $) in 2020, derived from World Bank DataBank on June 1, 2023.

SPRINGER NATURE

GPSR Compliance

The European Union's (EU) General Product Safety Regulation (GPSR) is a set of rules that requires consumer products to be safe and our obligations to ensure this.

If you have any concerns about our products, you can contact us on ProductSafety@springernature.com

In case Publisher is established outside the EU, the EU authorized representative is:

Springer Nature Customer Service Center GmbH
Europaplatz 3
69115 Heidelberg, Germany

The manufacturer's authorised representative in the EU is Springer Nature Customer Service Centre GmbH, Europaplatz 3, 69115 Heidelberg, Germany. If you have any concerns regarding our products, please contact ProductSafety@springernature.com

Printed and bound by CPI Group (UK) Ltd, Croydon, CR0 4YY
01/10/2026
02225907-0012